# Mathematical Methods of Quantum Physics

*Essays in Honor of Professor Hiroshi Ezawa*

**Proceedings of the 2nd Jagna International Workshop**

Jagna, Philippines, 4-8 January 1998

*Edited by*

**Christopher C. Bernido**
University of the Philippines
and Central Visayan Institute,
Philippines

**Koichi Nakamura**
Meiji University, Japan

**M. Victoria Carpio-Bernido**
Central Visayan Institute,
Philippines

**Keiji Watanabe**
Meisei University, Japan

**GORDON AND BREACH SCIENCE PUBLISHERS**
Australia • Canada • China • France • Germany • India • Japan
Luxembourg • Malaysia • The Netherlands • Russia
Singapore • Switzerland

Amsteldijk 166
1st Floor
1079 LH Amsterdam
The Netherlands

---

**British Library Cataloguing in Publication Data**

A catalogue record for this book is available from the British Library.

ISBN 90-5699-211-2

# Contents

## STOCHASTIC PROCESSES

## QUANTUM FIELD THEORY AND STATISTICAL MECHANICS

# FUNCTIONAL INTEGRALS AND THEIR APPLICATIONS

# SPIN SYSTEMS AND COHERENT STATES

# GAUGE FIELD THEORY AND SUPERSYMMETRY

## THERMO FIELD DYNAMICS

# Preface

The 2nd Jagna International Workshop: Mathematical Methods of Quantum Physics was held on January 4–8, 1998 at the Research Center for Theoretical Physics (RCTP), Central Visayan Institute in Jagna, Bohol, the Philippines. The Workshop had a double purpose: (a) to bring together physicists and mathematicians for an intense exchange of ideas and perspectives on significant achievements, unsolved problems, and current trends related to the theme of the Workshop, and (b) to celebrate the 65th birthday of Professor Hiroshi Ezawa, a member of the RCTP Scientific Advisory Board, whose works have revolved around quantum and mathematical physics.

The venue for the Workshop – the RCTP, located in the coastal town of Jagna in the island province of Bohol – allowed an informal and friendly atmosphere. Since only a small group was invited, there was remarkably close interaction between the 33 speakers and participants from 8 countries. Young physicists and graduate students had much chance to interact with the experts as well as to establish new contacts. There were 23 scientific sessions: five 45-minute lectures and 17 half-hour presentations of papers where various mathematical tools – from algebraic methods to functional integrals – were discussed and employed in wide-ranging topics from gauge field theories to statistical physics. In addition, the schedule allowed a roundtable discussion session where comments on possible directions for breakthroughs in physics and mathematics were exchanged. Impressions and suggestions were also shared on the status and development of physics research in the Philippines. This proceedings of the Workshop, published as a *Festschrift* in honor of Professor Hiroshi Ezawa, should reflect the scientific flavor of the Jagna meeting. For this we thank all the contributors for the prompt submission of their camera-ready manuscripts.

We also wish to express our gratitude to the lecturers and participants who made the meeting a memorable experience, and to the Sponsors, Patrons, and Donors for their generous and enthusiastic support.

The Editors

## Major Sponsors

Department of Science and Technology
Abdus Salam International Centre for Theoretical Physics
Alexander von Humboldt-Stiftung
Philippine Charity Sweepstakes Office
De La Salle University
Philippine Convention and Visitors Corporation

## Sponsors

Municipality of Jagna, Bohol
Consuelo Limestone Corporation
Philippine Sinter Corporation
Dr Corazon C. Bernido
Mr and Mrs L. Stanford Hoghe
Attorney Stephen C. Bernido
Drs Raul and Cristina V. Fabella
Carpio Family of Naga
Dr Thomas L. Pe
The Class of 1989 – National Institute of Physics,
University of the Philippines:
*M. C. Carpio, J. Castro-Nolasco, R. Cruz, M. Fernandez*
*D. J. Jalandoni-Wieczorek, G. Labanda, F. Sandejas,*
*and N. Urmeneta*
Carpio, General and Cortes Law Office
Geo-Seven International Corp.
Bleustar Manufacturing and Marketing Corp: ADVAN

## Donors

Gila Tan-Flueckiger, Dr Edgar Genio, Dr Julita Lambating
Dr Dan E. Zarraga, Fely Tiempo-Ramos, FILBARS

## Workshop Directors

Christopher C. Bernido          Koichi Nakamura
M. Victoria Carpio-Bernido     Keiji Watanabe

The Workshop organizers are grateful

to Dean Danilo M. Yanga, Professor Jose A. Magpantay, Professor Ludwig Streit, Professor Frederik W. Wiegel, Professor Swee Cheng Lim, Attorney Jose Maria Carpio, Dr Greg Labanda, and Mrs Flora Tinampay, for their invaluable assistance in generating financial support for the Workshop;

to Mayor Ocmeja-Tsurumi, Vice-Mayor Cafe and the Jagna Sangguniang Bayan, Mrs Consuelo C. Bernido, Justice and Mrs Bernardo Salas, Mrs Perfecta Salas-Du, Attorney Jesus Acebes, Mrs Camila Pajo, Dr Alice Balane-Lim, Mr and Mrs Celestino Ranoco, Ms Edna Luna, Dr and Mrs Cecilio Cero, Engineer Casenas, and Engineer Jesus Acedillo for technical assistance and accommodation arrangements for participants of the Workshop;

to Cora Bernido, Teena and Vigile Fabella, Loreste and Emmanuel Bernido, James and Pablo Carpio, Mrs Macaria Casenas and family, Tita Bagaipo, and Mona Espiritu, for their untiring help in the preparations for, and conduct of, the Workshop; and

to the CLC and CVI staff, especially, Amancia Acedo, Jovencia Acera, Virgilia Carcallas, Clemente Jotojot, Daisy Kang, and Marcial Revuelto; and local staff, Apolinar Galendez, Felicidario Neuda, and Gonzala Orias, for cheerfully coping with the demands of assisting at an international event.

Front Row (1-r): C. C. Bernido, K. Watanabe, Y. X. Gui, N. Sakai, T. Hida, H. Ezawa, J. A. Carpio, L. Streit, C. DeWitt-Morette, K. Nakamura, J. Klauder

Middle Row (1-r): M. V. Carpio-Bernido, D. M. Yanga, H. C. S. Lam, I. Ojima, A. Arai, T. Nakamura, A. Mann, J. A. Magpantay, R. Quiroga, K. Yasue, D. Garcia, I. Timonera, J. Bornales

Back Row (1-r): T. Ichinose, J. Arafune, J. Q. Liang, M. Hirokawa, T. Dennis, F. W. Wiegel, F. Basco, E. Gravador, J. Guevarra

# Opening Address

## Honorable Rene Relampagos
*Governor*
*Province of Bohol*

Greetings...

One of the greatest, if not the greatest, among the Greek philosophers, Plato, once said that the noblest among the leaders of the people is the "philosopher king". In other words, an ideal leader must be one who has wisdom and exercises such wisdom in leading his followers. Wisdom, of course, does not necessarily mean a degree or a diploma; for indeed, you could find the most profound wisdom from the lives of even the lowliest farmer in the hinterlands.

Nevertheless, I find it relevant to state that before I entered politics, I was a college instructor in philosophy at a prominent local college here in our province. But that does not make me a philosopher king. What I am merely trying to say is that, with our basic knowledge in philosophy, the leaders in Bohol are trying to exercise wisdom in local governance.

In fact, the local leaders here in Jagna spearheaded by Mayor Ocmeja-Tsurumi as well as the leaders of the nongovernment organizations here, particularly the local academic institutions led by Dr Bernido, specifically the Research Center for Theoretical Physics of the Central Visayan Institute, have demonstrated wisdom in having been able to sponsor this one-of-a-kind international workshop here in our province. I understand this is already the second Workshop. But to be able to gather together the expert minds in the international field of theoretical physics in one venue is a feat unequalled in the province, if not, in the entire Philippines.

We, in Bohol, have started to get used to international visitors as we have increasingly noted that more and more national and international conferences are being held in our province every month during our last three years in office. But still, I wish to congratulate the leaders here in Jagna, both in the government and nongovernment sectors, for having been able to organize this workshop here in the municipality of Jagna where our distinguished visitors could savor this town's pristine beauty and the warmth and friendliness of its people.

Let me also express my sincerest appreciation to our expert friends here from abroad for having come to our beloved province. You are most welcome here. Do please come to our province again and again. Perhaps, we should adopt all of you as sons and daughters of Bohol. Then, you will have to come back and make Bohol your second home. And when you go back to your respective original home countries, you could tell your friends and relatives about your wonderful experience here; and you shall have served as our ambassadors of goodwill. As you can see, tourists usually come here for various reasons. They have heard about the world-famous Chocolate Hills, the Philippine Tarsier which is the world's smallest primate found in Bohol, or our world class dive sites and white sandy beaches, or the delicious *calamay* delicacy here in Jagna. Whatever may be your initial reasons for coming, we intend to make you come back because we have set our singular vision in making our province the prime eco-cultural tourist destination of the Philippines in the next five years. Not just any tourist destination, but an ecologically-sound tourist destination. And not just ecologically-sound, but also culturally-relevant. Meaning, we intend to share with you the best of our local culture; just as we also respect yours. Perhaps, we may not have known also that you may be part of families of businessmen or investors back there at your home country. Then we are encouraging you to come back and bring your investments here, particularly in the agro-industrial sector. Bohol, as you know, has a great potential in agriculture. And we are known to have the thickest remaining forest in this part of our country. That is also the reason why, while the rest of the country is suffering from water shortage, we here have enough water supply for the next 200 years to come. With this backdrop of our vision for Bohol's development, let me also mention the fact that our province can boast of its 92 per cent literacy rate; one of the highest in the region.

It may be fitting to mention also that as a former college professor in the field of philosophy, I have learned to appreciate the relevance of Quantum Physics on the cultivation of international understanding and brotherhood. In my humble understanding,

> Quantum Theory has revealed a basic oneness of the universe. It shows that we cannot decompose the world into independently existing smallest units. Quantum Theory teaches that as we penetrate into matter, nature does not show us any isolated 'basic building blocks', but rather appears as a complicated web of relations between the various parts of the whole.

You may, or you may not, totally agree with me in taking the perspective of Fritjof Capra's New Physics. But I am sure you would not deny me the privilege of labelling this international workshop as an avenue, *par excellence,* for the promotion of cross-cultural exchange and understanding.

It is in this context that I welcome you all to this small, but certainly beautiful, province of ours. As I have said earlier, we would love to see you here again and again. May you all have a most enlightening workshop and a wonderful experience here in Jagna and around the province.

Good morning and have a nice day!

## Opening Remarks

## A Glimpse Beyond Paradigms

J. Antonio M. Carpio
*Benefactor*

A blessed, happy and prosperous New Year, 1998, to all of you!

It is indeed a rare privilege to be with active, leading minds in theoretical physics. Scientists who come all the way from China in the East; Canada in the North; Germany and The Netherlands in the heartland of the Old World; Israel in the Middle East; in the Far East, from Japan – where the honoree, Professor Hiroshi Ezawa, comes from – and from the Philippines; and in the New World, from the United States of America, and, special mention, although registered as coming from the USA, by consanguinity, she comes from France – Professor Cecile DeWitt-Morette.

Knowing I would be before practitioners of a different discipline, I did a bit of reading about the world you work in. Because to me, it is as unknown as whether or not life – as we know it – existed on the planet Mars. I chanced upon a thin book, 172 pages, Thomas S. Kuhn's historical essay on the *Structure of Scientific Revolutions*. Being an historical approach, I thought it would be quite digestible even for a layman like me. It's funny, but I must confess that the first time I read it, I thought I could see what Kuhn was trying to say. Yet on repeated readings, I could not find the things I knew I saw. So I stopped thinking about it. And slept on it.

From what I recall, I can glimpse a few things: (1) you and I have something in common in that we are puzzle-solvers. You solve puzzles of the physical world; we, lawyers, try to find solutions to human problems of justice and law and order; (2) the fact that a question is a puzzle presupposes that there is a solution and there are rules of the game; (3) these are contained in conceptual and instrumental frameworks called paradigms; (4) the only permanent thing about paradigms is that they are not permanent; hence (5) disobedience is a virtue, in that scientific revolutions – and progress – come about because some scientists, hearing a different drummer, march to a beat different from that sounded by existing paradigms.

For us laymen, we readily see this demonstrated in the age-old stock knowledge which included the naked-eye perceptions that the earth is flat, that it does not move and that it is the center around which the sun revolves. Then, one day, that paradigm burst like a bubble. Because it could no longer contain the new discoveries that it was designed to define and delimit. Columbus, confirmed by Cano, and then Copernicus had shown that the earth is round, that it rotates around its axis, and that it revolves around the sun. A revolution! A new scientific structure had come of age. And like new wine, must be contained in a new wineskin. A new paradigm. Because ideas in ferment are far more powerful than fermenting wine.

There is another paradigm, in another discipline that, in a shadowy way, is still with us – the theory of evolution. As a story goes, however, the apes objected. They emphatically, angrily, and vehemently denied that man descended from them! Their argument, their syllogism runs like this – *Nemo dat quod non habet.* But not a single ape was ever, is ever immoral. *Ergo,* man with all his immorality to man could never have descended from them. [Incidentally, if evolution is from the simple to the complex, why do we still have the one-celled organisms today? Why do modern apes no longer evolve into *homo sapiens?* Why should evolution suddenly, unceremoniously stop somewhere in the misty past, long, long before recorded time?] But that is a different discipline. Let's go back to physics.

I suggest that a paradigm is people. Not books, essays, or other publications. For while knowledge is compiled – whether accurately or not so accurately – and passed from generation unto generation through the written word, these are only passive records gathering dust in libraries. Scientific minds reacting with one another – thesis and anti-thesis; hard-headedness is but the other side of commitment – are the source of concepts and computations proclaiming new discoveries. And <u>scientists are people.</u>

Hence, I venture to say that this Workshop, regardless of how extensive or intensive you designed it to be, is a paradigm. The 1998 Jagna paradigm on 'Math in Quantum Physics'. May it be new wine in new wineskin.

I remember a ditty which runs something like this – "once I saw a little mountain,...and what do you think I did?...I climbed the little mountain,... and what do you think I saw?...I saw another mountain...and what do you think I did?... I climbed the other mountain,..." It goes on *ad infinitum.*

This, to me, is the situation you are in. You have come to share your specific solutions to present puzzles. But, I bet you have already 'seen another mountain' and have already decided – no way to go but 'climb the new mountain'. So welcome to Jagna, by the Mayana mountain and the Bohol Sea, in the island province of Bohol, the Philippines.

In the language of another discipline, yours is a sacred quest. For in seeking, in myriad ways, to know *all that there is to know,* yours is a glimpse beyond present paradigms, an intimation of immortality, a foreview of the infinite – the Alpha and the Omega of all reality.

I wish you success in this endless endeavor. *Vaya con Dios!*

# STOCHASTIC PROCESSES

# WHITE NOISE ANALYSIS AND QUANTUM DYNAMICS

*Dedicated to Professor Hiroshi Ezawa*

TAKEYUKI HIDA

Department of Mathematics, Meijo University
Nagoya, 468-8502 Japan

## 1   Introduction.

White noise analysis has extensively developed in these years as we can see in the monographs [6], [7] and [9], and it is now the time to discuss some future directions of the theory.

The present short note plans to present two topics as the proposed directions. One is the interpretation of the notion of the idealized elementary random variables (abbr. i.e.r.v.'s), and the other is the variational calculus for random fields, let them be denoted by $X(C)$, that are functionals of white noise and are indexed by a manifold $C$ which runs through the Euclidean space taken to be the parameter space of the basic white noise. The introduction of the notion of i.e.r.v. was suggested by Professor John R. Klauder when we met at Osaka on the occasion of the 20th International Colloquium on Group Theoretical Methods in Physics. With the spirit of taking this name the background of the white noise calculus has been well established. Namely, the system of i.e.r.v. can be thought of as a collection of atomic random variables that describes the fluctuation generating the random phenomena in question.

Some more details of these two topics will be discussed in the following three sections. It should be noted that these topics have close connections with quantum dynamics.

## 2   System of i.e.r.v.'s.

The most typical example of an i.e.r.v. is certainly a white noise. An actual realization is just to take the time derivative of a Brownian motion $B(t)$. Namely, $\dot{B}(t) = \frac{dB(t)}{dt}$ is a white noise. The system $\mathbf{B} = \{\dot{B}(t)\}$ is Gaussian in distribution, although each member $\dot{B}(t)$ is an infinitesimal random variable with infinite variance. The main reason why such a system is thought of as a typical example of a system of i.e.r.v.'s is that the system is Gaussian as is mentioned above and it is a collection of independent, identically distributed random variables.

A function (in general, nonlinear function) of white noise is called a white noise functional. We prefer to call it a functional rather than a function, since $\dot{B}(t), t \in R$, is viewed as a generalized stochastic process and not simply as a collection of infinitesimal random variables.

Let $\mu$ be the probability distribution of the white noise $\{\dot{B}(t), t \in R\}$. The $\mu$ is supported by a space of generalized functions on $R$. We choose a space $E^*$ that supports the measure $\mu$ ; the space $E^*$ is the dual space of some nuclear subspace of the real Hilbert space $L^2(R)$. Then, we are given a complex Hilbert space $(L^2) = L^2(E^*, \mu)$. A member $x$ of the space $E^*$ equipped with the measure $\mu$ is a sample (generalized) function of the $\dot{B}(t)$ and $\varphi(x)$ is a realization of a white noise functional with finite variance with respect to $\mu$. Thus the $(L^2)$ is just the basic space on which we are going to discuss the white noise analysis.

Before we develop the analysis we have to make an important remark. Again we use the intuitive notation $\dot{B}(t)$. The collection **B** does not mean simply the set of continuously many independent random variables, but it should be understood that "$\dot{B}(t)$ is associated with the infinitesimal time interval $dt$". Such an understanding is in agreement with the following facts:

**(1)** the definition of the differential operator with respect to $\dot{B}(t)$ uses the Fréchet derivative,

**(2)** exponentials of the differential operator define the shift of the variable $x$.

These will be discussed in Section 3.

Since we take, still in terms of an intuitive notation, the $\dot{B}(t)$'s to be the variables of white noise functionals, it is quite natural to have polynomials in them, although to form polynomials is not permitted in the usual analysis since $\dot{B}(t)$ is a generalized function of $t$. In fact, it was a clever trick to apply a renormalization technique to define a generalized white noise functional having had modified the polynomials (see [5]).

**Remark.** Professor L. Streit and the present author tried to reformulate Feynman's path integral in a more visualized manner by using white noise analysis. In that course we are much stimulated by the important result by Professor H. Ezawa et al who established the Feynman-Kac average. Note that the fluctuation of the trajectories is expressed as a Brownian motion, so that the term of the kinetic energy should involve the square of $\dot{B}(t)$ in our setup.

Later a beautiful method of introducing the space of generalized white noise functionals was introduced by Kubo and Takenaka. With the help of the second quantization method, the following Gel'fand triple has been introduced:

$$(S) \subset (L^2) \subset (S)^*$$

The space $(S)$ is that of test functionals and $(S)^*$ is the space of generalized white noise functionals. It may be said that the above triple is an infinite dimensional analogue of the case of the Schwartz distribution.

The space $(S)^*$ is quite fitting for our purpose, however, based on this much larger spaces are often introduced depending on the purpose of application.

## 3  Differential operators.

As a basic tool for the actual calculus we introduce a representation of white noise (generalized) functionals. It is done by the so called $S$-transform which is defined by

$$S \; : \; \varphi \; \to \; (S\varphi)(\xi) = \int \varphi(x + \xi) d\mu(x).$$

The differential operator $\partial_t$ is defined by

$$\partial_t \varphi(x) = S^{-1} \left( \frac{\delta}{\delta \xi(t)} (S\varphi) \right)(x),$$

where $\frac{\delta}{\delta \xi(t)}$ is the Fréchet derivative.

It is known that the differential operator $\partial_t$ has a rich domain involving the test functionals and it is a derivation. In terms of quantum dynamics the differential operator is an annihilation operator.

It is easy to see that exponentials of the differential operators can be defined:

$$\exp\left[ a \partial_t \right],$$

$$\exp\left[ \int a(t) \partial_t dt \right].$$

**Theorem 1.**    The exponential operator acts as the shift of the variable of white noise functionals.

Outline of the proof. Let a white noise functional be taken to be

$$\varphi(x) = \exp[< x, \eta >].$$

Then, we have

$$\exp\left[a\partial_t\right]\varphi(x) = \sum \frac{a^k}{k!}\partial_t^k\varphi(x)$$
$$= \sum \frac{a^k\eta(t)}{k!}\varphi(x)$$
$$= \exp\left[< x, \eta > +a\eta(t)\right].$$

Hence, the operator $A = \exp\left[\int f(t)\partial_t dt\right]$ with $f \in L^2(R)$ is applied to the functional $\varphi$ given above, then we must have

$$A\varphi(x) = \exp[< f, \eta >]\varphi(x) = \varphi(x + f).$$

Since the algebra generated by exponential functionals are dense in the domain of differential operators of the form $A$, we conclude the proof.

It is noted that the definition and applications of the differential operators are meaningful when they are associated with $dt$.

## 4   Variational calculus.

As a significant future direction of white noise analysis, we propose variational calculus for random fields $X(C)$ indexed by a manifold $C$. There are many topics on random fields, among others we wish to discuss the innovation approach in this note. Assuming that the given field $X(C)$ is a functional of some white noise, our problem is to find the white noise by taking various variations $\delta X(C)$. Such a question arises in many applications in physics and biology, in such cases as only observed data of random phenomena are given but input information is unknown. Some results have been obtained although we are far from a general theory.

We should like to emphasize the importance of the random field from the viewpoint of information theory and of innovation theory. The main reason is that as $C$ deforms, more information is carried by $X(C)$ compared to the case where a stochastic process does as $t$ goes by. As for the choice of a manifold $C$ we require many geometric properties; in addition to the regularity we assume some convexity of $C$ for convenience. Also the way of deformation of $C$ must be much restricted.

An exploring work was done in [10] in line with the innovation approach. The basic idea is referred to that literature, and we now mention only the result.

**Theorem 2.** ([10]). Let $X(C)$ be given by

$$X(C) = \int_{(C)^n} F(C; u_1, u_2, ..., u_n) : x(u_1)x(u_2)...x(u_n) : du^n,$$

where : : means the Wick product and where $(C)$ denotes the domain enclosed by $C$. Assume further that the above expression is a canonical representation. Then, $x(s)$ with $s \in C$ is formed as the innovation from the variations $\delta X(C)$ and its conditional expectations given by $X(C')$, where $C'$ runs through the inside of $C$.

A random field $X(C)$ expressed above is a homogeneous polynomial in $x$ of degree $n$. A linear combination of homogeneous polynomials is simply a polynomial. It is easy to have a generalization of Theorem 2, although the assertion is much weaker, in the following manner.

**Proposition.** Let $X(C)$ be a polynomial which is a functional of $x(s)$ with $s \in (C)$. Then, the innovation in the weak sense is obtained from the variations $\delta X(C)$.

The proof easily follows by using the idea of the proof of Theorem 2.

We have hopes that the innovation approach to random fields would develop since the theory has profound probabilistic meaning and has many interesting applications.

**Acknowledgement.** The author is grateful to Professor C. Bernido who is the organizer of the exciting International Jagna Conference and to the support given by the research project "Quantum Information Theoretical Approach to Life Science" at Meijo University.

# References

[1] P. Lévy, Sur la variation de la distribution de l'electricite sur un conducteur la surface se deforme. Bull de la Societe Mathematique de France. 46. (1918), 36-68

[2] P. Lévy, Processus stochastiques et mouvement brownien. Gauthier-Villars, 1948.

[3] P. Lévy, Problèmes concréts d'analyse fonctionnelle. Gauthier-Villars, 1951.

[4] H. Ezawa, J. R. Klauder and L. A. Shepp, A path space picture for Feynman-Kac averages. Ann. of Physics, 88 (1974), 588 - 620.

[5] T. Hida, Analysis of Brownian functionals. Carleton Math. Notes #13, 1975.

[6] T. Hida, H.-H. Kuo, J. Potthoff and L. Streit, White noise. - An infinite dimensional calculus. Kluwer Academic Pub. 1993.

[7] H.-H. Kuo, White Noise Distribution Theory. CRC Press. 1996.

[8] N. Obata, White Noise Calculus and Fock space. Lecture Notes in Math. no. 1577. Springer-Verlag, 1994.

[9] K. Saitô, A group generated by the Lévy Laplacian and the Fourier- Mehler transform. Stochastic analysis on infinite dimensional spaces. ed. H. Kunita and H.-H., Kuo, Longman Sci. & Tech. 1994, 274-288.

[10] Si Si, Innovation of some random fields. to appear in the Proceedings of the International Conference on Probability and its Applications, KAIST, Taejon, 1998.

# SOME OBSERVATIONS ON STOCHASTIC VS. QUANTUM DYNAMICS - FOLLOWING EZAWA, KLAUDER & SHEPP

LUDWIG STREIT

GFM - Universidade de Lisboa
CCM - Universidade da Madeira
BiBoS - Universität Bielefeld

## 1 Ezawa and (some of) the consequences

My first acquaintance with Hiroshi Ezawa coincided with my first acquaintance with his work on a path space picture for Euclidean quantum dynamics [1], it was also the moment of my first acquaintance with Japan. The impact of the latter two is reflected in most of my subsequent work, the impact of my acquaintance with Hiroshi is of course more personal, less visible, but those who know him will no doubt understand my delight in having known him for more than two decades, marred only by the fact that one meets too infrequently - here my thanks go to all the Bernidos for the present opportunity.

The message of Ezawa, Klauder, and Shepp in [1] was contrary to the prevalent method of imposing quantum dynamics on trajectories by killing them; more naturally, quantum dynamics should translate itself into specific dynamical evolutions of the paths.

The following section is intended to review some of the work that this observation has triggered. Of course on the Bernido estate sums over histories tend to be Feynmanian, so the third section will be devoted to these, in particular since some of my earliest steps in this direction were taken at Gakushuin University.

## 2 Distorted Brownian Paths

### 2.1 Quantum Mechanics

In the mid-seventies the prevalent path space approach to quantum dynamics such as encoded in

$$H = -\frac{1}{2m}\Delta + V \tag{1}$$

was via Brownian paths

$$x(t) = x(0) + \sqrt{\frac{\hbar}{m}}B(t)$$

9

i.e. the paths are killed with probability $V(x(\tau)) d\tau$. Instead of this, Ezawa, Klauder, and Shepp proposed to deform the paths according to the dynamics $V$

$$E(F) = \int F[\Re x] d\mu$$

with $y = \Re x$ obeying a stochastic differential equation

$$dy = dx + \beta_V(y) dt, \tag{2}$$

the quantum dynamics is now encoded in the drift term $\beta$. $\beta$ depends on the potential $V$, and in fact more directly on the ground state of the dynamics [3]. Standard existence theorems for such equations require no-more-than-linear growth of $\beta$ which translates itself into no-more-than-quadratic growth of $V$, an unfortunate limitation, but I still recall John Klauder pointing out in his 1975 Kyoto lecture that this seemed not so much a shortcoming of the approach but rather a lack of appropriate mathematical tools.

This was indeed the case, and in the following years the following pattern emerged of a ground state representation for Schrödinger theory

| | *Schrödinger Representation* | *Ground State Repn.* |
|---|---|---|
| State Space $\mathfrak{H}$ | $\mathfrak{H} = L^2(R^n, d^n x)$ | $\mathfrak{H}\check{} = L^2(R^n, d\nu)$ |
| Ground State $\Omega$ | $\Omega(x) = <x\|\Omega>$ | $\Omega\check{}(x) = 1$ |
| States $\psi$ | $\psi(x) = <x\|\psi>$ | $\psi\check{}(x) = \frac{<x\|\psi>}{<x\|\Omega>}$ |
| Hamiltonian $H$ | $H = \nabla_x^* \nabla_x + V(x)$ | $H\check{} = \nabla_x^* \nabla_x$ |
| Energy Form $\varepsilon(\psi)$ | $\varepsilon(\psi) = (\nabla\psi, \nabla\psi) + (\psi, V\psi)$ | $\varepsilon(\psi\check{}) = \int (\nabla\psi\check{})^2 d\nu$ |
| = Dirichlet Form for: | Brownian m. with killing | distorted Brownian m. |

Via the choice of the "ground state measure"

$$d\nu(x) = <x\|\Omega>^2 d^n x \tag{3}$$

the two representations, whenever both exist, are unitarily equivalent. Note the universal form of the Hamiltonian $H = \nabla^* \nabla$; for any ground state measure $\nu$, it is the latter which contains exclusively the dynamical information.

The Dirichlet form $\varepsilon$ gives rise to a diffusion process which solves the stochastic differential equation of Ezawa, Klauder, and Shepp;

$$dy = \beta(y)dt + dB,$$

where the drift $\beta$ is given in terms of the ground state wave function:

$$\beta(x) = \nabla_x \log \Omega^2(x) \tag{4}$$

This is dynamics in terms of the ground state. What is its scope? It turns out that we are confronted with a vast extension of Schrödinger theory. While formally we can recuperate the potential from the ground state through the eigenvalue relation

$$V(x) = \frac{\Delta_x \Omega(x)}{\Omega(x)}, \tag{5}$$

the ground state representation extends to $\Omega(x)$, and hence to perfectly welldefined dynamics, for which $V$ will not be a valid perturbation of the free Hamiltonian, or will not even exist! Ground states are always smoother than the corresponding potentials, so that the former may survive in limiting cases where the latter fail to exist. All this is treated in detail in [3], here we only mention zero range, multiparticle "pseudopotentials" as an example which fits nicely into this scheme.

Of course various questions come to mind, such as

- Whether, and in what sense, will similar ground states produce similar Hamiltonians?

$$\Omega_\varepsilon \to \Omega \overset{?}{\Longrightarrow} H_\varepsilon \to H$$

In [4], [5] one finds criteria which ensure the strong resolvent convergence of the Hamiltonians.

- What if $\Omega$ has nodes which would appear to cause problems in (4) and in (5)? This phenomenon is in fact closely related to the work of Ezawa, Klauder, and Shepp in their companion paper [2]. Depending on the nature of the nodes, these may or may not constitute barriers for the diffusion, and for the quantum mechanical motion; these barriers could be described completely in [6].

- Thinking of Brown as he observed random motions in his microscope one might wonder about the following: assume there is a black spot in the field of vision and we can observe the diffusion particles only as they move outside, and as they enter and leave the region under the black spot. Can we find out in this way what the evolution is in the invisible region? An affirmative answer is given in [7].

## 2.2   Quantum Field Theory

### 2.2.1   The Vacuum Density

For field theories the analogue of (3) would be

$$dv(\varphi) = \rho(\varphi) d^\infty\varphi \tag{6}$$

a measure on field configurations, with vacuum density $\rho$. Unfortunately, $d^\infty\varphi$ refuses to exist; but even a more modest attempt

$$dv(\varphi) = \rho(\varphi) dv_0(\varphi) \tag{7}$$

- where now $\rho$ would be the density of the physical vacuum with respect to the (well defined) free vacuum measure $v_0$ - will not work: Haag's theorem about inequivalence of free and interacting fields excludes the existence of such a density. In fact the saga of constructive quantum field theory was in essence the quest for inequivalent measures $v$

$$\langle\Omega|\phi(x_1)\dots\phi(x_n)|\Omega\rangle \overset{!}{=} \int dv(\phi)\phi(x_1)\dots\phi(x_n) \tag{8}$$

on infinite dimensional spaces.

What however if, guided by singular measures arising from generalized functions such as e.g. Dirac's $\delta$−function, one would look for measures arising from positive *generalized* functions in infinite dimensional analysis? T. Hida's program of "White Noise Analysis" [8] provides such a framework with Gaussian white noise as independent coordinates. Gaussian white noise gives of course rise to a Gaussian measure $\mu$ and we might begin our quest for vacuum densities in the corresponding $L^2$ space

$$L^2(d\mu) \equiv (L^2)$$

In the light of the above however, this will not be rich enough, we shall need to embed $(L^2)$ in a space of generalized functions of white noise (cf. [8] for details of such a construction):

$$(S) \subset L^2(d\mu) \subset (S)^*.$$

Fortunately, by a theorem of Kondratiev and of Yokoi, and just as in finite dimensional distribution theory, all positive generalized functions turn out to be measures, hence candidates for vacuum densities:

$$dv(\phi) = \rho(\phi)dv_0(\phi) \text{ with } \rho \in (S)^*? \tag{9}$$

Indeed, physical vacuum densities (Euclidean and Minkowski, $P(\varphi)_2$, Sine-Gordon, etc.) are Hida distributions:

$$\langle 0|F(\varphi)|0\rangle =< \rho, F >$$

for suitable positive $\rho \in (S)^*$, cf. [9]-[13]. Proofs use the "Froehlich bounds" on moments

$$\langle 0 \,|\varphi^n(f)|\, 0 \rangle = O((n!)^{1/2} \,|f|_p^n)$$

In particular such an estimate holds for canonical Bose fields if they obey a $\phi$-bound

$$\pm\varphi(f) \leq aH + b|f|_p^2 + c, \text{ with } a, b, c \geq 0. \tag{10}$$

### 2.2.2  Dynamics in terms of the Vacuum

Quantum field dynamics in terms of the ground state goes back to the work of Coester and Haag [24], and of Araki [23], in 1960.

Our goal was, as in the quantum mechanical case, to define the Hamiltonian as a Dirichlet form defined on smooth white noise functionals, and in terms of a generalized vacuum density $\rho \in (S)^*$, as follows

$$\varepsilon(\Psi) = \langle \Psi \,|H|\, \Psi \rangle = < \rho, |\nabla\Psi|^2 > \qquad D(\varepsilon) = (S)$$

This program was indeed carried out in [11], [12], including the realization of the fields as infinite dimensional diffusions.

What about the field theory counterpart(s) of the stochastic evolution equation (2) proposed by Ezawa, Klauder, and Shepp? This comes up in the stochastic quantization program with all its difficulties, but a Gaussian "toy model" is quite tractable [14].

## 3   Feynman Integrals

To compute (time ordered) quantum mechanical expectations, Feynman proposes an average over paths

$$\langle F \rangle_T = \int d^\infty x(t) e^{\frac{i}{\hbar}S} F[x].$$

These objects are intrinsically much trickier than the ones discussed before. There is no measure $I$ here, not even a complex one, for which we might have

$$\langle F \rangle_T = \int d^\infty x(t) e^{\frac{i}{\hbar}S} F[x] = < I, F >$$

But what if there were a generalized function in $(S)^*$, not necessarily a measure, such that

$$\langle F \rangle_T = \int d^\infty x(t) e^{\frac{i}{\hbar}S} F[x] = < I, F >, \ I \in (S)^*?$$

This indeed works beautifully. The basic ideas are

- Brownian paths

$$x(t_0 + \tau) = x_0 + \left(\frac{\hbar}{m}\right)^{1/2} B(\tau) \tag{11}$$

(Note that the "velocity" of Brownian motion is white noise $\omega$: $B(\tau) = \int_0^\tau \omega(s)ds$.)

- the free Feynman integrand should then be

$$I_0(x, t \mid x_0, t_0) \tag{12}$$

$$= N \exp\left(\frac{i+1}{2} \int_{\mathbf{R}} \omega^2(\tau) \, d\tau\right) \delta(x(t) - x). \tag{13}$$

where the exponential contains the $\frac{i}{\hbar}$ times the free action, while the $\delta$-function serves to pin down the final position of the paths; the extra seemingly spurious real term in the exponent actually is crucial to compensate the Gaussian fall-off of the white noise measure.

- To tackle this expression it is helpful to calculate its "$T$-transform" defined for general(ized) white noise functionals $\Phi$ as follows

$$T\Phi(\Phi) = E\left(\Phi(\omega)e^{i\int \Phi(t)\omega(t)dt}\right),$$

i.e. we should calculate an (infinite dimensional) Gauss-Fourier transform.

This can be done rather straightforwardly for $I_0$ and gives, with $m = \hbar = 1$,

$$TI_0(\Phi) = \frac{1}{(2\pi i \, |t - t_0|)^{\frac{d}{2}}} \exp\left[-\frac{i}{2} \int_{\mathbf{R}} \xi^2(\tau) \, d\tau \right. \tag{14}$$

$$\left. - \frac{1}{2i \, |t - t_0|} \left(\int_{t_0}^t \xi(\tau) \, d\tau + x - x_0\right)^2\right],$$

In fact

$$TI_0(\xi) \equiv K_0^{(\xi)}(x, t | x_0, t_0) \tag{15}$$

has a physical meaning, it obeys Schrödinger equation

$$\left(i\partial_t + \frac{1}{2}\triangle_d - \dot{\xi}(t) \cdot x\right) K_0^{(\xi)}(x, t | x_0, t_0) = 0 \tag{16}$$

with the initial condition

$$\lim_{t \searrow t_0} K_0^{(\xi)}(x, t | x_0, t_0) = \delta(x - x_0).$$

## 3.1    The interactions

The big question is: Which potentials are admissible to be included in this framework? Over the recent years various classes of admissible potentials have been identified [15]-[22]. The most recent, and rather surprising, class is given in the form

$$V(x) = \int_{\mathbf{R}^d} e^{\alpha \cdot x} dm(\alpha). \tag{17}$$

where $m$ is any complex measure with

$$\int_{\mathbf{R}^d} e^{C|\alpha|} d|m|(\alpha) < \infty, \quad \forall C > 0. \tag{18}$$

**Example 1** Dirac measure $m(\alpha) := g\, \delta_a(\alpha)$, $g \in \mathbf{R}$., i.e. $V(x) = g\, e^{ax}$. Likewise , e.g. $\sinh(ax)$, $\cosh(ax)$.

**Example 2** Morse potential $V(x) := g(e^{-2ax} - 2\gamma e^{-ax})$ with $g, a, x \in \mathbf{R}$ and $\gamma > 0$

**Example 3**    A Gaussian density gives $V(x) = g e^{bx^2}$ with $b \in \mathbf{R}$.

**Example 4** entire functions of arbitrary high order of growth are in this class. $m(\alpha) := \Theta(\alpha) \exp\left(-k\alpha^{1+1/n}\right)$ with $n$, $k > 0 \Rightarrow V$ is entire of order $1 + n$.

The construction of the Feynman integrand is perturbative

$$I = I_0 \cdot \exp\left(-i \int_{t_0}^{t} V(x(\tau))\, d\tau\right) \tag{19}$$

$$= \sum_{n=0}^{\infty} \frac{(-i)^n}{n!} \int_{[t_0,t]^n} \int_{\mathbf{R}^{dn}} I_0 \cdot \prod_{j=1}^{n} e^{\alpha_j \cdot x(\tau_j)} \prod_{j=1}^{n} dm(\alpha_j)\, d^n\tau \tag{20}$$

**Theorem 5** *Let $V$ be as above. Then*

$$I := \sum_{n=0}^{\infty} \frac{(-i)^n}{n!} \int_{[t_0,t]^n} \int_{\mathbf{R}^{dn}} I_0 \cdot \prod_{j=1}^{n} e^{\alpha_j \cdot x(\tau_j)} \prod_{j=1}^{n} dm(\alpha_j)\, d^n\tau \tag{21}$$

*exists as a generalized white noise functional.*

The proof can be found in [22]. The main issue from the physics point of view is of course whether the corresponding Feynman integral solves the Schroedinger equation as in the free case (16):

**Theorem 6** *Let V be as above. Then the T-transform of I solves the Schrödinger equation for all $x, x_0, t_0 < t$*

$$\left( i\frac{\partial}{\partial t} + \frac{1}{2}\Delta_d - gV(x) - x \cdot \dot{\xi}(t) \right) K^{(\xi)}(x, t \mid x_0, t_0) = 0. \tag{22}$$

*with initial condition*

$$\lim_{t \searrow t_0} K^{(\xi)}(x, t \mid x_0, t_0) = \delta(x - x_0) \tag{23}$$

The proof is by verification, and in fact the result extends to time dependent potentials. We shall not reproduce here the explicit, but longish form of the propagator for the general case. Instead, because of its interest in applications we present a particular example in more detail.

## 3.2   The Morse Potential

Its Hamilton operator is

$$H := -\frac{1}{2}\Delta + g\left(e^{-2ax} - 2\gamma e^{-ax}\right) \tag{24}$$

**Remark 1** *H is essentially self-adjoint for $g \geq 0$ and it is not essentially self-adjoint for $g < 0$.*
*The Green function, the eigenvectors and the discrete eigenvalues are not analytic in $g$.*

The propagator, with $\xi \equiv 0$, is in this case

$$K(x, t \mid x_0, t_0) \tag{25}$$

$$= K_0(x, t \mid x_0, t_0)$$

$$\cdot \sum_{n=0}^{\infty} \frac{(-ig)^n}{n!} (t - t_0)^n \sum_{j_1,\ldots,j_n=1}^{2} (-2\gamma)^{2n - \sum_{k=1}^{n} j_k} \tag{26}$$

$$\int_{[0,1]^n} \exp\left\{-a \sum_{l=1}^{n} j_l \left(\sigma_l x + (1 - \sigma_l) x_0\right)\right\}$$

$$\exp\left\{-\frac{i}{2} (t - t_0) a^2 \sum_{l=1}^{n} \sum_{k=1}^{n} j_k j_l \left[\sigma_j \sigma_k - \sigma_j \wedge \sigma_k\right]\right\} d^n \sigma.$$

It is not hard to verify that, in spite of the above remark, this is a convergent series! In fact it is implicit in our construction (21) that the propagators for all the potentials (17) admit convergent perturbation series *for their propagators*. It is an easy exercise to verify that this is not the case for the corresponding "Euclidean" heat equations.

Indications are that once again new tools are called for: a better understanding of oscillatory integrals is needed to supplant the intuition carried over from the perturbative analysis of self-adjoint operators.

* * *

Here ends this account of some work which owes much to Hiroshi Ezawa, with the hope that paths will cross again, for inspiration and for good friendship.

**Acknowledgments:** This work has been done under a project of PRAXIS XXI with support from FEDER. Its presentation was made possible by a travel grant of the Alexander-von-Humboldt-Foundation and by the hospitality of the RCTP.

## References

[1] H. Ezawa H , . Klauder J. R, . Shepp L. A (1974) A Path Space Picture for Feynman-Kac Averages *Ann. Phys.* **88**, 588-620.

[2] Ezawa H., Klauder J. R., Shepp L. A. (1975) Vestigial Effects of Singular Potentials in Diffusion Theory and Quantum Mechanics *J. Math. Phys.* **16**, 783-799

[3] Albeverio S. , Hoegh-Krohn R., Streit L. (1977) Energy Forms, Hamiltonians, and Distorted Brownian Paths *J. Math. Phys.* **18**, 907

[4] Albeverio S. , Hoegh-Krohn R., Streit L. (1980) Regularization of Hamiltonians and Processes J. *Math. Phys.* **21**, 1636.

[5] Albeverio S., Kusuoka S., Streit L. (1986) Convergence of Dirichlet Forms and Associated Schrödinger Operators S. *J. Funct. Anal.* **68**, 130.

[6] "Capacity and Quantum Mechanical Tunneling" (with S. Albeverio, M. Fukushima, W. Karwowski) - *Comm. math. Phys.* **81**, 501 (1981).

[7] "An Inverse Problem for Stochastic Differential Equations" (with S. Albeverio, Ph. Blanchard, S. Kusuoka) - *J. Stat. Phys.* **57**, 347 (1989).

[8] Hida, T., Kuo, H.-H., Potthoff, J. and Streit, L. (1993) *White Noise - An Infinite Dimensional Calculus.* Dordrecht: Kluwer-Academic 1993.

[9] Hida, T., Potthoff, J. and Streit, L. (1988) Dirichlet Forms and White Noise Analysis *Comm. math. Phys.* **116**, 235.

[10] Albeverio S., Hida, T., Potthoff, J., Streit, L. (1989) The vacuum of the Hoegh-Krohn model as a generalized white noise functional *Phys. Lett.* **B 217**, 511.

[11] Albeverio S., Hida, T., Potthoff, J., Roeckner M., Streit, L. (1990) Dirichlet Forms in Terms of White Noise Analysis I - Construction and QFT Examples *Rev. Math. Phys.* **1**, 291

[12] Albeverio S., Hida, T., Potthoff, J., Roeckner M., Streit, L. (1990) Dirichlet Forms in Terms of White Noise Analysis II - Construction of Infinite Dimensional Diffusions and QFT Examples *Rev. Math. Phys.* **1**, 313

[13] T., Potthoff, J., Streit, L. (1993) Invariant states on random and quantum fields: $\phi$ bounds and white noise analysis. *J. Funct. Anal.* **111**, 295-311.

[14] Hida T., Streit L. (1977) On Quantum Theory in Terms of White Noise *Nagoya Math. J.* **68**, 21.

[15] Hida T., Streit L. (1983) Generalized Brownian Functionals and the Feynman Integral *Stoch. Proc. Appl.* **16**, 55.

[16] de Faria M. , Potthoff J., Streit L. (1991) The Feynman Integrand as a Hida Distribution *J. Math. Phys.* **32**, 2123

[17] Khandekar D. C., Streit L. (1992) Constructing the Feynman Integrand *Annalen d. Physik* **1**, 49.

[18] Lascheck A., Leukert P., Streit L., Westerkamp W. (1993) Quantum Mechanical Propagators in Terms of Hida Distributions ), *Rep. Math. Phys.* **33**, 221-232.

[19] Lascheck A., Leukert P., Streit L., Westerkamp W. (1994) More about Donsker's delta function, *Soochow Math. J.* **20**, 401-418.

[20] Streit L. The Feynman Integral. Answers and Questions. Proc. Opening Conference RCTP , Bohol, Philippines

[21] Grothaus M., Khandekar D. C., Silva J. L., Streit L. (1997) The Feynman Integral for time-dependent anharmonic oscillators *Journ. Math. Phys.* **38**(6), 3278-3299.

[22] Kuna T., Streit L., Westerkamp W. (1997) Feynman Integrals for a Class of Exponentially Growing Potentials. UMa preprint 19/97, to appear in *J. Math. Phys.*

[23] Araki H. (1960) Hamiltonian Formalism and the Canonical Commutation Relations in Quantum Field Theory. *Journal of Math. Physics* 1, 492-504.

[24] Coester F., Haag R. (1960) Representation of States in a Field Theory with Canonical Variables. *Physical Review*, **117**,1137-1145.

# ON A KUBO NOISE ASSOCIATED WITH A MULTIDIMEN-SIONAL STATIONARY CURVE IN A HILBERT SPACE

Dedicated to Professor Hiroshi Ezawa on his 65th birthday

YASUNORI OKABE

Department of Mathematical Engineering and Information Physics
Graduate School and Faculty of Engineering, University of Tokyo
Bunkyo-Ku, Tokyo 113, JAPAN

Abstract    This is a theory for a multi-dimensional stationary curve moving in a Hilbert space. The main purpose of this paper is to derive a Kubo noise associated with it and to show Kubo's fluctuation-dissipation theorem based upon the second KMO-Langevin equation with the Kubo noise its random force.

## 1.  INTRODUCTION

The time evolution of Brownian motion with the Alder-Wainwright effect can be governed by Stokes-Boussinesq-Langevin equation which gives a precise description of the time evolution of Ornstein-Uhlenbeck's Brownian motion considered by A. Einstein and P. Langevin[1,2,3,4,5,6,7,8,9]. It has been remained to clarify mathematically a physical structure of a random force in the above Stokes-Boussinesq-Langevin equation to which Kubo's linear response theory[10,11,12] is applied.

With an aim to clarify a mathematical structure of a random force in Kubo's linear response theory, we have shown[13] that the random force in Kubo's linear response theory can be mathematically clarified as a Kubo noise. And we have shown[14] that we can introduce two kinds of random forces—a white noise and a Kubo noise—as a random force in Stokes-Boussinesq-Langevin equation and then found that the one-dimensional stochastic process as the solution to each Stokes-Boussinesq-Langevin equation has both the stationarity and the T-positivity.

As a natural generalization of Stokes-Boussinesq-Langevin equation, we have characterized[13] analytically a mathematical structure of any one-dimensional stationary process with T-positivity and then shown that the time evolution of such a stationary process can be written by two kinds of equivalent KMO-Langevin equations: one is the first KMO-Langevin equation whose random force is a white noise and the other is the second KMO-Langevin equation whose random force is a Kubo noise. By a Kubo noise associated with the one-dimensional stationary curve $\mathbf{A} = (A(t); t \in \mathbf{R})$ in a Hilbert space $\mathcal{H}$, we mean an $\mathcal{H}$-valued stationary tempered distribution $\mathbf{I} = (I(\varphi); \varphi \in \mathcal{S}(\mathbf{R}))$[15] such that

$$(1) \qquad A(\varphi) = \lim_{\epsilon \downarrow 0} \frac{1}{\sqrt{2\pi}} \int_0^\infty e^{-\epsilon t} R(t) I(\varphi(\cdot + t)) dt,$$

where $R = (R(t); t \in \mathbf{R})$ is the correlation matrix function for the stationary curve $\mathbf{A}$ and $(A(\varphi); \varphi \in \mathcal{S}(\mathbf{R}))$ is the $\mathcal{H}$-valued stationary tempered distribution defined by

$$(2) \qquad A(\varphi) = \int_{\mathbf{R}} \varphi(t) A(t) dt.$$

Moreover, we have proved[13] two kinds of fluctuation-dissipation theorems based upon the above KMO-Langevin equations. In particular, a mathematical and physical embodyment for Kubo's fluctuation-dissipation theorem can be given based upon the second KMO-Langevin equation as the time evolution of the stationary curve.

On the other hand, we have discussed[14] a modeling problem whether either of a white noise and a Kubo noise in Stokes-Boussinesq-Langevin equation is adequate as a physical random force for Brownian motion with the Alder-Wainwright effect and stated a conjecture that comes from the fluctuation-dissipation theorem based upon the first Stokes-Boussinesq-Langevin equation.

Apart from a qualitative property of T-positivity, we have treated[16] any one-dimensional stationary curve $\mathbf{A}$ in a Hilbert space. Under some conditions for the covariance function $R$ of the stationary curve $\mathbf{A}$, we have proved a representaion theorem for the complex mobility function by using the theory[17] of Mori's Brownian motion which can be applied only for the stationary process whose sample paths are differentiable. By introducing the Kubo noise associated with the stationary curve $\mathbf{A}$, we have derived the second KMO-Langevin equation with Kubo's fluctuation-dissipation theorem.

We have treated[18] any multi-dimensional stationary curve $\mathbf{A}$ in a Hilbert space and proved a representation theorem for the complex mobility matrix function under certain general conditions for the covarince matrix function.

As a continuation of the previous paper[18], we shall in this paper introduce the Kubo noise associated with the multi-dimensional stationary curve $\mathbf{A}$ under certain stronger conditions than those in the previous paper[18]. Moreover, we shall derive the second KMO-Langevin equation and then prove Kubo's fluctuation-dissipation theorem.

## 2. COMPLEX MOBILITY MATRIX FUNCTION

Let $\mathcal{H}$ be any Hilbert space over complex field $\mathbf{C}$ and $\mathbf{A} = (A(t); t \in \mathbf{R})$

any $d$-dimensional stationary curve in $\mathcal{H}$ with covariance matrix function $R = (R(t); t \in \mathbf{R})$ such that

(3)                           $R(s - t) = (A(s), \,{}^t A(t))$        $(s, t \in \mathbf{R})$.

We define the complex mobility matrix function $[R]$ defined on $\mathbf{C}^+$ by

(4)                           $[R](\zeta) \equiv \dfrac{1}{2\pi} \displaystyle\int_0^\infty e^{i\zeta t} R(t) dt.$

We shall recall briefly the results in the previous paper[18]. We assume the following conditions:

(R.1)                           $R(0) \in \mathrm{GL}(d; \mathbf{C})$

(R.2)                           $D \equiv \lim_{\eta \downarrow 0}[R](i\eta)$ exists.

Under these conditions (R.1) and (R.2), we have proved the following theorem.

**Theorem 2.1**[18]    *There exists a triple $[\alpha, \beta, \kappa]$ such that*

(i)       $[R](\zeta) = Z(\zeta)^{-1} \dfrac{\alpha}{\sqrt{2\pi}}$      $(\zeta \in \mathbf{C}^+),$

*where*

(ii)       $\alpha$ *is a positive definite $d \times d$-matrix $(\in \mathrm{GL}(d; \mathbf{C}))$*

(iii)       $\beta$ *is a $d \times d$-matrix $(\in \mathrm{GL}(d; \mathbf{C}))$*

(iv)       $\kappa = (\kappa_{pq})_{p,q=1,\cdots,d}$ *is a $d \times d$-matrix valued signed measure on $\mathbf{R}$ such that*

(a)       $\sum_{p,q=1}^d c_p \overline{c_q} \kappa_{pq}$ *is a nonnegative Borel measure on $\mathbf{R}$   for $\forall c_p \in$ $\mathbf{C}$ $(1 \leq p \leq d)$*

(b)        $\int_{\mathbf{R}} \frac{1}{1+\lambda^2} \sum_{p,q=1}^{d} c_p \overline{c_q} \kappa_{pq}(d\lambda) < \infty \quad for\ \forall c_p \in \mathbf{C}\ (1 \le p \le d)$

(c)        $\kappa^* = \kappa$

(d)        $\lim_{\epsilon \downarrow 0} \int_{\mathbf{R}} \frac{1}{\pi} \frac{\epsilon}{\lambda^2+\epsilon^2} \kappa(d\lambda) = \frac{1}{\sqrt{2\pi}} (\beta\alpha + (\beta\alpha)^*)$

(e)        $\lim_{\eta \downarrow 0}(\lim_{\epsilon \downarrow 0} \int_{\mathbf{R}} (\frac{1}{\lambda-i\eta} - \frac{1}{\lambda-i\epsilon}) \kappa(d\lambda)) = 0$

(v)        $Z(\zeta) = (\beta - i\zeta - i\zeta \lim_{\epsilon \downarrow 0} \int_{\mathbf{R}} \frac{1}{(\lambda-\zeta-i\epsilon)(\lambda-i\epsilon)} \kappa(d\lambda)) \cdot (\sqrt{2\pi}\alpha)^{-1} \in$
GL$(d; \mathbf{C})$   $for\ \forall \zeta \in \mathbf{C}^+$

(vi)        $\sup_{\eta>0} \int_{\mathbf{R}} \mathrm{trace}(Z(\xi+i\eta)^{-1}\alpha + (Z(\xi+i\eta)^{-1}\alpha)^*)d\xi < \infty.$

The triple $[\alpha, \beta, \kappa]$ can be obtained from the correlation matrix function $R$ as follows.

**Theorem 2.2[18]**

(i)        $\alpha = \frac{R(0)}{\sqrt{2\pi}}$

(ii)        $\beta = R(0)D^{-1}$

(iii)        $\kappa(d\lambda) = \frac{1}{(2\pi)^2} R(0)\lim_{\eta \downarrow 0}([R]\,(\xi+i\eta)^{-1} + ([R]\,(\xi+i\eta)^{-1})^*)R(0)d\xi$
$in\ \mathcal{D}'(\mathbf{R})$

## 3.   KUBO NOISE and FLUCTUATION-DISSIPATION THEOREM

Similarly as in the previous paper[18], we shall assume the following conditions (R.3) and (R.4) concerning the complex mobility matrix function.

(R.3)        There exists a null set $\Lambda$ in $\mathbf{R} - \{0\}$ such that

$$[R]\,(\xi + i0) \equiv \lim_{\epsilon \downarrow 0}[R](\xi + i\epsilon) \text{ exists for } \forall \xi \in \Lambda.$$

(R.4)        There exist a positive number $c_1$ and a positive integer $m_1$ such that

$$\|\,[R]\,(\zeta)\| \geq c_1(1 + |\zeta|^{m_1})^{-1} \qquad \text{for } \forall \zeta \in \mathbf{C}^+.$$

We denote by $(E(\xi); \xi \in \mathbf{R})$ the decomposition of identity associated with the stationary curve $\mathbf{A}$:

(5)                          $$A(t) = \int_{\mathbf{R}} e^{-it\xi} dE(\xi)A(0).$$

By using this decomposition of identity, we can define a random tempered distribution $\mathbf{I} = (I(\varphi); \varphi \in \mathcal{S}(\mathbf{R}))$ by

(6)                  $$I(\varphi) \equiv \frac{1}{\sqrt{2\pi}} \int_{\mathbf{R}} \hat{\varphi}(\xi)[R](\xi + i0)^{-1} dE(\xi)A(0).$$

We note that the tempered stationary distribution $(A(\varphi); \varphi \in \mathcal{S}(\mathbf{R}))$ defined in (2) can be rewritten into

(7)                          $$A(\varphi) = \int_{\mathbf{R}} \hat{\varphi}(\xi) dE(\xi)A(0).$$

Similarly as in Theorem 6.1 in the previous paper[16], we can show

**Theorem 3.1**  *Let us suppose that there exist a positive number $c_2$ and a positive integer $m_2$ such that*

$$\|[R](\zeta)\| \leq c_2(1 + |\zeta|^{m_2}) \qquad (\zeta \in \mathbf{C}^+).$$

*Then the random tempered distribution $\mathbf{I}$ defined in (6) becomes the Kubo noise associated with the stationary curve $\mathbf{A}$.*

**Theorem 3.2**  *Under the same condition as in Theorem 3.1, the spectral measure of the tempered stationary distribution $\alpha\mathbf{I}$ is $\kappa$:*

$$(\alpha I(\varphi), \, {}^t(\alpha I(\varphi))) = \int_{\mathbf{R}} |\hat{\varphi}(\xi)|^2 \kappa(d\xi) \qquad (\varphi \in \mathcal{S}(\mathbf{R})).$$

Moreover, similarly as in Theorem 7.1 in the previous paper[16], we can derive the second KMO-Langevin equation describing the time evolution of $\mathbf{A}$.

**Theorem 3.3**  *As $\mathcal{H}$-valued tempered distributions, we have*

$$(8) \qquad\qquad \dot{A} = -\beta A - \lim_{\epsilon \downarrow 0} \gamma_\epsilon * \dot{A} + \alpha I,$$

*where $\gamma_\epsilon$ is the tempered distribution defined by*

$$(9) \qquad \gamma_\epsilon \equiv \frac{1}{2\pi} \textit{Fourier transform of } \Big( \int_{\mathbf{R}} \frac{1}{(\lambda - \cdot - i\epsilon)(\lambda - i\epsilon)} \kappa(d\,\lambda) \Big).$$

By virtue of Theorem 3.3, we can interprete Theorem 2.2 in section 2 from the pointview of fluctuation-dissipation theorem. Since $\beta$ becomes a part of drift coefficients in the second KMO-Langevin equation (8), we find that relation (ii) in Theorem 2.2 is the very Einstein relation. On the other hand, it follows from Theorems 3.2 and 3.3 that relation (iii) in Theorem 2.2 is the very Kubo's second kind of fluctuation-dissipation theorem[10,11,12].

## 4.  KUBO NOISE (2)

In this section, under some different conditions from (R.3) and (R.4), we shall derive the Kubo noise associated with the multi-dimensional stationary curve **A**. For that purpose, we define a matrix function $\Theta$ deined on $\mathbf{C}^+$ by

$$(10) \qquad\qquad \Theta(\zeta) \equiv [R](\zeta)([R](\zeta)^*)^{-1} \qquad (\zeta \in \mathbf{C}^+)$$

and for each $\xi \in \mathbf{R}$ a matrix function $\Theta_\xi$ defined on $\mathbf{R}$ by

$$(11) \qquad\qquad \Theta_\xi(\eta) \equiv \Theta(\xi + i\eta) \qquad (\eta \in \mathbf{R}).$$

Differently from (R.3) and (R.4), we shall assume the following conditions (R.5)–(R.9):

(R.5) $\qquad\qquad \Theta(\xi) \equiv \lim_{\eta \downarrow 0} \Theta(\xi + i\eta)$ exists.

(R.6) $\qquad\qquad \Theta(\xi;t) \equiv \lim_{\eta \downarrow 0} \Theta_\eta(\xi + i\eta)$ exists for $\forall \xi \in \mathbf{R}, \forall t \in \mathbf{R}$.

It is to be noted that $\Theta(\xi;0) = R(\xi)$ for $\forall \xi \in \mathbf{R}$.

(R.7) $\qquad\qquad \Theta(\xi;t) = R(\xi) \qquad$ for $\forall \xi \in$ supp $\kappa, \forall t \in \mathbf{R}$.

(R.8) $\qquad \exists C > 0$ such that $\left\| \frac{\Theta(\xi_1) - \Theta(\xi_2)}{\xi_1 - \xi_2} \right\| \leq C \quad a.e. \xi_1, \xi_2 \ (\xi_1 \neq \xi_2).$

(R.9)        $\sum_{p,q=1}^{d} c_p \overline{c_q} \kappa_{pq}(\mathbf{R}) < \infty$  for $\forall c_p \in \mathbf{C} \ (1 \le p \le d)$.

Under the above conditions (R.1),(R.2),(R.5)–(R.9), we shall show the existence of Kubo noise associated with the stationary curve $A$. For each positive $\eta$, we define a random tempered distribution $I_\eta = (I_\eta(\varphi); \varphi \in \mathcal{S}(\mathbf{R}))$ by

(12)        $I_\eta(\varphi) \equiv \dfrac{1}{\sqrt{2\pi}} \int_{\mathbf{R}} P_\eta * (\hat{\varphi}(\cdot)[R](\cdot + i\eta)^{-1}(\xi)) dE(\xi) A(0),$

where $P_\eta$ is the Cauchy kernel given by

(13)                        $P_\eta(t) \equiv \dfrac{1}{\pi} \dfrac{\eta}{t^2 + \eta^2} \quad (t \in \mathbf{R}).$

**Theorem 4.1**  *Under conditions* (R.1),(R.2),(R.5)–(R.9), *there exists a stationary random tempered distribution* $\mathbf{I} = (I(\varphi); \varphi \in \mathcal{S}(\mathbf{R}))$ *such that for*

$$s - \lim_{\eta \downarrow 0} I_\eta(\varphi) = I(\varphi) \quad \text{for } \forall \varphi \in \mathcal{S}(\mathbf{R})$$

*and* $\mathbf{I}$ *becomes the Kubo noise associated with the stationary curve* $A$.

**Theorem 4.2**  *Under the same conditions as in Theorem 4.1, the spectral measure of the tempered stationary distribution* $\alpha\mathbf{I}$ *is* $\kappa$:

$$(\alpha I(\varphi), \ ^t(\alpha I(\varphi))) = \int_{\mathbf{R}} |\hat{\varphi}(\xi)|^2 \kappa(d\xi) \qquad (\varphi \in \mathcal{S}(\mathbf{R})).$$

## ACKNOWLEDGEMENTS

This research was partially supported by Grant-in-Aid for Science Research (B) No. 07459007 and Grant-in-Aid for Exploratory Research No. 08874007, the Ministry of Education, Science, Sports and Culture, Japan and by Promotion Work for Creative Software, Information-Technology Promotion Agency, Japan.

## REFERENCES

1. Stokes, G.G. (1850) On the effect of the internal friction of fluids on the motion of pendulums. *Trans. Cambridge Phylosop. Soc.*, **9**.
2. Boussinesq, J. (1885) Sur la résistance qu'oppose un liquide indéfini en repos, sans pesanteur, au mouvement varié d'une sphère solide.

*Compte Rendus de l'Académie des Sciences,* **100**, 935-937.

3. Einstein, A. (1905) *Über* die von der molekularkinetischen Theorie der Wärme geforderte Bewegung von in ruhenden Flüssigkeiten suspendierten Teichen. *Brudes Ann.,* **17**, 549-560.

4. Langevin, P. (1908) Sur la théorie du movement brownien. *Comptes Rendus,* **146**, 530-533.

5. Uhlenbeck, G.E. and Ornstein, L.S. (1930) On the theory of the Brownian motion. *Phys. Rev.,* **36**, 823-842.

6. Alder, B.J. and Wainwright T.E. (1967) Velocity autocorrelations for hard spheres. *Phys. Rev. Lett.,* **18**, 988-990.

7. Alder, B.J. and Wainwright T.E. (1970) Decay of the velocity autocorrelation function. *Phys. Rev.,* A1, 18-21.

8. Kubo. R. (1979) Irreversible Processes and Stochastic Processes, RIMS. Kyoto, 50-93 (in Japanese).

9. Oobayashi, K., Kohno, T. and Utiyama, H (1983) Photon correlation spectroscopy of the non-Markovian Brownian motion of spherical particles. *Phys. Rev.,* A**27**, 2632-2641.

10. Kubo. R. (1957) Statistical mechanical theory of irreversible processes I, general theory and simple applications to magnetic and conduction problems. *J. Phys. Soc. Japan,* **12**, 570-586.

11. Kubo. R. (1966a) The Fluctuation-Dissipation Theorem and Brownian Motion. *1965 Tokyo Summer Lectures in Theoretical Physics,* edited by R.Kubo (Part 1, Many-Body Theory), Syokabo (Tokyo) and Benjamin (New York), 1-16.

12. Kubo. R. (1966b) The fluctuation-dissipation theorem. *Reports on Progress in Physics,* **29**, 255-284.

13. Okabe, Y. (1986a) On KMO-Langevin equations for stationary Gaussian processes with T-positivity. *J. Fac. Sci. Univ. Tokyo,* Sect IA **33**, 1-56.

14. Okabe, Y. (1986b) On the theory of the Brownian motion with Alder-Wainwright effect. *J. Statist. Phys.,* **45**, 953-981.

15. Itô, K. (1954) Stationary random distributions. *Mem. Coll. Sci. Univ. Kyoto,* **28**, 209-223.

16. Okabe, Y. (1986c) KMO-Langevin equation and fluctuation-dissipation theorem (I). *Hokkaido Math. J.,* **15**, 163-216.

17. Mori, H. (1965) Transport, collective motion and Brownian motion. *Progr. Theor. Phys.,* **33**, 423-455.

18. Okabe, Y. (1986d) KMO-Langevin equation and fluctuation-dissipation theorem (II). *Hokkaido Math. J.,* **15**, 317-355.

# QUANTUM FIELD THEORY
# AND
# STATISTICAL MECHANICS

# OPERATOR ALGEBRA METHODS IN QUANTUM PHYSICS

HUZIHIRO ARAKI

Department of Mathematics, Faculty of Science and Technology
The Science University of Tokyo
Noda-shi, Chiba-ken, 278-8510 Japan

**Abstract**    The significance of operator algebra methods in quantum physics is reviewed for theoretical physicists by examples in application to quantum field theory and quantum statistical mechanics.

## 1   Introduction

In quantum mechanics of a system of a finite degree of freedom, such as atoms and molecules, an observable is represented by an operator $Q$ on a Hilbert space $\mathcal{H}$, a (pure) state of the system is represented by a vector $\overline{\Psi}$ (of a unit length: $\|\overline{\Psi}\| = 1$), a mixture state such as Gibbs canonical ensemble by a density matrix $\rho$ on $\mathcal{H}$ (a positive operator of the unit trace: $\rho \geq 0$, $tr\, \rho = 1$ : in particular $\rho = |\overline{\Psi} >< \overline{\Psi}|$ in the case of a pure state represented by a vector $\overline{\Psi}$ as above ) and the expectation value $\Psi(Q)$ (the averaged measured value) of an observable $Q$ in a state $\overline{\Psi}$ or $\rho$ is given by

$$\Psi(Q) = (\overline{\Psi}, Q\overline{\Psi}) \qquad \text{(for a pure state)} \qquad (1.1)$$
$$= tr\,(\rho Q) \qquad \text{(for a mixture state)} \qquad (1.2)$$

where $(\xi, \eta)$ denotes the inner product of two vectors $\xi$ and $\eta$ in $\mathcal{H}$ and $trA = \sum_n (\xi_n, A\xi_n)$ for any orthonormal basis $\{\xi_n\}$ of $\mathcal{H}$.

Furthermore, the dynamics of the system is specified by a selfadjoint operator H, called the Hamiltonian, via the group of unitary operators

$$U(t) = \exp(iHt) \qquad (1.3)$$

with the real parameter t representing time, which induces the time change of an observable $Q$ in the Heisenberg picture by

$$\alpha_t(Q) = U(t)\, Q\, U(t)^*, \qquad (1.4)$$

the time change of a state in the Schrodinger picture by

$$\overline{\Psi}_t = U(t)^* \overline{\Psi} \qquad \text{(for a pure state)} \qquad (1.5)$$
$$\rho_t = U(t)^* \rho U(t) \qquad \text{(for a mixture state)} \qquad (1.6)$$

29

and the time change of the expectation value by

$$\Psi_t(Q) = \Psi(\alpha_t(Q))$$
$$= (\overline{\Psi}_t, Q\overline{\Psi}_t) \qquad \text{(for a pure state)} \qquad (1.7)$$
$$= tr(\rho_t Q) \qquad \text{(for a mixture state)}. \qquad (1.8)$$

For a system of an infinite degree of freedom, such as an infinitely extended system of a non-zero density in quantum statistical mechanics and a system of quantized fields, the above stated basic formulation of quantum theory has to be somewhat modified . We shall illustrate a necessity of modifications in terms of some examples.

To start with, let us consider the Hamiltonian operator H which is used to describe the time change of the system. For the Heisenberg model describing the interaction of spins on lattice sites, the Hamiltonian is given by

$$H_V = J \sum H_{ij}, \qquad H_{ij} = \sigma_x^{(i)}\sigma_x^{(j)} + \sigma_y^{(i)}\sigma_y^{(j)} + \sigma_z^{(i)}\sigma_z^{(j)} \qquad (1.9)$$

where $\sigma_\alpha^{(i)}, \alpha = x, y, z$ describes the spin $\frac{1}{2}$ at the lattice site $i$, $H_{ij}$ is the interaction of spins at the lattice sites $i$ and $j$, which are limited to neighbouring sites in the Heisenberg model, and the sum is over all pairs of neighbouring sites inside a region $V$. ( $J$ is a coupling constant.) If $V$ is finite, then the sum is a well-defined bounded operator but for an infinite region $V$, the sum does not converge and does not make sense. It represents the interaction energy in a region $V$. The energy of a uniform system is the energy density times the volume and hence it is infinite for an infinitely extended system (unless it is zero), which is essentially the reason for nonexistence of the operator $H_V$ for an infinite $V$.

On the other hand, the time change of an observable $Q$ can be given by

$$\alpha_t(Q) = \lim_{V \to \infty} e^{iH_V t} Q e^{-iH_V t}. \qquad (1.10)$$

While $H_V$ does not have a limit, (1.10) has a limit for a large class of infinite systems. The time change $\alpha_t$ of observables is taken to be an automorphism of the algebra of observables,as will be explained below, and it is not a unitary transformation in general. The distinction between automorphisms and unitary transformations is one of the points of modification.

In the Heisenberg approach to quantum mechanics, one starts with the canonical commutation relations (CCRs).

$$[q_j, q_k] = [p_j, p_k] = 0, \qquad [q_j, p_k] = i\delta_{jk} \qquad (1.11)$$

among coordinates $q_j$ and momenta $p_j$ of particles $j = 1, 2, ...N$. Under a mild assumption , these operators are realized by the so-called Schrödinger representation on the $L_2$ space over $\mathbb{R}^N$ (for a simplicity of notation, we are using one-dimensional space for each particle instead of the usual 3 dimensions):

$$(q_j\overline{\Psi})(x_{1,\dots,}x_N) = x_j\overline{\Psi}(x_{1,\dots,}x_N), \tag{1.12a}$$

$$(p_j\overline{\Psi})(x_{1,\dots,}x_N) = -i\frac{\partial}{\partial x_j}\overline{\Psi}(x_{1,\dots,}x_N). \tag{1.12b}$$

Namely, under a mild condition, the CCRs (1.11) have a unique irreducible representation (1.12). This uniqueness of representation no longer holds for $N = \infty$. Thus one has to choose suitable representation(s) out of an extremely large variety of representations for an infinite system. This is another point of modification.

The last situation should not be taken negatively as physically irrelevant complications. For example, it can be used positively in obtaining a deeper understanding of anticommutativity of Fermion fields at a space-like distance in contrast to commutativity for Boson fields, as will be explained in the last section. According to the special theory of relativity, two events occuring in space-like separated regions should not influence each other. Thus two observables which can be measured in space-like separated regions should be simultaneously observable and operators representing these two observables should commute with each other according to quantum theory. Bose fields commute at a space-like distance and conform with this principle. On the other hand, Fermion fields are not observables due to the following reason. Under $360^0$ rotation, there should not be any observable effect. However, Fermion fields change sign under $360^0$ rotation and hence are not observables. Thus their anticommutativity is not a contradiction with the above principle. However, one is curious why they should satisfy the anticommutativity. A deeper understanding of this can be obtained by studying different but mutually related representations of the observable algebra.

## 2  Operator Algebras

Any set $\mathfrak{A}$ of bounded linear operators (on a Hilbert space $\mathcal{H}$) which is closed under the following operations is called a (concrete) C*-algebra.

  (1) Linear combination:
   $Q_1, Q_2 \in \mathfrak{A}, \quad c_1, c_2 \in \mathbb{C} \Longrightarrow c_1Q_1 + c_2Q_2 \in \mathfrak{A}$
   ($\mathbb{C}$ denotes the set of all complex numbers.)
  (2) Multiplication:    $Q_1, Q_2 \in \mathfrak{A} \Longrightarrow Q_1Q_2 \in \mathfrak{A}$
  (3) Adjoint:   $Q \in \mathfrak{A} \Longrightarrow Q^* \in \mathfrak{A}$
  (4) Limit in norm: $\lim\limits_{n\to\infty} \|Q_n - Q\| = 0, \quad Q_n \in \mathfrak{A} \Longrightarrow Q \in \mathfrak{A}$.

   ($\|Q\| = \sup\limits_{\xi \neq 0} \frac{\|Q\xi\|}{\|\xi\|}$ where the norm $\|\xi\|$ of $\xi \in \mathcal{H}$ is $\|\xi\| = (\xi, \xi)^{1/2}$.)

In the algebraic approach to quantum theory, one assumes that the set of observables is a C*-algebra. (One may restrict the terminology "observable" to selfadjoint elements of a C*-algebra. Then one deals with a Jordan algebra.)

A representation $\pi$ of a C*-algebra $\mathfrak{A}$ on a Hilbert space $\mathcal{L}$ is a map $A \in \mathfrak{A} \to \pi(A)$ : a bounded linear operator on $\mathcal{L}$, preserving the algebraic operation (1), (2), (3) on $\mathfrak{A}$ (i.e. a * homomorphism):

$$\pi(c_1 Q_1 + c_2 Q_2) = c_1 \pi(Q_1) + c_2 \pi(Q_2) ,$$
$$\pi(Q_1 Q_2) = \pi(Q_1)\pi(Q_2),$$
$$\pi(Q)^* = \pi(Q^*) .$$

It is automatically continuous and, more strongly, contractive:$\|\pi(Q)\| \leqq \|Q\|$. If $\pi(Q) = 0$ occurs only for $Q = 0$, then the representation is called faithful and $\|\pi(Q)\| = \|Q\|$ for all $Q \in \mathfrak{A}$.

An abstract C*-algebra can be defined, without referring to any Hilbert space, as a * Banach algebra whose norm satisfies $\|Q^*Q\| = \|Q\|^2$ for any $Q \in \mathfrak{A}$. Then there is a concrete C*-algebra which is "the same" as the given abstract C*-algebra.

A state $\Psi$ on a C*-algebra is defined to be any mapping $Q \in \mathfrak{A} \to \Psi(Q) \in \mathbb{C}$ (a functional) satisfying the following properties.

(1) Linear:     $\Psi(c_1 Q_1 + c_2 Q_2) = c_1 \Psi(Q_1) + c_2 \Psi(Q_2)$ .

(2) Positive:     $\Psi(Q^*Q) \geqq 0$.

(3) Normalized: $\|\Psi\| = 1$ where $\|\Psi\| = \sup\limits_{Q \neq 0} |\Psi(Q)| / \|Q\|$.

If $\mathfrak{A}$ is unital (i.e. $1 \in \mathfrak{A}$), the condition (3) is equivalent to

(3)' $\Psi(1) = 1$.

A fundamental result due to Gelfand, Naimark and Segal (GNS) is as follows.

**Theorem 1** *For any state $\Psi$ of a C*-algebra, there exists a Hilbert space $\mathcal{L}_\Psi$, a representation $\pi_\Psi$ on $\mathcal{L}_\Psi$ and a unit vector $\Omega_\Psi \in \mathcal{L}_\Psi$ satisfying*

$$\Psi(Q) = (\Omega_\Psi, \pi(Q)\Omega_\Psi) . \qquad (2.1)$$

An additional condition that $\{\pi(Q)\Omega_\Psi; Q \in \mathfrak{A}\}$ is dense in $\mathcal{L}_\Psi$ can be satisfied (called cyclicity) and the triplet $(\mathcal{L}_\Psi, \pi_\Psi, \Omega_\Psi)$ satisfying cyclicity is unique up to a unitary equivalence (i.e., if there is another such triplet $(\mathcal{L}'_\Psi, \pi'_\Psi, \Omega'_\Psi)$, then there exists a unitary map $u$ from $\mathcal{L}_\Psi$ to $\mathcal{L}'_\Psi$ satisfying $u\Omega_\Psi = \Omega'_\Psi$, $u\pi_\Psi(Q)u^* = \pi'_\Psi(Q)$).

This connects the above abstract definition of a state with the quantum mechanical state (1.1). The state $\Psi$ is pure if and only if the representation $\pi$ is irreducible in the sense that there is no subspace of $\mathcal{L}_\Psi$, other than $\mathcal{L}_\Psi$ and $O$, which is mapped into itself by all $\pi(Q), Q \in \mathfrak{A}$. (For the mixture state $\rho$ in section 1, let $\Psi_i$ $(i = 1, 2, ...)$ be an orthonormal set of all eigenvectors of $\rho$ with non-zero eigenvalues $\lambda_i$. Then $\mathcal{L}_\rho$ is the direct sum $\sum_i^\oplus \mathcal{H}$, $\pi(Q)$ is the direct sum $\sum_i^\oplus Q$, and $\Omega_\rho = \sum_i^\oplus (\lambda_i)^{1/2}\Psi_i$.)

An automorphism $\alpha$ of a C*-algebra $\mathfrak{A}$ is a map $Q \in \mathfrak{A} \to \alpha(Q) \in \mathfrak{A}$ satisfying the following conditions:

(1) * homomorphism: $\alpha(c_1 Q_1 + c_2 Q_2) = c_1\alpha(Q_1) + c_2\alpha(Q_2)$,

$$\alpha(Q_1 Q_2) = \alpha(Q_1)\alpha(Q_2), \ \alpha(Q^*) = \alpha(Q)^*.$$

(2) Bijective: $\alpha(\mathfrak{A}) = \mathfrak{A}$ and $\alpha(Q) = 0$ holds only if $Q = 0$.

The time translation $\alpha_t$ is a one-parameter group of automorphisms of $\mathfrak{A}$, the group property being $\alpha_{t_1}\alpha_{t_2} = \alpha_{t_1+t_2}$. It is further assumed (and proved in concrete cases) that $\alpha_t(Q)$ is continuous in t for each $Q : \lim_{t\to 0} \|\alpha_t(Q) - Q\| = 0$.

A pair $(\mathfrak{A}, \alpha_t)$ is called a C*-dynamical system.

A von Neumann algebra $M$ is a concrete C*-algebra on a Hilbert space $\mathcal{H}$ which contains the identity operator 1 of $\mathcal{H}$ and is weakly closed in the sense that

$(4)_W \quad Q_\nu \in M, \ (\xi, Q_\nu \eta) \to (\xi, Q\eta)$ for all $\xi, \eta \in \mathcal{H} \implies Q \in M$.

If $\|Q_n - Q\| \to 0$, then $Q_n$ weakly approaches to $Q$ in the above sense. Thus a von Neumann algebra can be defined by (1), (2), (3) at the beginning of this section together with $(4)_W$ above and (5) $1_\mathcal{H} \in M$. (The last condition is not essential.)

For a state $\Psi$ of a C*-algebra $\mathfrak{A}$, one can obtain the smallest von Neumann algebra containing the representing (GNS cyclic) algebra $\pi_\Psi(\mathfrak{A})$ on $\mathcal{L}_\Psi$ by adding to $\pi_\Psi(\mathfrak{A})$ all weak limits of operators in $\pi_\Psi(\mathfrak{A})$. It is a convenient tool to study the state $\Psi$ and is denoted by $\pi_\Psi(\mathfrak{A})$" because of the following reason: the <u>commutant</u> $M'$ of a von Neumann algebra $M$ (or any set of bounded linear operators) is the set of all bounded linear operators $A$ which commute with every element $B$ of $M$: $[A, B] \equiv AB - BA = 0$. It is automatically a von Neumann algebra (if $M$ is any selfadjoint set, i.e., if $B \in M$ implies $B^* \in M$). The commutant of the commutant is called the <u>double commutant</u> and is denoted by $M$". It is the smallest von Neumann algebra containing $M$ if $M$ is a selfadjoint set. In particular, $\pi_\Psi(\mathfrak{A})$ is the double commutant of $\pi_\Psi(\mathfrak{A})$.

The center $Z(M) = M \cap M'$ of a von Neumann algebra $M$ is the set of all $B \in M$ commuting with all elements of $M$. If it is trivial, being the set of complex multiples of the identity operator 1, then $M$ is called a <u>factor</u>. The center being commutative, one can diagonalize all elements of $Z(M)$ and decompose $M$ into an integral over factors (in a rather delicate sense). Murray and von Neumann, the initiators of von Neumann algebras, introduced the classification of factors into type I $(I_n, n = 1, 2, ..., \infty)$, II $(II_1$ and $II_\infty)$ and III. The same terminology applies to a general von Neumann algebra if its central decomposition (explained above) consists of factors of one type.

If $\pi_\Psi(\mathfrak{A})$" is a factor of type I, then $\pi_\Psi(\mathfrak{A})$" is the set of all bounded linear operators on $\mathcal{L}_\Psi$ and we are back to the case of the quantum mechanics formulation in Section 1. This is the case if and only if $\Psi$ is a pure state of $\mathfrak{A}$.

On the other hand, if $\Psi$ is an equilibrium state of a quantum statistical system, we have type III von Neumann algebras for $\pi_\Psi(\mathfrak{A})$".

It is appropriate to emphasize at this point the change in mathematical definition of states in the operator algebraic approach in comparison with the usual quantum mechanics, namely a state is the (linear, positive, normalized) functional on $\mathfrak{A}$ giving the expectation value of the observables, rather than a (unit) vector on a Hilbert space as in the usual quantum mechanics. Such a formulation, necessitated by the existence of many inequivalent representations, has a number of merits:

(1) The expectation value is more directly related to measurements compared with vectors in a Hilbert space.

(2) It includes all possible states arising from vectors in many inequivalent representations. Especially, the situation where the relevant representations are all inequivalent for different time can be dealt with in this formulation of states, but not in the usual method of quantum mechanics based on a selfadjoint Hamiltonian operator.

(3) Pure states described by vectors and mixture states described by the density matrices (the latter containing the former) can be viewed in a unified manner. (From a linear functional point of view, the mixture of two states $\Psi_1$ and $\Psi_2$ (as linear functionals on $\mathfrak{A}$) is a convex combination $\Psi = \lambda\Psi_1 + (1-\lambda)\Psi_2$, $0 \leq \lambda \leq 1$ : For $A \in \mathfrak{A}$,

$$(\lambda\Psi_1 + (1-\lambda)\Psi_2)(A) = \lambda\Psi_1(A) + (1-\lambda)\Psi_2(A).$$

A state $\Psi$ is a pure state when the mixture decomposition $\lambda\Psi_1 + (1-\lambda)\Psi_2$ holds only for trivial cases: $\lambda = 0, \lambda = 1$ or $\Psi_1 = \Psi_2 = \Psi$.)

(4) The new formulation is still essentially within the framework of quantum theory due to the GNS theorem above.

## 3   Quantum Statistical Mechanics.

Consider a quantum spin system on a $\nu$-dimensional lattice $\mathbb{Z}^\nu$, where $\mathbb{Z}$ is the set of integers and a lattice site is labelled by $\mathbf{n} = (n_1, ..., n_\nu)$, $n_k \in \mathbb{Z}$. At each lattice site $\mathbf{n}$, an alegbra $\mathfrak{A}_\mathbf{n}$ of all $d \times d$ matrices describes the $(d-1)/2$ spin system at the site $\mathbf{n}$. If $d = 2$, for example, an element of $\mathfrak{A}_\mathbf{n}$ is a linear combination of Pauli spin operators $\sigma_x^{(\mathbf{n})}, \sigma_y^{(\mathbf{n})}, \sigma_z^{(\mathbf{n})}$ and the identity 1. Elements of $\mathfrak{A}_\mathbf{m}$ and $\mathfrak{A}_\mathbf{n}$ commute for distinct sites $\mathbf{m}$ and $\mathbf{n}$. Their union for all sites $\mathbf{n} \in \mathbb{Z}^\nu$ generates a unique C*-algebra $\mathfrak{A}$ which is taken to be the algebra of observables for this system. For a subset $I$ of $\mathbb{Z}^\nu$, $\mathfrak{A}(I)$ denotes the C*- subalgebra of $\mathfrak{A}$ generated by all $\mathfrak{A}_\mathbf{n}$, $\mathbf{n} \in I$.

The translation of spin operators along the lattice such as

$$\tau_\mathbf{m}(\sigma_\alpha^{(\mathbf{n})}) = \sigma_\alpha^{(\mathbf{n}+\mathbf{m})} \tag{3.1}$$

(more generally $\tau_\mathbf{m}$ maps $\mathfrak{A}_\mathbf{n}$ onto $\mathfrak{A}_{\mathbf{n}+\mathbf{m}}$ as a *-isomorphism) determines the group of automorphisms $\tau_\mathbf{m}$ of $\mathfrak{A}$ ($m \in \mathbb{Z}^\nu$ ).

The interaction between spins is described by a potential $\Phi$ which assigns a selfadjoint operator $\Phi(I) \in \mathfrak{A}(I)$ for each finite subset $I$ of $\mathbb{Z}^\nu$ . For example, for a two point set $I = \{\mathbf{n}^a, \mathbf{n}^b\}$, $\Phi(I)$ is the interaction energy between spins at lattice sites $\mathbf{n}^a$ and $\mathbf{n}^b$. We assume the following properties:

(1) $\Phi(\phi) = 0$. ($\phi$ is the empty set.)

(2) $\Phi(I + m) = \tau_{\mathbf{m}}\Phi(I)$ (translation invariance).

(3) $\Phi(I) = 0$ if $diam(I) = \max\{|\mathbf{n}_i - \mathbf{n}_j|; i, j \in I\} > r$ (finite range).

The condition (3) is taken for simplicity and can be weakened.

The energy in a finite subset $I$ in $\mathbb{Z}^\nu$ (the Hamiltonian for spins in $I$ as a closed system) is defined to be

$$U(I) = \sum_{\Lambda \subset I} \Phi(\Lambda). \tag{3.2}$$

The dynamics of the system is defined by

$$\alpha_t(A) = \lim_{I \nearrow \mathbb{Z}^\nu} e^{itU(I)} A e^{-itU(I)} \tag{3.3}$$

which can be proven to exist as a limit in the norm of $\mathfrak{A}$.

We note that the time derivative at $t = 0$ can be given by

$$\overset{\bullet}{A} = \frac{d}{dt}\alpha_t(A)|_{t=0} = i \sum_{\Lambda} [\Phi(\Lambda), A] \tag{3.4}$$

where the sum on the right hand side becomes a finite sum if $A \in \mathfrak{A}(I)$ for some finite subset $I$ of $\mathbb{Z}^\nu$ because $[\Phi(\Lambda), A] = 0$ if $\Lambda$ is more than $r$ distance away from $I$. Formally, $\overset{\bullet}{A} = i[H, A]$ where $H = \sum \Phi(\Lambda)$, which does not converge.

A comprehensive general theory about the dynamical system $(\mathfrak{A}, \alpha_t)$, has been established. We mention a few results below. ($\beta = (kT)^{-1}$ with absolute temperature T.)

An equilibrium state $\varphi$ can be characterized by the following Kubo-Martin Schwinger (KMS) boundary condition: For any pair $A, B \in \mathfrak{A}$, there exists a function $F^{AB}(z)$ of a complex variable $z$ for $0 \leq \beta^{-1}(\mathrm{Im}\ z) \leq 1$ which is continuous on the whole strip region and analytic inside, satisfying

$$F^{AB}(t) = \varphi(A\alpha_t(B)), \quad F^{AB}(t + i\beta) = \varphi(\alpha_t(B)A). \tag{3.5}$$

For a finite region $I \subset \mathbb{Z}^\nu$, the Gibbs ensemble state

$$\varphi_I^G(A) = tr(e^{-\beta U(I)}A)/tr(e^{-\beta U(I)}) \tag{3.6}$$

satisfies the KMS condition above for

$$\alpha_t^I(A) = e^{itU(I)} A e^{-itU(I)} \tag{3.7}$$

instead of $\alpha_t$, and any accumulation point of $\varphi_I^G$ as $I \nearrow \mathbb{Z}^\nu$ (in van Hove limit, where the surface to volume ratio of $I$ tends to 0) satisfies the KMS condition for $\alpha_t$.

For high temperature (i.e. for small $\beta$), a state satisfying the KMS condition (called a KMS state) is unique and is the limit of the finite volume Gibbs state $\varphi_I^G$ as $I \nearrow \mathbb{Z}^\nu$.

Even if the interaction potential $\Phi$ is assumed to be translation invariant, a KMS state need not be translation invariant ($\varphi(\tau_{\mathbf{m}}(A)) = \varphi(A)$) if there are many KMS states. However, there exists always a translation invariant KMS state. For any translation invariant state $\varphi$, one can define its energy density

$$e(\varphi) = \lim_{I \nearrow \mathbb{Z}^\nu} | I |^{-1} \varphi(U(I)) \tag{3.8}$$

(which is proved to converge) and the entropy density

$$s(\varphi) = \lim_{I \nearrow \mathbb{Z}^\nu} | I |^{-1} (-tr_I(\rho_\varphi^I \log \rho_\varphi^I)) \tag{3.9}$$

(which is also proved to converge) where $\rho_\varphi^I \in \mathfrak{A}(I)$ is uniquely defined by

$$\varphi(A) = tr_I(\rho_\varphi^I A) \quad \text{for all } A \in \mathfrak{A}(I). \tag{3.10}$$

Then the pressure function

$$P(\beta\Phi) = \lim_{I \nearrow \mathbb{Z}^\nu} | I |^{-1} \log \varphi(e^{-\beta U(I)}) \tag{3.11}$$

(which is proved to exist) satisfies the variational equality

$$P(\beta\Phi) = \sup_\varphi (s(\varphi) - \beta e(\varphi)) \tag{3.12}$$

where $\varphi$ runs over all translation invariant states. Translation invariant KMS states are characterized as those states $\varphi$ giving the supremum value

$$P(\beta\Phi) = s(\varphi) - \beta e(\varphi) \tag{3.13}$$

This shows the equivalence of the KMS condition and the thermodynamical variational principle as far as translation invariant states are concerned. A local variational principle, equivalent to the KMS condition for a general state can also be formulated.

Another class of physically interesting states of the C*-dynamical system are ground states $\varphi$ which can be defined as follows (among some alternative equivalent definitions). Let $(\mathcal{L}_\varphi, \pi_\varphi, \Omega_\varphi)$ be the GNS triplet. $\varphi$ is the ground state if there exists a positive selfadjoint operator $H_\varphi$ such that

(1) $\pi_\varphi(\alpha_t(A)) = e^{itH_\varphi} \pi_\varphi(A) e^{-itH_\varphi}$,

(2) $H_\varphi \Omega_\varphi = 0$.

(This means that $H_\varphi$ is an energy operator in the representation space $\mathcal{L}_\varphi$ in some sense and $\Omega_\varphi$ is its eigenvector with the lowest eigenvalue.)

Operator algebra methods can be used to solve a concrete system. The XY-model on a one-dimensional lattice is described by the Hamiltonian

$$H = -J \sum_{j \in \mathbb{Z}} \{(1+\gamma)\sigma_x^{(j)}\sigma_x^{(j+1)} + (1-\gamma)\sigma_y^{(j)}\sigma_y^{(j+1)} + \lambda(\sigma_z^{(j)} + \sigma_z^{(j+1)})\}$$

(or in terms of the potential, $\Phi(\{j\}) = -2\lambda J\sigma_z^{(j)}$, $\Phi(\{j, j+1\}) = -J\{(1+\gamma)\sigma_x^{(j)}\sigma_x^{(j+1)} + (1-\gamma)\sigma_y^{(j)}\sigma_y^{(j+1)}\}$, and $\Phi(I) = 0$ for all other $I$).

By a general theorem on one-dimensional systems, the KMS state is unique for any $\beta$. It can be explicitly described.

For the ground states, an interesting phase diagram can be written.

(i) For $|\lambda| \geqq 1$ and for $|\lambda| < 1, \gamma = 0$, the ground state is unique.

(ii) For $|\lambda| < 1, \gamma \neq 0$, there are exactly two pure ground states and a general ground state is their mixture (an average with weight $\lambda$ and $1 - \lambda, 0 \leqq \lambda \leqq 1$) except the case (iii).

(iii) For $(\lambda, \gamma) = (0, \pm 1)$, there are continuously many pure ground states. However, there are exactly 4 irreducible, mutually non-equivalent (i.e., disjoint) representations of $\mathfrak{A}$ containing ground state vectors and a general ground state is a mixture of mixture states on these 4 representations. Thus, $\pi_\varphi(\mathfrak{A})$" for any ground state $\varphi$ in this model is of type I.

Except for the H-shaped phase boundary (i.e., except for $|\lambda| \models 1$ as well as for $|\lambda| < 1, \gamma = 0$), the ground states $\varphi(A)$ (the unique one in case (i) and appropriately chosen 2 pure ground states in cases (ii) and (iii)) as a function of $\gamma$ and $\lambda$ is real analytic for each $A$ belonging to a finite region, has an energy gap above the ground state vector (in GNS representation) and has an exponentially decreasing long distance correlation. On the other hand, at the phase boundary, all these properties do not hold.

One can also analyse all positive energy representations (those representations satisfying the condition (1) for ground states but not necessarily the condition (2)). There is a unique irreducible positive energy representation in the interior of the region (i) (i.e. for $|\lambda| > 1$), there are exactly 4 irreducible positive energy representations in the regions (ii) and (iii), two of which are ground state GNS representations while the other two are soliton representations, containing ground state vectors only in case (iii). In all these cases, all positive energy representations are type I, while in the case of phase boundary, there are continuously many positive energy irreducible representations as well as factor representations of type III, which can be called infrared representations.

The main operator algebra techniques used in obtaining these results are connected with two points of modification of usual quantum theory described in Section 1.

As for the type of von Neumann algebras in the general theory of a quantum spin system on a lattice, the type of $\pi_\varphi(\mathfrak{A})''$ for an equilibrium state for $\beta \neq 0$ is always type III, and the representations $\pi_\varphi$ of $\mathfrak{A}$ for different $\beta$ are different (i.e. disjoint).

## 4  Local Quantum Theory

A general theory for quantum fields can be developed without mentioning concrete quantum fields by introducing the concept of local observables as follows. We start with the C*-algebra of all observables, denoted $\mathfrak{A}$.

For each space-time region $D$, we consider the subset $\mathfrak{A}(D)$ of $\mathfrak{A}$ consisting of all observables which can be measured within the region $D$. The shape of $D$ may be restricted to simple forms such as the double cone

$D = \{x; a - x \text{ and } x - b \text{ are positive time-like}\}$

for $a - b$ positive time-like. (For quantum fields $A_i(x)$, we are considering $A(f) = \int A_i(x)f(x)dx$ for those $f$ vanishing outside of $D$, or more precisely bounded operators obtained from $A(f)$ such as spectral projections if $A(f)$ is selfadjoint, for example.)

We assume the following basic and natural axioms.

(1) Isotony: If $D_1 \supset D_2$, then $\mathfrak{A}(D_1) \supset \mathfrak{A}(D_2)$

(2) Covariance: there exists a continuous representation of the inhomogenouse Lorentz group $\mathcal{P}$ (special theory of relativity) by automorphisms $\alpha_g, g \in \mathcal{P}$, of $\mathfrak{A}$ satisfying

$$\alpha_g(\mathfrak{A}(D)) = \mathfrak{A}(gD) \tag{4.1}$$

where $gD = \{gx; x \in D\}$. ($g = (a, \Lambda)$ with a vector $a$ representing space-time translation and a homogeneous Lorentz transformation $\Lambda$.)

(3) Locality: If $D_1$ and $D_2$ are mutually space-like, then operators in $\mathfrak{A}(D_1)$ and $\mathfrak{A}(D_2)$ commute:

$$\mathfrak{A}(D_1)' \supset \mathfrak{A}(D_2). \tag{4.2}$$

Here, $\mathfrak{A}(D_1)'$ denotes the commutant of $\mathfrak{A}(D_1)$ (see Section 2).

The C*-subalgebra of $\mathfrak{A}$ generated by $\mathfrak{A}(D)$ also satisfies these properties. Hence, we may replace $\mathfrak{A}(D)$ by the C*-subalgebra of $\mathfrak{A}$ generated by $\mathfrak{A}(D)$. We assume therefore that $\mathfrak{A}(D)$ is a C*-subalgebra of $\mathfrak{A}$ .

A vacuum state is defined to be a state which has the lowest energy (ground state) in any coordinate frame of special relativity. This then implies the invariance of any vacuum state $\omega$ under $\mathcal{P}$ : $\omega(\alpha_g(A)) = \omega(A), g \in \mathcal{P}, A \in \mathfrak{A}$ and there exists a continuous unitary representation $U_\omega(g), g \in \mathcal{P}$ of the group $\mathcal{P}$ in the

GNS representation space $\mathcal{H}_\omega$ for the state $\omega$ which implements the action $\alpha$ of the group $\mathcal{P}$ on $\mathfrak{A}$ by

$$U_\omega(g)\pi_\omega(A)U_\omega(g)^* = \pi_\omega(\alpha_g(A)) \tag{4.3}$$

and leaves the representative vector $\Omega_\omega$ of $\omega$ invariant: $U_\omega(g)\Omega_\omega = \Omega_\omega$. Furthermore, the ground state condition implies that the generator $\mathbf{P} = (\mathbf{P}^0, \mathbf{P}^1, \mathbf{P}^2, \mathbf{P}^3)$ of the translations

$$U((a,1)) = e^{i(P,a)} \quad ((P,a) = P^0 a^0 - P^1 a^1 - P^2 a^2 - P^3 a^3) \tag{4.4}$$

has a spectrum in the positive time-like cone:

$$Spec\, P \subset V_+ \equiv \{P; (P,P) = (P^0)^2 - \sum_{i=1}^{3}(P^i)^2 \geq 0,\, P^0 \geq 0\}. \tag{4.5}$$

We assume the existence of vacuum states. Then it follows that there exist pure vacuum states and their mixtures are dense in all vacuum states. For any pure vacuum state $\omega$, $\pi_\omega(\mathfrak{A})$ is irreducible and $\Omega_\omega$ is the unique vector invariant under $U((a,1))$ (uniqueness up to a phase factor).

With this general set-up, one can develop the scattering theory for particles where a particle of mass $m$ and spin $j$ is defined as the irreducible invariant subspace of $U(\mathcal{P})$ where it is the representation of $\mathcal{P}$ with mass $m$ and spin $j$. As a result, the S-matrix of the scattering is already determined by the above seemingly general structure.

To deal with Fermions, it will be convenient to find a representation of $\mathfrak{A}$, in which the universal covering group $\widetilde{\mathcal{P}}$ of $\mathcal{P}$ has a continuous unitary representation $U(g), g \in \widetilde{\mathcal{P}}$ which implements the action $\alpha$ of $\mathcal{P}$ in the sense of (4.3) for this representation and has an irreducible invariant subspace where it is the representation of $\mathcal{P}$ for the relevant Fermion. Such a representation of $\mathfrak{A}$ has to be different from the vacuum state representation $\pi_\omega$ and we will now discuss how one can obtain such a representation.

One proposes to study all representations $\pi$ of $\mathfrak{A}$ satisfying the following 2 conditions.

(1) The representation space of $\pi$ is obtained from that of $\pi_\omega$ by some strictly localized excitation. If the excitation is in a finite (double cone) region $D$, then the representations $\pi$ and $\pi_\omega$ of the subalgebra $\mathfrak{A}$ for the region

$$D' = \{x; x - y \text{ space-like for all } y \in D\}, \tag{4.6}$$

which should not feel any disturbance from the excitation in the region $D$ (by the special theory of relativity — so-called Einstein causality), are to be unitarily equivalent.

(2) The local excitation of (1) is to be transportable in space-time.

Under the mathematical (and plausible) assumption of the following Haag duality on the vacuum representation and an additional technical assumption due to Borchers, one can completely analyze the above type of representations.

$$\text{Haag duality}: \pi_\omega(\mathfrak{A}(D))' = \pi_\omega(\mathfrak{A}(D'))'' \tag{4.7}$$

where $\mathfrak{A}(D')$ is the C*-subalgebra of $\mathfrak{A}$ generated by all $\mathfrak{A}(D_1)$, $D_1 \subset D'$ (i.e. $D_1$ which is space-like to $D$).

As a result of such an analysis, one arrives at the following structure which contains all desirable representations of $\mathfrak{A}$ as well as operators connecting different representations of $\mathfrak{A}$, similar to Fermion (and non-observable Boson) fields.

One has ($\alpha$) an irreducible field algebra $\mathcal{F}$, acting on a Hilbert space $\widetilde{\mathcal{H}}$ containing the vacuum representation space $\widetilde{\mathcal{H}}_\omega$ of $\mathfrak{A}$ and generated by its subalgebras $\mathcal{F}(D)$ for local regions $D$ which satisfy the properties (1) and (2) for $\mathfrak{A}(D)$, where the action $\alpha$ of $\mathcal{P}$ is now to be replaced by an action $\tilde{\alpha}$ of $\widetilde{\mathcal{P}}$ satisfying $\alpha_{\tilde{g}}(\mathcal{F}(D)) = \mathcal{F}(gD)$ ($\tilde{g} \in \widetilde{\mathcal{P}}$ covers $g \in \mathcal{P}$), ($\beta$) a compact group G of automorphisms of $\mathcal{F}$, called the gauge group, such that G-invariant elements of $\mathcal{F}$ are exactly a representation $\tilde{\pi}$ of $\mathfrak{A}$ and G-invariant elements of $\mathcal{F}(D)$ coincide with $\tilde{\pi}(\mathfrak{A}(D))$, ($\gamma$) continuous unitary representations of G and $\widetilde{\mathcal{P}}$ in $\widetilde{\mathcal{H}}$, implementing their action on $\mathcal{F}$ (i.e. $u_g B u_g^* = g(B)$ for $g \in G$ and $= \alpha_g(B)$ for $g \in \widetilde{\mathcal{P}}$) and leaving $\Omega_\omega$ invariant, such that the spectral condition (of positive energy) (4.5) is satisfied for the translation part of $\widetilde{\mathcal{P}}$.

The element $u \in \widetilde{\mathcal{P}}$ representing 360° rotation (and hence corresponds to identity of $\mathcal{P}$) satisfies $u^2 = 1$ and hence splits $\mathcal{F}$ into even and odd parts: For any $F \in \mathcal{F}$, $F_\pm = \frac{1}{2}(F \pm \tilde{\alpha}_u(F))$ satisfies $\tilde{\alpha}_u(F_\pm) = \pm F_\pm$ and $F = F_+ + F_-$. $F_+$ and $F_-$ are called Bose and Fermi parts of $F$. We can arrange the above structure such that for any mutually space-like regions $D_1$ and $D_2$, the Bose and Fermion parts of operators on $\mathcal{F}(D_1)$ and $\mathcal{F}(D_2)$ satisfy the normal commutation relations, namely Fermion parts mutually anticommute and other pairs commute. Thus we obtain a deeper understanding of the connection of spin and statistics without assuming anticommutation relations or any commutation relations other than the commutativity of observables localized in space-like separated two regions.

To illustrate the above structure, let us consider quantum field theory of one Dirac Fermion (for example free Dirac fields). The algebra of local observables, $\mathfrak{A}(D)$, is the C*-algebra generated by $\Psi_\alpha(f)^*\Psi_\beta(g)$ where $\Psi_\alpha(f)$ denotes $\int \Psi_\alpha(f)f(x)dx$, $\alpha$ and $\beta$ are arbitrary and $f$ and $g$ vanish outside of $D$. The vacuum representation space is the total Fermion number $O$ space. The field algebra $\mathcal{F}(D)$ is generated by $\Psi_\alpha(f)$ with $f$ vanishing outside of $D$. The gauge group G is the group $\mathbb{T}$ of complex numbers $e^{i\theta}$ of modulus 1 and its action is $g(\Psi_\alpha(f)) = e^{-i\theta}\Psi_\alpha(f)$, the gauge group of the first kind. The main point is that there is no explicit information about $\mathcal{F}$ and G in the algebras $\mathfrak{A}(D)$ and $\mathfrak{A}$. However we can recover these objects out of $\mathfrak{A}(D)$ and $\mathfrak{A}$ and prove the anticommutation relation of Fermi fields at a space-like distance.

The merit of the approach above is that one can find out all possibilities without any arbitrary assumption. One obtains possibilities of para Bose and para Fermi statistics and an infinite statistics. The significance of the last one is not necessarily well-understood but can be excluded under a reasonable assumption. For 4-dimensional space-time, these are the only possibilities.

For a lower dimensional case such as 2-dimensional space-time, one has a more complicated situation, obtaining braid statistics.

The above analysis is due to Doplicher, Haag and Roberts. There are some more general analyses due to Buchholz and Fredenhagen treating all finite energy excitation with separated one particle spectrum. In this case, the excitation is not restricted to a strictly localized region but it is restricted to a stringlike region extending to infinity. For further details, see the reference below.

## REFERENCES

[1] For quantum statistics and operator algebras, see, O. Bratteli and D.W. Robinson: *Operator Algebras and Quantum Mechanics I & II.* (Springer Verlag).

[2] For local quantum theory, see, R. Haag: *Local Quantum Physics* (Springer Verlag).

# IRREVERSIBILITY FROM A REVERSIBLE EQUATION

HIROSHI EZAWA

Department of Physics, Gakushuin University
Mejiro, Toshima-ku, Tokyo 171-8588, Japan

KEIJI WATANABE

Department of Physics, Meisei University,
Hino, Tokyo 191-8506, Japan

KOICHI NAKAMURA

Division of Natural Science, Meiji University, Izumi Campus
Eifuku, Suginami-ku, Tokyo 168-8555, Japan

Abstract    A simple example is taken to show without misty approximations that irreversibility can follow from a reversible equation. The example is a Bose gas having an interaction quadratic in the field operators. It is shown that the particle-number density exhibits diffusion superposed with an oscillatory damping when looked at a macroscopic scales in space and time. Some additional remarks are given.

Keywords    irreversibility, nonequilibrium statistical mechanics, Bose gas, diffusion, soluble models, Bogolubov Hamiltonian

## 1. INTRODUCTION

The irreversible processes caused by temperature inhomogeneity, say, heat conduction, thermal diffusion, remain to be understood despite the long history and many attempts[1,2]. For the irreversible processes caused mechanically, say, electric conduction, we have Kubo's theory of linear response[3], which is based on the perturbation theory in Hamiltonian dynamics. But, the heat flow is not related to any forces in such a direct way.

We have been using the Thermo Field Dynamics[4], but in this paper we shall put forward a simpler idea. A system in thermal equilibrium can be described by a density matrix $\hat{\rho} = e^{-\beta(\hat{\mathcal{H}} - \mu\hat{N})}$ (not normalized) of grand canonical ensemble, with the inverse temperature $\beta = 1/(k_B T)$, Hamiltonian $\hat{\mathcal{H}}$ and the particle-number operator $\hat{N}$ of the system. The thermal process is caused not by a mechanical perturbation, but by the distortion of the state. If so, how can the process be irreversible when it is dictated by a reversible equation of motion?

## 1.1 Local equilibrium

When we say that the temperature is inhomogeneous, it varies only on a macroscopic scale much larger than any scales of atomic motion in the system. Therefore, we can speak of a local equilibrium; otherwise the temperature would not make sense.

The system in a local equilibrium state would be represented by a density matrix,

$$\hat{\rho}(\boldsymbol{x}) = e^{-\int \beta(\mathbf{x})\{\hat{\mathcal{H}}(\mathbf{x}) - \mu \hat{N}(\mathbf{x})\}d^3 x}, \tag{1.1}$$

(not normalized) with the Hamiltonian density $\hat{\mathcal{H}}(\boldsymbol{x})$ and the particle-number density $\hat{N}(\boldsymbol{x})$, and with the inverse local temperature $\beta(\boldsymbol{x})$ varying extremely slowly in atomic scales; the chemical potential $\mu$ is assumed to be constant. We write

$$\beta(\boldsymbol{x}) = \beta_0 + \beta_a(\boldsymbol{x}), \qquad \int \beta_a(\boldsymbol{x}) d^3 x = 0 \tag{1.2}$$

and assume that $|\beta_a(\boldsymbol{x})| \ll \beta_0$ throughout the space. Let us also write

$$\hat{\mathcal{K}}(\boldsymbol{x}) := \hat{\mathcal{H}}(\boldsymbol{x}) - \mu \hat{N}(\boldsymbol{x}) \tag{1.3}$$

for the sake of brevity, and call its space integral "Hamiltonian" $\hat{\mathcal{K}}$.

Physically, we expect that the inhomogeneity will be smeared out in time. We would take these expressions as the initial conditions at $t = 0$, and let the system evolve by the Hamiltonian equation of motion. We would not look at the evolution on microscopic scales of time and space; it would appear chaotic if we did. We would be interested only in the macroscopic features of the evolution.

## 1.2 Zubarev's No Go theorem

Zubarev wrote in his well-known textbook[5]:

> In the local equilibrium state (1.1), there are no dissipative processes, i.e., no thermal conduction, diffusion nor viscosity.

We shall construct a simple counter example. But, let us first see how he argues. With respect to the space-reflection, he points out, $\hat{\mathcal{H}}(\boldsymbol{x})$ and $\hat{N}(\boldsymbol{x})$ in (1.1) are even, while the current densities of energy and particle number are odd. Consequently, he says, the averages of these operators in the state (1.1) vanish. Q.E.D. Note that he is concerned only with the current without delving into any dissipative features.

Let us now look at his argument closely, taking an example of a Bose gas (mass $1/2$) in a box $V$, whose Hamiltonian density is given by

$$\hat{\mathcal{H}}(\boldsymbol{x}) = \frac{1}{V} \sum_{pp'} (\boldsymbol{p}' \cdot \boldsymbol{p}) \hat{a}_{p'}^{\dagger} \hat{a}_p e^{i(p-p')\cdot x} + \frac{1}{V^2} \sum_{pp'qq'} \tilde{v}(\boldsymbol{q} - \boldsymbol{q}') \hat{a}_{p'}^{\dagger} \hat{a}_{q'}^{\dagger} \hat{a}_p \hat{a}_q e^{-i(p'+q'-p-q)\cdot x} \tag{1.4}$$

and the number density by the first term without $(p' \cdot p)$, where $\hat{a}_k$ and $\hat{a}_k^\dagger$ are the annihilation and creation operators of a particle [1] with momentum $k$; the interaction potential $v(x)$ is reflection symmetric and so is its Fourier transform $\tilde{v}(k)$.

The space reflection, $\hat{\mathcal{P}}^{-1}\hat{a}_k\hat{\mathcal{P}} = \hat{a}_{-k}$, reverses the direction of momentum $k$, so that it acts on the first term of $\hat{\mathcal{H}}(x)$ as

$$\hat{\mathcal{P}}^{-1}\hat{\mathcal{H}}_1(x)\hat{\mathcal{P}} = \frac{1}{V}\sum_p (p' \cdot p)\,\hat{a}_{-p'}^\dagger \hat{a}_{-p} e^{i(p'-p)\cdot x} = \hat{\mathcal{H}}_1(-x),$$

and similarly on the second term of $\hat{\mathcal{H}}(x)$ as well as on $\hat{N}(x)$. Therefore,

$$\hat{\mathcal{P}}^{-1}\hat{\rho}\hat{\mathcal{P}} = \exp\left[-\int d^3x\beta(-x)\left\{\hat{\mathcal{H}}(x) - \mu\hat{N}(x)\right\}\right]. \tag{1.5}$$

Thus, the density matrix has changed and Zubarev's argument breaks down.

This conclusion does not negate Zubarev's statement that the average of current densities vanish. Counter examples will be given in the following sections.

## 2. NUMBER AND CURRENT DENSITIES

### 2.1   Perturbation expansion

Let us expand the density matrix (1.1) in powers of $\beta_a$. Put

$$\hat{\rho}(\lambda) = \hat{\rho}_0(\lambda)\hat{\eta}(\lambda), \quad \text{with} \quad \hat{\rho}_0(\lambda) := \exp\left[-\lambda\beta_0\hat{\mathcal{K}}\right]. \tag{2.1}$$

Then,

$$\frac{d}{d\lambda}\hat{\eta}(\lambda) = -\hat{\rho}_0(\lambda)^{-1}\left\{\int \beta_a(x)\hat{\mathcal{K}}(x)d^3x\right\}\rho_0(\lambda)\hat{\eta}(\lambda),$$

so that, by iteration to the first order in $\beta_a$, with the initial condition $\hat{\eta}(0) = 1$,

$$\hat{\rho} := \hat{\rho}(1) = e^{-\beta_0\hat{\mathcal{K}}}\left\{1 - \int_0^1 d\lambda \int \beta_a(x)e^{\lambda\beta_0\hat{\mathcal{K}}}\hat{\mathcal{K}}(x)e^{-\lambda\beta_0\hat{\mathcal{K}}}d^3x\right\}. \tag{2.2}$$

### 2.2   Time development

Let us remark on the time development. In general, the average at time $t$ of an observable $\hat{A}$ is given by $Z^{-1}\text{Tr}[\rho\hat{A}(t)]$ where $\hat{A}(t) := e^{i\hat{\mathcal{H}}t}\hat{A}e^{-i\hat{\mathcal{H}}t}$.

Since the total number of particles $\hat{N}$ commutes with the total Hamiltonian $\hat{\mathcal{H}}$,

$$e^{-i\hat{\mathcal{K}}t} = e^{-i(\hat{\mathcal{H}}-\mu_0\hat{N})t} = e^{i\mu_0\hat{N}t}e^{-i\hat{\mathcal{H}}t}.$$

If further $\hat{N}$ commutes with $\hat{A}$, then

$$\hat{A}(t) = e^{i\hat{\mathcal{H}}t}\hat{A}e^{-\hat{\mathcal{H}}t} = e^{i\hat{\mathcal{H}}t}e^{-i\mu_0\hat{N}t}\hat{A}e^{i\mu_0\hat{N}t}e^{-i\hat{\mathcal{H}}t} = e^{i\hat{\mathcal{K}}t}\hat{A}e^{-i\hat{\mathcal{K}}t}. \tag{2.3}$$

---

[1] In this paper we use the unit system in which $\hbar = 1$ and mass of the particle $= 1/2$. In this system, $[\hbar] = [ML^2T^{-1}]$ as well as $[M]$ has no dimension, the time has the dimension $[L^2]$.

By (2.2), therefore

$$\mathrm{Tr}[\hat{\rho}\hat{A}(t)] = \mathrm{Tr}\Big[e^{-\beta_0\hat{\mathcal{K}}}\Big\{1 - \int_0^1 d\lambda \int \beta_a(\boldsymbol{x})e^{\lambda\beta_0\hat{\mathcal{K}}}\hat{\mathcal{K}}(\boldsymbol{x})e^{-\lambda\beta_0\hat{\mathcal{K}}}d^3\mathbf{x}\Big\}e^{i\hat{\mathcal{K}}t}\hat{A}e^{-i\hat{\mathcal{K}}t}\Big],$$

which can be rewritten as

$$\mathrm{Tr}[\hat{\rho}\hat{A}(t)] = \mathrm{Tr}\Big[e^{-\beta_0\hat{\mathcal{K}}}\Big\{1 - \int_0^1 d\lambda \int \beta_a(\boldsymbol{x})\hat{\mathcal{K}}(\boldsymbol{x})d^3\mathbf{x}\Big\}e^{i\hat{\mathcal{K}}(t+i\lambda\beta_0)}\hat{A}e^{-i\hat{\mathcal{K}}(t+i\lambda\beta_0)}\Big],$$

by shifting the $\lambda$ dependence of the "Hamiltonian" density $\hat{\mathcal{K}}(\boldsymbol{x})$ to $\hat{A}$ using the invariance of trace under cyclic permutation of operators. Thus, with $\tau = t + i\lambda\beta_0$,

$$\mathrm{Tr}[\hat{\rho}\hat{A}(t)] = \mathrm{Tr}\Big[e^{-\beta_0\hat{\mathcal{K}}}\Big\{1 - \int_0^1 d\lambda \int \beta_a(\boldsymbol{x})\hat{\mathcal{K}}(\boldsymbol{x})d^3\mathbf{x}\Big\}\hat{A}(\tau)\Big]. \tag{2.4}$$

## 2.3 Current density of particles

The number and the current densities at $t = 0$ are given by

$$\hat{j}_0(\boldsymbol{x}) = \hat{\phi}^\dagger(\boldsymbol{x})\hat{\phi}(\boldsymbol{x}), \qquad \hat{j}_l(\boldsymbol{x}) = -i\Big\{\hat{\phi}^\dagger(\boldsymbol{x})\frac{\partial\hat{\phi}(\boldsymbol{x})}{\partial x_l} - \frac{\partial\hat{\phi}^\dagger(\boldsymbol{x})}{\partial x_l}\hat{\phi}(\boldsymbol{x})\Big\} \qquad (l = 1, 2, 3),$$

in terms of the field operator $\hat{\phi}(\boldsymbol{x}) = (1/\sqrt{V})\sum_{\mathbf{p}}\hat{a}_{\mathbf{p}}e^{i\mathbf{p}\cdot\mathbf{x}}$. Or,

$$\hat{j}_0(\boldsymbol{x}) = \frac{1}{V}\sum_{\mathbf{p},\mathbf{p}'}\hat{a}_{\mathbf{p}'}^\dagger\hat{a}_{\mathbf{p}}e^{i(\mathbf{p}-\mathbf{p}')\cdot\mathbf{x}}, \qquad \hat{j}_l(\boldsymbol{x}) = \frac{1}{V}\sum_{\mathbf{q}'\mathbf{q}}(q_l + q_l')\hat{a}_{\mathbf{q}'}^\dagger\hat{a}_{\mathbf{q}}e^{i(\mathbf{q}-\mathbf{q}')\cdot\mathbf{x}}. \tag{2.5}$$

The average of the current $\hat{j}_l(\boldsymbol{x}, t)$ with respect to the density matrix (2.2),

$$\langle\hat{j}_l(\boldsymbol{x}, t)\rangle = \frac{1}{Z}\mathrm{Tr}[\hat{\rho}\hat{j}_l(\boldsymbol{x}, t)] \qquad (Z := \mathrm{Tr}\,\hat{\rho}),$$

is, in our perturbation expansion, the sum of two terms, one zeroth order in $\beta_a$,

$$\langle\hat{j}_l(\boldsymbol{x}, t)\rangle^{(0)} = \frac{1}{Z^{(0)}}\mathrm{Tr}\Big[e^{-\beta_0\hat{\mathcal{K}}}\hat{j}_l(\boldsymbol{x}, t)\Big] \qquad (Z^{(0)} := \mathrm{Tr}\,e^{-\beta_0\hat{\mathcal{K}}}), \tag{2.6}$$

and the other in the first order,

$$\langle\hat{j}_l(\boldsymbol{x}, t)\rangle^{(1)}$$
$$= \frac{1}{Z^{(0)2}}\mathrm{Tr}\Big[e^{-\beta_0\hat{\mathcal{K}}}\int_0^1 d\lambda \int d^3x'\beta_a(\boldsymbol{x}')e^{\lambda\beta_0\hat{\mathcal{K}}}\hat{\mathcal{K}}(\boldsymbol{x})e^{-\lambda\beta_0\hat{\mathcal{K}}}\Big] \cdot \mathrm{Tr}\Big[e^{-\beta_0\hat{\mathcal{K}}}\hat{j}_l(\boldsymbol{x}, t)\Big]$$
$$- \frac{1}{Z^{(0)}}\mathrm{Tr}\Big[e^{-\beta_0\hat{\mathcal{K}}}\int_0^1 d\lambda \int d^3x'\beta_a(\boldsymbol{x}')\hat{\mathcal{K}}(\boldsymbol{x}')\hat{j}_l(\boldsymbol{x}, \tau)\Big]\Big\}. \tag{2.7}$$

We shall now show that (2.6) and the first term in (2.7) vanish provided that the space reflection $\hat{\mathcal{P}}$ commutes with $\hat{\mathcal{K}}$. The current density is transformed as

$$\hat{\mathcal{P}}^{-1}\hat{j}_l(\boldsymbol{x}, t)\hat{\mathcal{P}} = \frac{1}{2V}\sum_{\mathbf{q}'\mathbf{q}}(q_l + q_l')\,e^{i\hat{\mathcal{K}}t}\hat{a}_{-\mathbf{q}'}^\dagger\hat{a}_{-\mathbf{q}}e^{-i\hat{\mathcal{K}}t}e^{i(\mathbf{p}-\mathbf{p}')\cdot\mathbf{x}} = -\hat{j}_l(-\boldsymbol{x}, t). \tag{2.8}$$

But, since (1) $[\hat{P}, \hat{K}] = 0$ by assumption, we have

$$\text{Tr}\Big[e^{-\beta_0 \hat{K}} \hat{P}^{-1} \hat{j}_l(\boldsymbol{x}, t)\hat{P}\Big] = \text{Tr}\Big[\hat{P}^{-1} e^{-\beta_0 \hat{K}} \hat{j}_l(\boldsymbol{x}, t)\hat{P}\Big]$$

and since (2) the trace is invariant under cyclic permutation of operators,

$$\text{Tr}\Big[\hat{P}^{-1} e^{-\beta_0 \hat{K}} \hat{j}_l(\boldsymbol{x}, t)\hat{P}\Big] = \text{Tr}\Big[e^{-\beta_0 \hat{K}} \hat{j}_l(\boldsymbol{x}, t)\hat{P}\hat{P}^{-1}\Big] = \text{Tr}\Big[e^{-\beta_0 \hat{K}} \hat{j}_l(\boldsymbol{x}, t)\Big],$$

we find

$$\langle \hat{j}_l(\boldsymbol{x}, t)\rangle^{(0)} = -\langle \hat{j}_l(-\boldsymbol{x}, t)\rangle^{(0)} \tag{2.9}$$

using (2.8). But, this quantity is independent of $\boldsymbol{x}$,

$$\langle \hat{j}_l(\boldsymbol{x}, t)\rangle^{(0)} = \langle \hat{j}_l(0, t)\rangle^{(0)}, \tag{2.10}$$

and hence $\langle \hat{j}_l(\boldsymbol{x}, t)\rangle^{(0)} = 0$, implying that the first term in (2.7) vanishes.

To see (2.10), we note that the space translation $e^{i\hat{P}\cdot\boldsymbol{x}}$ commutes with $\hat{K}$,

$$\text{Tr}\Big[e^{-\beta_0 \hat{K}} e^{-i\hat{P}\cdot\boldsymbol{x}} \hat{j}_l(0, t) e^{i\hat{P}\cdot\boldsymbol{x}}\Big] = \text{Tr}\Big[e^{-i\hat{P}\cdot\boldsymbol{x}} e^{-\beta_0 \hat{K}} \hat{j}_l(0, t) e^{i\hat{P}\cdot\boldsymbol{x}}\Big],$$

and the trace is invariant under cyclic permutation of operators.

Thus, the contribution first order in $\beta_a(\boldsymbol{x})$ turns out to be given by

$$\langle \hat{j}_l(\boldsymbol{x}, t)\rangle^{(1)} = -\frac{1}{Z^{(0)}}\text{Tr}\Big[e^{-\beta_0 \hat{K}} \int_0^1 d\lambda \int d^3x' \beta_a(\boldsymbol{x}')\hat{K}(\boldsymbol{x}')\hat{j}_l(\boldsymbol{x}, \tau)\Big]. \tag{2.11}$$

The space reflection,

$$\langle \hat{P}^{-1} \hat{j}_l(\boldsymbol{x})\hat{P}\rangle = -\frac{1}{Z^{(0)}}\text{Tr}\Big[e^{-\beta_0 \hat{K}} \int_0^1 d\lambda \int d^3x' \beta_a(-\boldsymbol{x}')\hat{K}(\boldsymbol{x}')\hat{P}^{-1}\hat{j}_l(\boldsymbol{x}, \tau)\hat{P}\Big],$$

implies by the transformation properties (2.8) and $\hat{P}^{-1}\hat{K}(\boldsymbol{x})\hat{P} = \hat{K}(-\boldsymbol{x})$ that

$$\langle \hat{j}_l(\boldsymbol{x}, t)\rangle^{(1)} \quad \text{is} \quad \left\{ \begin{array}{c} \text{even} \\ \text{odd} \end{array} \right\} \quad \text{if} \quad \beta_a(\boldsymbol{x}) \quad \text{is} \quad \left\{ \begin{array}{c} \text{odd} \\ \text{even}. \end{array} \right. \tag{2.12}$$

We can deduce this much from the symmetry of the current density under the space reflection. The current can be non-vanishing only when the temperature is inhomogeneous. That there is in fact a case of nonvanishing current will be shown by an example in §3.

## 2.4 Number density

The average of the particle-number density (2.5) is given, up to the first order in $\beta_a(\boldsymbol{x})$ by the sum of

$$\langle \hat{j}_0(\boldsymbol{x}, t)\rangle^{(0)} = \frac{1}{Z^{(0)}}\text{Tr}\Big[e^{-\beta_0 \hat{K}} \hat{j}_0(\boldsymbol{x})\Big], \tag{2.13}$$

$\langle \hat{j}_0(\boldsymbol{x}, t) \rangle^{(1)}$

$$= \frac{1}{Z^{(0)2}} \text{Tr}\left[e^{-\beta_0 \hat{\mathcal{K}}} \int_0^1 d\lambda \int d^3 x' \beta_a(\boldsymbol{x}') e^{\lambda \beta_0 \hat{\mathcal{K}}} \hat{\mathcal{K}}(\boldsymbol{x}') e^{-\lambda \beta_0 \hat{\mathcal{K}}}\right] \cdot \text{Tr}\left[e^{-\beta_0 \hat{\mathcal{K}}} \hat{j}_0(\boldsymbol{x}, t)\right]$$

$$- \frac{1}{Z^{(0)}} \text{Tr}\left[e^{-\beta_0 \hat{\mathcal{K}}} \int_0^1 d\lambda \int d^3 x' \beta_a(\boldsymbol{x}') \hat{\mathcal{K}}(\boldsymbol{x}) \hat{j}_0(\boldsymbol{x}, \tau)\right]. \tag{2.14}$$

The zeroth order contribution is independent of $\boldsymbol{x}$ and $t$ similarly to (2.10):

$$\langle \hat{j}_0(\boldsymbol{x}, t) \rangle^{(0)} = \langle \hat{j}_0(0, 0) \rangle^{(0)} := \rho_0, \tag{2.15}$$

giving a uniform background. The first term in (2.14) is zero for the odd $\beta_a(\boldsymbol{x})$:

$$\text{Tr}\left[e^{-\beta_0 \hat{\mathcal{K}}} \int d^3 x' \beta_a(\boldsymbol{x}') e^{\lambda \beta_0 \hat{\mathcal{K}}} \hat{\mathcal{K}}(\boldsymbol{x}') e^{-\lambda \beta_0 \hat{\mathcal{K}}}\right]$$

$$= \text{Tr}\left[e^{-\beta_0 \hat{\mathcal{K}}} \int d^3 x' \beta_a(\boldsymbol{x}') e^{\lambda \beta_0 \hat{\mathcal{K}}} \hat{\mathcal{K}}(0) e^{-\lambda \beta_0 \hat{\mathcal{K}}}\right] = 0.$$

Thus, the part of the particle density that may vary with $\boldsymbol{x}$ and $t$ is given by

$$\langle \hat{j}_0(\boldsymbol{x}, t) \rangle^{(1)} = -\frac{1}{Z^{(0)}} \text{Tr}\left[e^{-\beta_0 \hat{\mathcal{K}}} \int_0^1 d\lambda \int d^3 x' \beta_a(\boldsymbol{x}') \hat{\mathcal{K}}(\boldsymbol{x}') \hat{j}_0(\boldsymbol{x}, \tau)\right]. \tag{2.16}$$

We can check that the density gradient, if any, is due to the temperature gradient:

$$\frac{\partial}{\partial x_l} \langle \hat{j}_0(\boldsymbol{x}, t) \rangle^{(1)} = -\frac{1}{Z^{(0)}} \text{Tr}\left[e^{-\beta_0 \hat{\mathcal{K}}} \int_0^1 d\lambda \int d^3 x' \frac{\partial \beta_a(\boldsymbol{x}')}{\partial x_l'} \hat{\mathcal{K}}(\boldsymbol{x}') \hat{j}_0(\boldsymbol{x}, \tau)\right]. \tag{2.17}$$

We see a general tendency that the particle-number density is higher at the places where the temperature is lower.

## 3. CASE OF FREE FIELD

We wish to illustrate that a temperature gradient can induce particle current with an example of the free field, for which the Hamiltonian density $\hat{\mathcal{H}}(\boldsymbol{x})$ is given by (1.4) with $\tilde{v}(\boldsymbol{k}) = 0$. For $\hat{\mathcal{K}}(\boldsymbol{x}) = \hat{\mathcal{H}}(\boldsymbol{x}) - \mu \hat{N}(\boldsymbol{x})$, we have

$$\int \beta_a(\boldsymbol{x}) \hat{\mathcal{K}}(\boldsymbol{x}) d^3 x = \frac{1}{V} \sum_{p'p} \tilde{\beta}_a(\boldsymbol{p}' - \boldsymbol{p}) \{(\boldsymbol{p}' \cdot \boldsymbol{p}) - \mu\} \hat{a}_{p'}^\dagger \hat{a}_p,$$

where the Fourier transform is defined by $\tilde{f}(\boldsymbol{k}) = \int d^3 x f(\boldsymbol{x}) e^{-i \boldsymbol{k} \cdot \boldsymbol{x}}$.

By (2.11), the average of the current density (2.5) can be written as

$$\langle \hat{j}_l(\boldsymbol{x}, t) \rangle = -\frac{1}{Z^{(0)} V^2} \sum_{p'pq'q} \tilde{\beta}_a(\boldsymbol{p}' - \boldsymbol{p})[(\boldsymbol{p}' \cdot \boldsymbol{p}) - \mu](q_l + q_l')$$

$$\times \text{Tr}\left[e^{-\beta_0 \hat{\mathcal{K}}} \hat{a}_{p'}^\dagger \hat{a}_p \hat{a}_{q'}^\dagger \hat{a}_q\right] e^{i(\boldsymbol{q} - \boldsymbol{q}') \cdot \boldsymbol{x} - i(\omega_q - \omega_{q'}) \tau}. \tag{3.1}$$

The trace in (3.1) is non-vanishing either for $p' = p$ and $q' = q$, or for $p = q'$ and $p' = q$. However, the former makes $\tilde{\beta}_a = 0$ giving no contribution to the current, and only the latter counts. In effect, we have

$$\frac{1}{Z^{(0)}} \mathrm{Tr}\left[e^{-\beta_0 \check{K}} \hat{a}_{p'}^\dagger \hat{a}_p \hat{a}_{q'}^\dagger \hat{a}_q\right] = f(\omega_{p'})[f(\omega_p) + 1]\delta_{p',q}\delta_{p,q'},$$

where

$$f(\omega_p) := \frac{1}{e^{\beta_0(\omega_p - \mu)} - 1}$$

with the chemical potential being determined by the density $\rho_0$ of the particles,

$$\rho_0 = \frac{N}{V} = \frac{1}{(2\pi)^3} \int \frac{1}{e^{\beta_0(\omega_p - \mu)} - 1} d^3p.$$

At high temperature approximation such that $e^{\beta_0(\omega - \mu)} \gg 1$, we have

$$f(\omega_p) = e^{\beta_0 \mu} e^{-\beta_0 \omega_p}, \quad f(\omega_p) + 1 \sim 1$$

so that, writing the particle mass $m$ and $\hbar$ explicitly,

$$\frac{N}{V} = \frac{e^{\beta_0 \mu}}{(2\pi\hbar)^3} \int e^{-\beta_0 p^2/(2m)} d^3p = \left(\frac{m}{2\pi\hbar^2\beta_0}\right)^{3/2} e^{\beta_0 \mu} \tag{3.2}$$

and

$$\langle \hat{j}_l(\boldsymbol{x}, t)\rangle^{(1)} = -\frac{e^{\beta_0 \mu}}{V^2} \sum_{p'p} \{(\boldsymbol{p}' \cdot \boldsymbol{p}) - \mu\}\tilde{\beta}_a(\boldsymbol{p}' - \boldsymbol{p})(p'_l + p_l)e^{i(\mathbf{p}' - \mathbf{p})\cdot\mathbf{x}}e^{-i\tau(\omega_{p'} - \omega_p)}e^{-\beta_0 p'^2}.$$

Put $p' = P + k/2$, $p = P - k/2$, then, in the limit of $V \to \infty$,

$$\langle \hat{j}_l(\boldsymbol{x}, t)\rangle^{(1)} = -\frac{e^{\beta_0 \mu}}{(2\pi)^6} \int_0^1 d\lambda \int d^3k \tilde{\beta}_a(\boldsymbol{k})e^{i\mathbf{k}\cdot\mathbf{x}}e^{-\beta_0 k^2/4}$$

$$\times \frac{1}{-i\tau'} \int d^3P \left(P^2 - \frac{k^2}{4} - \mu\right) e^{-\beta_0 P^2} \frac{\partial}{\partial k_l} e^{-2i\tau' P \cdot k}, \tag{3.3}$$

where

$$\tau' := \tau - \frac{i}{2}\beta_0 = t + i\left(\lambda - \frac{1}{2}\right)\beta_0.$$

We begin with the $P$-integration. The integration over the angle between $P$ and $k$, followed by the partial differentiation $\partial/\partial k_l$ yields

$$\langle j_l(\boldsymbol{x}, t)\rangle^{(1)} = -\frac{e^{\beta_0 \mu}}{(2\pi)^6} \int_0^1 d\lambda \int d^3k \tilde{\beta}_a(\boldsymbol{k})e^{i\mathbf{k}\cdot\mathbf{x}}e^{-\beta_0 k^2/4}\frac{2\pi}{\tau'}\frac{ik_l}{k^2}\left(\frac{i}{2k\tau'}I_1 + I_2\right)$$

where

$$I_\nu := \int_{-\infty}^{\infty} P^\nu \left(P^2 - \frac{k^2}{4} - \mu\right) e^{-\beta_0 P^2} e^{2i\tau' Pk} dP \quad (\nu = 1, 2). \tag{3.4}$$

By shifting the path of integration on the complex $P$-plane in the standard way, we carry out the Gaussian integration. The result can be written as

$$\langle \hat{j}_l(\boldsymbol{x}, t)\rangle^{(1)} = \frac{1}{(2\pi)^3} \int d^3y \frac{\partial}{\partial x_l} G(|\boldsymbol{x} - \boldsymbol{y}|, t) \beta_a(\boldsymbol{y}) \qquad (3.5)$$

by defining

$$G(x, t) := \int_{-1/2}^{1/2} d\lambda' \tau' \int d^3k \frac{e^{\beta_0 \mu}}{2^{1/2}(2\pi)^{3/2}\beta_0^{7/2}} \left\{ \frac{5}{2} - \left( \frac{\tau'^2}{\beta_0} + \frac{\beta_0}{4} \right) k^2 - \beta_0 \mu \right\}$$
$$\times \exp\left[ i\boldsymbol{k} \cdot \boldsymbol{x} - \left( \frac{\tau'^2}{\beta_0} + \frac{\beta_0}{4} \right) k^2 \right], \qquad (3.6)$$

where $\lambda' = \lambda - 1/2$ and $\tau' = t + i\lambda'\beta_0$. In (3.5), we notice that $-\partial/\partial x_l$ can be replaced by $\partial/\partial y_l$ and then shifted on $\beta_a$ by integration by parts; the result exhibits that the current is due to the temperature gradient.

By shifting again the path of integration on the complex $k$-plane, we obtain

$$G(x, t) = \frac{e^{\beta_0 \mu}}{4\beta_0^2} \int_{-1/2}^{1/2} d\lambda' \tau' \left( \frac{1}{\tau'^2 + (\beta_0/2)^2} \right)^{3/2} \left( 1 - \beta_0\mu + \frac{1}{4} \frac{\beta_0 x^2}{\tau'^2 + (\beta_0/2)^2} \right)$$
$$\times \exp\left[ -\frac{1}{4} \frac{\beta_0 x^2}{\tau'^2 + (\beta_0/2)^2} \right]. \qquad (3.7)$$

After a long time $t \gg \hbar/(k_B T)$, we have

$$G(x, t) = \frac{e^{\beta_0 \mu}}{4\beta_0^2 t^2} \left\{ 1 - \beta_0\mu + \frac{\beta_0}{4}\left(\frac{x}{t}\right)^2 \right\} \exp\left[ -\frac{\beta_0}{4}\left(\frac{x}{t}\right)^2 \right] \int_{-1/2}^{1/2} d\lambda' \exp\left[ i\frac{\beta_0^2 x^2}{2t^3}\lambda' \right],$$

so that

$$G(x, t) = \frac{e^{\beta_0 \mu}}{4\beta_0^2 t^2} \left\{ 1 - \beta_0\mu + \frac{\beta_0}{4}\left(\frac{x}{t}\right)^2 \right\} \exp\left[ -\frac{\beta_0}{4}\left(\frac{x}{t}\right)^2 \right] \frac{\sin \chi(x, t)}{\chi(x, t)}, \qquad (3.8)$$

where, recalling $m$ and $\hbar$

$$\chi(x, t) := \frac{\beta_0^2}{4t}\left(\frac{x}{t}\right)^2 = \beta_0^2 \frac{\hbar}{t} \frac{m}{2}\left(\frac{x}{t}\right)^2.$$

The factor $\dfrac{\sin \chi}{\chi}$, standing for a quantum effect, is 1 after a macroscopic time. Then,

$$\frac{\partial}{\partial x_l} G(x, t) = -\frac{1}{2^3} \frac{x_l}{t^4} \left\{ \frac{1}{4}\left(\frac{x}{t}\right)^2 - \mu \right\} e^{-\beta_0\{(x/t)^2/4 - \mu\}} \qquad (3.9)$$

and hence

$$\langle j_1(\boldsymbol{x}, t)\rangle = -\frac{1}{(4\pi)^3} \frac{1}{t^4} \int e^{-\beta_0\{(y/t)^2/4 - \mu\}} \beta_a(\boldsymbol{x} - \boldsymbol{y}) y_l \left\{ \frac{1}{4}\left(\frac{y}{t}\right)^2 - \mu \right\} d^3y, \quad (3.10)$$

which can be interpreted in terms of classical statistics as "ballistic" behavior as expected for free particles (see Appendix).

## 4. A MODEL OF INTERACTING BOSE GAS

To illustrate that diffusive behaviour can result from the interaction of particles even when their equation of motion is reversible, we consider a simple model[6,7] of a Bose gas (particle mass: 1/2) with a quadratic "Hamiltonian,"

$$\hat{\mathcal{K}} = \hat{\mathcal{H}} - \mu_0 \hat{N} := \sum_p p^2 \hat{a}_p^\dagger \hat{a}_p + \frac{N_0}{2V} \sum_{p \leq K} v_0 (\hat{a}_p^\dagger \hat{a}_{-p}^\dagger + 2\hat{a}_p^\dagger \hat{a}_p + \hat{a}_p \hat{a}_{-p}). \tag{4.1}$$

The density of the "Hamiltonian" may be taken as

$$\hat{\mathcal{K}}(x) = \frac{\partial \hat{\phi}^\dagger(x)}{\partial x_l} \frac{\partial \hat{\phi}(x)}{\partial x_l} + \frac{\rho_0}{2} \int v(r) \Big[ \hat{\phi}^\dagger(x + \tfrac{1}{2}r) \hat{\phi}^\dagger(x - \tfrac{1}{2}r)$$
$$+ \hat{\phi}^\dagger(x + \tfrac{1}{2}r) \hat{\phi}(x - \tfrac{1}{2}r) + \hat{\phi}^\dagger(x - \tfrac{1}{2}r) \hat{\phi}(x + \tfrac{1}{2}r) + \hat{\phi}(x + \tfrac{1}{2}r) \hat{\phi}(x - \tfrac{1}{2}r) \Big] d^3 r$$

where $v(r)$ is such that

$$\tilde{v}(k) = \int v(r) e^{-ik \cdot r} d^3 r = \begin{cases} v_0 & (k \leq K) \\ 0 & (k > K), \end{cases} \quad \text{or} \quad v(r) = 4\pi^2 v_0 K^3 j_1(Kr)/(Kr).$$

In terms of the Fourier transform, we find

$$\int v(r) \hat{\phi}^\dagger(x + \tfrac{1}{2}r) \hat{\phi}^{(\dagger)}(x - \tfrac{1}{2}r) d^3 r = \frac{1}{V} \sum_{p'p} \tilde{v}\Big(\frac{p' + p}{2}\Big) \hat{a}_{p'}^\dagger \hat{a}_{-p}^{(\dagger)} e^{i(p - p') \cdot x}$$

and similarly for their Hermitian conjugates. We assume for simplicity that $K$ is sufficiently large to make all the $\tilde{v}(k)$'s for thermally agitated particles to have the constant value $v_0$; we shall hereafter suppress the proviso $p \leq K$.

We wish to find the long-time behavior of the particle-number density (2.5),

$$\langle \hat{j}_0(x, t) \rangle^{(1)} = -\frac{1}{Z^{(0)}} \text{Tr} \left[ e^{-\beta_0 \hat{\mathcal{K}}} \int_0^1 d\lambda \int d^3 x' \beta_a(x') \hat{\mathcal{K}}(x') e^{i\hat{\mathcal{K}}\tau} \hat{j}_0(x) e^{-i\hat{\mathcal{K}}\tau} \right]. \tag{4.2}$$

The "Hamiltonian" (4.1) can be diagonalized as

$$\hat{\mathcal{K}} = \sum_p \omega_p \hat{\alpha}_p^\dagger \hat{\alpha}_p \quad (\omega_p := p\sqrt{p^2 + 2c^2}, \quad c^2 := \rho_0 v_0) \tag{4.3}$$

where the zero-point energy has been dropped, by a Bogolubov transformation

$$\hat{a}_p = \hat{\alpha}_p \cosh \theta_p - \hat{\alpha}_{-p}^\dagger \sinh \theta_p,$$

with

$$\sinh^2 \theta_p = \frac{p^2 + c^2}{2p\sqrt{p^2 + 2c^2}} - \frac{1}{2}, \quad \cosh \theta_p \sinh \theta_p = \frac{c^2}{2p\sqrt{p^2 + 2c^2}}. \tag{4.4}$$

For the factors in (4.2), we have

$$\int d^3x' \beta_a(x') \hat{K}(x') = \frac{1}{V} \sum_{p'p} \tilde{\beta}(p' - p)(p' \cdot p + c^2)$$

$$\times (\hat{\alpha}_{p'}^\dagger \cosh \theta_{p'} - \hat{\alpha}_{-p'} \sinh \theta_{p'})(\hat{\alpha}_p \cosh \theta_p - \hat{\alpha}_{-p}^\dagger \sinh \theta_p) + \cdots \qquad (4.5)$$

and

$$\hat{j}_0 = \frac{1}{V} \sum_{q'q} e^{i\hat{K}\tau}(\hat{\alpha}_{q'}^\dagger \cosh \theta_{q'} - \hat{\alpha}_{-q'} \sinh \theta_{q'})(\hat{\alpha}_q \cosh \theta_q - \hat{\alpha}_{-q}^\dagger \sinh \theta_q)e^{-i\hat{K}\tau}e^{i(q-q')\cdot x} + \cdots.$$

$$(4.6)$$

The thermal average turns out to be given by the sum:

$$\langle j_0(x,t) \rangle = \langle j_0(x,t) \rangle^{(s,-)} + \langle j_0(x,t) \rangle^{(s,+)} + \langle j_0(x,t) \rangle^{(c,-)} + \langle j_0(x,t) \rangle^{(c,+)},$$

where

$$\langle j_0(x,t) \rangle^{(\kappa,\pm)} = \frac{-1}{2(2\pi)^6} \int d^3k \, \tilde{\beta}(k) e^{ik\cdot x} \int d\lambda F^{(\kappa,\pm)}(k,\lambda) \qquad (4.7)$$

with

$$F^{(s,\pm)}(k,\tau) = \int d^3P \, g(p',p) \sinh(\theta_{p'} + \theta_p) \begin{cases} f(\omega_{p'})f(\omega_p)e^{-i(\omega_{p'}+\omega_p)\tau} \\ \{f(\omega_{p'}) + 1\}(f(\omega_p) + 1)e^{i(\omega_{p'}+\omega_p)\tau} \end{cases}$$

$$(4.8)$$

and

$$F^{(c,\pm)}(k,\tau) = \int d^3P \, h(p',p) \cosh(\theta_{p'} + \theta_p) \begin{cases} f(\omega_{p'})\{f(\omega_p) + 1\}e^{-i(\omega_{p'}-\omega_p)\tau} \\ \{f(\omega_{p'}) + 1\}f(\omega_p)e^{i(\omega_{p'}-\omega_p)\tau}, \end{cases} \qquad (4.9)$$

Here, $p' = P + k/2$, $p = P - k/2$, and

$$\left. \begin{aligned} g(p',p) \\ h(p',p) \end{aligned} \right\} := (p' \cdot p) \begin{Bmatrix} \sinh \\ \cosh \end{Bmatrix} (\theta_{p'} + \theta_p) \mp c^2(\cosh \theta_{p'} - \sinh \theta_{p'})(\cosh \theta_p - \sinh \theta_p),$$

$$(4.10)$$

of which $g$ is associated with the off-diagonal term $\alpha_{p'}^\dagger \alpha_{-p}^\dagger$ and $h$ with the "diagonal" $\alpha_{p'}^\dagger \alpha_p$ in the "Hamiltonian" density $\hat{K}(x)$ [cf. (4.5)]. We note

$$\frac{1}{Z^{(0)}} \text{Tr}\left[e^{-\beta_0 \hat{K}} \alpha_{p'}^\dagger \alpha_{-p}^\dagger \alpha_{-q'} \alpha_q\right] = f(\omega_{p'})f(\omega_p)(\delta_{p',-q'}\delta_{-p,q} + \delta_{p',q}\delta_{p,q'}). \qquad (4.11)$$

## 4.1   Evaluation of the integral $F^{(s,\pm)}$

As a key for the following calculations, the temperature inhomogeneity $\beta(x)$ is assumed to be varying on a macroscopic scale, so that the support of its Fourier transform is very small in comparison with the scales on which $\omega_p$ and the coefficients of the Bogoliubov transformation vary.

We begin the evaluation of (4.7) from the $P$-integration,

$$F^{(s,+)}(\boldsymbol{k},\tau) = \int P^2 dP \, \sinh[2\theta_P]\{f(\omega_P)\}^2 \int d\Omega_P g(\boldsymbol{p}',\boldsymbol{p}) e^{-i(\omega_{p'}+\omega_p)\tau}, \qquad (4.12)$$

where, in sinh and $f$, we have put $\boldsymbol{p} = \boldsymbol{p}' = \boldsymbol{P}$ taking advantage of the small support of $\tilde{\beta}(\boldsymbol{k})$. However, we cannot do so in $g$ and the exponent, because $g$ would vanish if we did and the exponent is proportional to $\tau$ that involves the macroscopicaly large variable $t$ . Instead, we expand to order $k^2$,

$$\omega_{p'} + \omega_p = 2\omega_p + \frac{P^2 + c^2}{2\omega_p}k^2 - \frac{c^4}{\omega_p^3}(\boldsymbol{P} \cdot \boldsymbol{k})^2 \qquad (4.13)$$

and put

$$g(\boldsymbol{p}',\boldsymbol{p}) = g_0(P) + g_1(P)(\boldsymbol{P} \cdot \boldsymbol{k}) + g_2(P)\boldsymbol{k} \cdot \boldsymbol{k} + g_3(P)\frac{(\boldsymbol{P} \cdot \boldsymbol{k})^2}{c^2};$$

finding, however, that $g_0 = g(\boldsymbol{P},\boldsymbol{P}) = 0$ by the Bogolubov transformation diagonalizing $\hat{\mathcal{K}}$ [cf, the remark after (4.10)], $g_1 = 0$ by the symmetry $g(\boldsymbol{p}',\boldsymbol{p}) = g(\boldsymbol{p},\boldsymbol{p}')$ from (4.10); we take in advance the later result into account that

$$g_3(P) = \cosh 2\theta_P \sinh^2 2\theta_P$$

vanishes at the value of $P$ that concerns us; see (4.24) below. Thus,

$$g(\boldsymbol{p}',\boldsymbol{p}) = g_2(P) \quad \text{with} \quad g_2(P) = -\frac{1}{2}\sinh 2\theta_P. \qquad (4.14)$$

The integration over the angles in (4.12),

$$I_0 = \int_0^\pi e^{i\alpha \cos^2 \gamma} \sin\gamma d\gamma = \frac{2}{\sigma}\int_0^\sigma e^{-v^2} dy,$$

is carried out by adding and subtracting $\int_1^\infty e^{-(-i\alpha)z^2} dz$ taking advantage of Im $\alpha = \lambda\beta_0 > 0$, where $\gamma$ is the angle between $\boldsymbol{P}$ and $\boldsymbol{k}$,

$$\alpha = \frac{c^4 P^2}{\omega_P^3}k^2\tau, \qquad \sigma := (-i\alpha)^{1/2}, \quad \text{and} \quad y = \sigma x. \qquad (4.15)$$

We note $-\pi/2 < \arg\sigma < 0$ and $|\sigma| \to \infty$ as $t \to \infty$. Then, asymptotically

$$I_0 \sim \frac{\sqrt{\pi}}{\sigma} - \frac{1}{\sigma^2}e^{-\sigma^2}. \qquad (4.16)$$

Thus,

$$F^{(s,+)}(\boldsymbol{k},\tau) \sim 2\pi \int_0^\infty P^2 dP \, \sinh[2\theta_P]\Big(f(P)\Big)^2 g_2(P)k^2$$

$$\times \left\{\frac{\sqrt{\pi}}{\sigma} - \frac{1}{\sigma^2}\exp\left[i\frac{c^4 P^2}{\omega_P^3}k^2\tau\right]\right\} \exp\left[-i\left(2\omega_P + \frac{P^2 + c^2}{2\omega_P}k^2\right)\tau\right]. \quad (4.17)$$

Similar calculations augmented by adding and subtracting $\int_1^{i\infty} e^{-i\alpha z^2} dz$ give

$$F^{(s,-)}(k,\tau) \sim 2\pi \int_0^\infty P^2 dP \, \sinh[2\theta_P]\Big(f(P)+1\Big)^2 g_2(P)k^2$$
$$\times \left\{ i\frac{\sqrt{\pi}}{\sigma} + \frac{1}{\sigma^2}\exp\left[-i\frac{c^4 P^2}{\omega_P^3}k^2\tau\right] \right\} \exp\left[ i\left(2\omega_P + \frac{P^2+c^2}{2\omega_P}k^2\right)\tau\right]. \quad (4.18)$$

If we change the variable of integration $P$ to $-P$ in (4.18) after analytically continuing the integrand to the complex $P$-plane with two cuts to cope with the singularity of $(P^2+2c^2)^{1/2}$; one extending along the imaginary axis from $i\sqrt{2}c$ to $i\infty$, and the other from $-i\sqrt{2}c$ to $-i\infty$, then $\omega_{-P} = -\omega_P$, and consequently

$$f(\omega_{-P}) + 1 = \frac{e^{-\beta_0 \hbar \omega_P}}{e^{-\beta_0 \hbar \omega_P} - 1} = -f(\omega_P), \quad \sinh\theta_{-P} = -\sinh\theta_P \quad (4.19)$$

so that for $F^{(s)} := F^{(s,+)} + F^{(s,-)}$,

$$F^{(s)}(k,\tau) \sim \int_{-\infty}^\infty dP\, G(P)$$
$$\times \left\{ \frac{\sqrt{\pi}}{\sigma} - \frac{1}{\sigma^2}\exp\left[i\frac{c^4 P^2}{\omega_P^3}k^2\tau\right] \right\} \exp\left[-i\left(2\omega_P + \frac{P^2+c^2}{2\omega_P}k^2\right)\tau\right], \quad (4.20)$$

where

$$G(P) := 2\pi P^2 \sinh[2\theta_P]\{f(\omega_P)\}^2 g_2(P)\, k^2.$$

### 4.1.1   Long-time behavior of $F_1^{(s)}$

The time scale we are interested in is macroscopic. To evaluate the integral (4.20) asymptotically for $t \to \infty$, we use the saddle-point method. Candidates for the saddle point $P_{s,1}$ for the part of the integral (4.20),

$$J_1 := -\frac{1}{\sigma^2}\int_{-\infty}^\infty dP\, G(P)\exp\left[-i\left\{2\omega_P + \left(\frac{P^2+c^2}{2\omega_P} - \frac{c^4 P^2}{\omega_P^3}\right)k^2\right\}\tau\right], \quad (4.21)$$

are determined by the condition of vanishing derivative of the exponent,

$$2\frac{d\omega_P}{dP} - \left\{ \left(\frac{P^2+c^2}{2\omega_P^2} - \frac{3c^4 P^2}{\omega_P^4}\right)\frac{d\omega_P}{dP} + \left(\frac{P}{\omega_P} - \frac{2c^4 P}{\omega_P^3}\right) \right\}k^2 = 0,$$

where

$$\frac{d\omega_P}{dP} = \frac{2(P^2+c^2)}{\sqrt{P^2+2c^2}}.$$

For small $k$, the main part of the root is given by $P = \pm ic$. Then, we put $P^2 = -c^2 + \xi$, finding $\xi = -(3/4)k^2$ up to order $k^2$, and hence $P_{s,1} = \pm ic\{1 + (3/8)(k^2/c^2)\}$ as candidates for the saddle points. Since

$$\frac{d^2\omega_P}{dP^2} = \frac{2P(P^2+3c^2)}{(P^2+2c^2)^{3/2}}.$$

takes the value $\pm 4i + O(k^2/c^2)$ at the saddle points, the exponent behaves like

$$2\omega_P + \left(\frac{P^2 + c^2}{2\omega_P} - \frac{c^4 P^2}{\omega_P^3}\right) k^2 = \pm i\left\{(2c^2 + k^2) + 4(P - P_{s,1})^2\right\}$$

in the neighborhod of $P_{s,1}$, so that

$$J_1 \sim -\frac{1}{\sigma^2} e^{\pm(2c^2 + k^2)\tau} \int_{-\infty}^{\infty} G(P) e^{\pm 4(P - P_{s,1})^2 \tau} dP. \tag{4.22}$$

For $\tau \sim t \to +\infty$, therefore, we should take the lower sign here

$$P_{s,1} = -ic\left(1 + \frac{1}{8}\frac{k^2}{c^2}\right)$$

for the saddle point which the path of integration should pass. Thus,

$$F^{(s)} \sim -\frac{1}{\sigma(P)^2} G(P_{s,1}) e^{-(2c^2 + k^2)\tau} \int_{-\infty}^{\infty} e^{-4\tau s^2} ds, \tag{4.23}$$

and we have only to carry out the expansion (4.14) around $P = P_{s,1}$ , obtaining

$$\sigma(P_{s,1})^2 = k^2 \tau, \qquad \sinh[2\theta_{P_{s,1}}] = i, \qquad G(P_{s,1}) = \pi c^2 k^2 f(\omega_{P_{s,1}})^2$$

and in particular

$$g_3(P_{s,1}) = \frac{1}{c^2} \cosh 2\theta_{P_{s,1}} \sinh^2 2\theta_{P_{s,1}} = 0, \tag{4.24}$$

which we have invoked before. Thus, we obtain

$$F^{(s)}(k, \tau)_1 \sim -\frac{1}{2}\left(\frac{\pi}{\tau}\right)^{3/2} f(\omega_{P_{s,1}})^2 e^{-(2c^2 + k^2)\tau}. \tag{4.25}$$

Note that the small factor $k^2$ of $G$ has been cancelled by $1/\sigma^2$. The factor $e^{-k^2\tau}$ is a welcome sign of the diffusive behavior. However, it is accompanied by $e^{-2c^2\tau}$, a damping, which is an unexpected feature.

### 4.1.2   Long-time behavior of $F_2^{(s)}$

We still have to evaluate the other part of the integral (4.20),

$$J_2 := \frac{\sqrt{\pi}}{\sigma} \int_{-\infty}^{\infty} dP\, G(P) \exp\left[-i\left(2\omega_P + \frac{P^2 + c^2}{2\omega_P}k^2\right)\tau\right]. \tag{4.26}$$

Candidates for the saddle point $P_{s,2}$ for this integral is given by the roots of

$$2\frac{d\omega_P}{dP} - \left(\frac{P^2 + c^2}{2\omega_P^2}\frac{d\omega_P}{dP} - \frac{P}{\omega_P}\right)k^2 = 0,$$

and hence

$$P_{s,2} = \pm ic \left( 1 + \frac{1}{8}\frac{k^2}{c^2} \right) \tag{4.27}$$

up to order $k^2$. Since

$$\omega_{P_{s,2}} = \pm ic \left( 1 + \frac{k^2}{8c^2} \right) \left( c^2 - \frac{k^2}{4} \right)^{1/2} = \pm ic + O(k^4),$$

we see the exponent in (4.26) behave as

$$-i \left( 2\omega_P + \frac{P^2 + c^2}{2\omega_P} k^2 \right) \tau = \pm 2\{c^2 + 2(P - P_{s,2})^2\}\tau$$

in a neighborhood of $P_{s,2}$. Therefore, we choose the saddle point $P_{s,2} = -ic \left( 1 + \frac{1}{8}\frac{k^2}{c^2} \right)$ to let the path of integration (4.26) pass, obtaining

$$F^{(s)}(\mathbf{k}, \tau)_2 \sim \frac{1}{2} k \left( \frac{\pi^2}{\tau} \right) f(\omega_{P_{s,1}})^2 e^{-2c^2 \tau}, \tag{4.28}$$

which implies a simple damping without any diffusive character.

### 4.1.3   Integration over $\lambda$

It remains to integrate (4.25) and (4.28) over $\lambda$. We may consider only the part,

$$f(\omega_{P_s})^2 \int_0^1 e^{-2ic^2\beta_0\lambda} d\lambda = -\frac{1}{2\beta_0 c^2} \cot \frac{\beta_0 c^2}{2}. \tag{4.29}$$

Thus, summarizing

$$\langle j_0(\mathbf{x}, t) \rangle^{(s)} = -\frac{1}{2(2\pi)^6} \frac{1}{4\beta_0 c^2} \cot \frac{\beta_0 c^2}{2} \left( \frac{\pi}{t} \right)^{3/2} e^{-2c^2 t} \int \left( e^{-k^2 t} - \sqrt{\pi k^2 t} \right) \beta_a(\mathbf{k}) d^3 k. \tag{4.30}$$

Two remarks are in order. (1) $e^{-k^2 t}$ leads to a diffusion. This does not contradict the basic reversibility of our model. If we reversed the time to look at the evolution towards the past, $t < 0$, then we must take the path of integration crossing the other saddle point as given by (4.27) with the upper sign, obtaining the diffusion towards the past. Our result is time symmetric in this sense, yet exhibiting the irreversible features. The same circumstance (but no damping) was found in our previous approach[8] with the thermo field dynamics using a more realistic model.

(2) The simple damping implied by (4.28) should call for an interpretation if it could be taken as an irreversible, behavior. Its amplitude has a factor $k\sqrt{t}$ extra to (4.25), $k$ being small but $t$ large.

## 4.2  Evaluation of the integral (4.9)

Since

$$h(P, P) = \omega_P, \quad \cosh[2\theta_P] = \frac{P^2 + c^2}{\omega_P}, \quad \omega_{P+k/2} - \omega_{P+k/2} = 2\frac{P^2 + c^2}{\omega_P}(P \cdot k)$$

(4.9) turns out to be

$$F^{(c,\pm)} = \int_0^\infty P^2 dP\,(P^2 + c^2)f(\omega_P)\{f(\omega_P) + 1\} \int d\Omega_P \exp\left[\mp 2i\frac{P^2 + c^2}{\sqrt{P^2 + 2c^2}}k\tau\cos\gamma\right].$$

The integration over angles gives the same result $F^{(c)}/2$ for both $F^{(c,\pm)}$:

$$\frac{1}{2}F^{(c)} = \pi\int_0^\infty P^2 dP\,\frac{\sqrt{P^2 + 2c^2}}{ik\tau}f(\omega_P)\{f(\omega_P) + 1\}\sum_{\pm}\pm\exp\left[\pm 2i\frac{P^2 + c^2}{\sqrt{P^2 + 2c^2}}k\tau\right]. \tag{4.31}$$

We use the saddle-point method after analytically continuing the integrand on the complex $P$-plane with two cuts, $[i\sqrt{2}c, \infty)$ and $(-\infty, -i\sqrt{2}c]$ as before. The saddle points are determined from the exponent by

$$\frac{d}{dP}\frac{P^2 + c^2}{(P^2 + 2c^2)^{1/2}} = \frac{P(P^2 + 3c^2)}{(P^2 + 2c^2)^{3/2}} = 0 \tag{4.32}$$

as $P_s = \pm i\sqrt{3}c + \epsilon$, two twins, each of which, being separated by one of the cuts, consists of the right and the left member with an infinitesimal $\epsilon = +0$ and $\epsilon = -0$, respectively. The other root $P = 0$ of (4.32) is discarded because the prefactor vanishes there. In the neighborhood of the saddle points $P_s$, the exponent behaves like

$$\frac{P^2 + c^2}{(P^2 + 2c^2)^{1/2}} = \frac{P^2 + c^2}{(P^2 + 2c^2)^{1/2}}\bigg|_{P=\pm i\sqrt{3}c+\epsilon} + \frac{3c^4}{(P^2 + 2c^2)^{5/2}}\bigg|_{P=\pm i\sqrt{3}c+\epsilon}(P - P_s)^2$$

times $\pm 2ik\tau$. We note

$$(P^2 + 2c^2)^{1/2}\big|_{P=\pm i\sqrt{3}c+\epsilon} = \begin{cases} \pm ic & (\epsilon = +0) \\ \mp ic & (\epsilon = -0). \end{cases}$$

Let us try the right members of the two twins, $P_{s+} = \pm i\sqrt{3}c + 0$. Then,

$$\pm 2ik\tau\frac{P^2 + c^2}{(P^2 + 2c^2)^{1/2}} = \mp 4ck\tau \pm \frac{6k\tau}{c}(P - P_{s+})^2, \tag{4.33}$$

which requires that the path of integration be deformed to run along the imaginary axis and the quadrant of a large circle to come back to the real axis. But, on the circle, the exponential behaves like

$$\left|\exp\left[\pm 2ik\tau\frac{P^2 + c^2}{(P^2 + 2c^2)^{1/2}}\right]\right| \sim \exp[\mp 2k\tau R\sin\phi] \quad (P := Re^{i\phi}) \tag{4.34}$$

requiring that $0 \leq \phi \leq \pi/2$ for $t > 0$. We remark that $|f(\omega_P)\{f(\omega_P) + 1\}|$ behaves on the circle as $e^{-R^2|\cos 2\phi|}$, not decreasing as $R \to \infty$ when $\phi = \pm\pi/4$, while helping suppress (4.34) when $\phi = 0, \pi$.

Thus, we take the path of integration passing the saddle (4.33),

$$\{iy + 0 : y \text{ runs from } 0 \text{ to } \pm R)\} \cup \{Re^{i\phi} : \phi \text{ runs from } \pm \pi/2 \text{ to } 0)\},$$

with $R \to \infty$, obtaining the same results for the two terms $\pm$ in (4.31):

$$F^{(c)} = -\sqrt{\frac{3}{2}}c^{7/2}\left(\frac{\pi}{k\tau}\right)^{3/2} f(\omega_{P_s})\{f(\omega_{P_s}) + 1\}e^{-4ck\tau}. \tag{4.35}$$

We remark that the double poles $P = \pm i\sqrt{2}c$ due to $f(\omega_P)\{f(\omega_P) + 1\}$ have the residues killed by the exponential.

In (4.35), $\lambda$ appears only in $e^{-4ck(t+i\lambda\beta_0)}$ and may be neglected because of the small factor $k$. Then,

$$\langle j_0(\boldsymbol{x}, t)\rangle^{(c)} = \frac{1}{2(2\pi)^6}\sqrt{\frac{3}{2^5}}\frac{c^{7/2}}{\sinh^2\frac{\sqrt{3}\beta_0 c}{2}} \int \beta_a(\boldsymbol{k})\left(\frac{\pi}{k\tau}\right)^{3/2} e^{i\boldsymbol{k}\cdot\boldsymbol{x}-4ckt}d^3\mathbf{k}, \tag{4.36}$$

which is to be added on (4.30).

Note that this result implies a kind of slow oscillatory diffusion as illustrated by

$$\left(\frac{\pi}{k\tau}\right)^{3/2}\int e^{i\boldsymbol{k}\cdot\boldsymbol{x}-4ckt}d^3\mathbf{k} = \frac{4\pi^{5/2}}{|\boldsymbol{x}|}\text{Im}\int_0^\infty \frac{e^{-(4ct-ix)k}}{\sqrt{k}}dk = \frac{8\pi^{5/2}}{|\boldsymbol{x}|}\frac{\sin\phi(|\boldsymbol{x}|, t)}{\{x^2 + (4ct)^2\}^{1/4}}, \tag{4.37}$$

where the sin of

$$\phi(|\boldsymbol{x}|, t) := \frac{1}{2}\tan^{-1}\frac{|\boldsymbol{x}|}{4ct}$$

is responsible for the oscillation mentioned.

## 5. CONCLUSION

We have shown with simple examples (1) that temperature gradient can induce particle current, and (2) that, in a system of interacting particles, diffusive behavior can arise from a reversible Hamiltonian dynamics. The damping and the slow oscillatory diffusion are interesting new features, the former being particularly intriguing. Our model for (2) is a mathematical example without a claim for physical reality.

## APPENDIX. CLASSICAL CURRENT DENSITY

For the ideal gas of particles (mass: $1/2$) having inverse temperature $\beta_0 + \beta_a(\boldsymbol{x})$ at $t = 0$, the number of particles in $d^3\mathbf{x}d^3\mathbf{v}$ at $(\boldsymbol{x}, \boldsymbol{v})$ is given by the Maxwell law :

$$dN(\boldsymbol{x}, \boldsymbol{v}) = \frac{1}{(4\pi)^3}\exp\left[-\{\beta_0 + \beta_a(\boldsymbol{x})\}\left\{\frac{v^2}{4} - \mu\right\}\right] d^3\mathbf{x}d^3\mathbf{v}. \tag{A.1}$$

After a time $t$, the current density at $x$ will be given by

$$j_l(x, t) = -\frac{1}{(4\pi)^3} \int e^{-\beta_0\{(v^2/4)-\mu\}} \beta_a(x - vt) v_l \left(\frac{v^2}{4} - \mu\right) d^3v,$$

to the first order in $\beta_a$. Change the variable of integration to $y = vt$, then

$$j_1(x, t) = -\frac{1}{(4\pi)^3} \frac{1}{t^4} \int e^{-\beta_0\{(y/t)^2/4-\mu\}} \beta_a(x - y) y_l \left\{\frac{1}{4}\left(\frac{y}{t}\right)^2 - \mu\right\} d^3y. \quad (A.2)$$

This is identical with (3.10), the long-time behavior from quantum statistics.

## REFERENCES

1.    Ichiyanagi, M. (1995) Conceptual development of non-equilibrium statistical mechanics in the early Days of Japan. *Physics Reports*, **262**, 227-310 .

2.    Nakano, H. (1993) Linear response theory – A historical perspective. *International Journal of Modern Physics*, **B7**, 2397-2467.

3.    Kubo, R. (1957) Statistical-mechanical theory of irreversible processes, I. General theory and simple application to magnetic and conduction phenomena. *Journal of the Physical Society of Japan*, **12**, 570-586.

4.    Umezawa, H., Matsumoto, H. and Tachiki, M. (1982) *Thermo Field Dynamics and Condensed States*. Amsterdam, North Holland.

5.    Zubarev, D.N. (1974) *Nonequilibrium Statistical Thermodynamics*, tr. by Shepherd, P.J. and ed. by Gray, P. and Shepherd, P.J., p. 294. New York - London, Consultants Bureau. p. 294. Original Russian edition was published in 1971.

6.    Bogoliubov, N.N. (1947) On the Theory of Superfluidity. *Journal of Physics (USSR)*, **11**, 23 - 32.

7.    Huang, K. and Yang, C.N. (1957) Quantum mechanical many-body problem with hard-sphere interaction. *Physical Review* **105** (1957), 767 - 775;

Huang, K., Yang, C.N. and Luttinger, J.M. Imperfect Bose gas with hard sphere interactions. *Physical Review* **105** (1957) 776 - 784;

Huang, K. (1959) Energy levels of a Bose-Einstein system of particles with attractive interactions. *Physical Review*, **115** (1959) 765 - 777.

8.    Ezawa, H., Watanabe, K. and Nakamura, K. (1991) Quantum Field Theory of Thermal Diffusion. in *Thermal Field Theories*, edited by Ezawa, H., Arimitsu, T., Hashimoto, Y., pp.79 - 94, Amsterdam, North Holland.

# SOLUTIONS OF COLORED YANG-BAXTER EQUATION

SHI-KUN WANG, HAI-TANG YANG

Institute of Applied Mathematics, Academia Sinica,
Beijing, 100080, P. R. China

Ke Wu

Institute of Theoretical Physics, Academia Sinica,
Beijing,100080, P. R. China

**Abstract:**   In this paper all seven-vertex type solutions of the colored
Yang-Baxter equation dependent on spectral as well as colored parameters
are given. It is proved that they are composed of six groups of basic solutions
up to five solution transformations. Moreover, all solutions can be classified
into two types called Baxter type and free-fermion type.

## 1. INTRODUCTION

The Yang-Baxter equation (YBE), first appeared in refs. [1-3] plays a prominent
role in many branches of physics. In the field theories, the YBE can determine completely the S-matrix for the two-body scattering amplitudes in the multi-particle
scattering processes [4]. In two dimensional integrable models of quantum fields
theories and statistical models, the YBE provides an essential and consistency condition in establishing the integrability and solving the models [5-7]. In conformal
field theory, the YBE is an important equation related to the KZB equation [8].
Motivated by the important role of the YBE, much attention has been directed to
the search for the solutions of YBE [9-13].

The colored Yang-Baxter equation (CYBE) dependent on spectral as well as
colored parameters is a generalization of the usual YBE. It has also attracted a lot
of research to find the solutions of CYBE [14-23]. The first solution of CYBE with
six-vertex can be found in ref. [14]. The eight-vertex solution of the CYBE has been
investigated previously in free fermion model in magnetic field by Fan C. and Wu
F. Y. [15], who first introduce a simple relation between the weight functions in the
model, so-called free-fermion condition. Based on the work and the condition, V.

V. Bazhanov and Yu. G. Stroganov obtained a eight-vertex solution of the CYBE [19]. J. Murakami gave another eight-vertex solution in discussing multi-variable invariants of links. The first and third authors in this paper gave all six-vertex and eight-vertex types solutions of CYBE and classified them into two kinds called Baxter and Free-Fermion types [22,23].

The main theme of this paper is to give and classify seven-vertex solutions of CYBE by a computer algebraic method. Moreover, it is proved from a theorem in ref. [24] that all seven-vertex type solutions of CYBE have indeed obtained in this paper. In section 1 we will introduce the symmetries or solution transformation for CYBE, then by the symmetries and a computer algebraic method to find the most simple system of equations, including differential equation, the solution set of which covers the one of CYBE. In section 2 we will show all solutions and classify them into two kinds called Baxter and Free-Fermion types. However, the details will be omitted due to the finite space for the paper.

## 2. THE SYMMETRIES FOR CYBE

The colored Yang-Baxter equation means the following matrix equation:

$$\check{R}_{12}(u,\xi,\eta)\check{R}_{23}(u+v,\xi,\lambda)\check{R}_{12}(v,\eta,\lambda) = \check{R}_{23}(v,\eta,\lambda)\check{R}_{12}(u+v,\xi,\lambda)\check{R}_{23}(u,\xi,\eta)$$

$$\check{R}_{12}(u,\xi,\eta) = \check{R}(u,\xi,\eta) \otimes E$$
$$\check{R}_{23}(u,\xi,\eta) = E \otimes \check{R}(u,\xi,\eta)$$

$$\tag{2.1}$$

where $\check{R}(u,\xi,\eta)$ is a matrix function of $N^2$ dimension, $E$ is the unit matrix of order $N$, $\otimes$ means the tensor product of two matrices, $u,v$ are spectral parameters and $\xi,\eta$ are colored parameters.

Let a matrix be the following form

$$\check{R}(u,\xi,\eta) = \begin{pmatrix} a_1(u,\xi,\eta) & 0 & 0 & a_7(u,\xi,\eta) \\ 0 & a_2(u,\xi,\eta) & a_6(u,\xi,\eta) & 0 \\ 0 & a_5(u,\xi,\eta) & a_3(u,\xi,\eta) & 0 \\ 0 & 0 & 0 & a_4(u,\xi,\eta) \end{pmatrix} \tag{2.2}$$

where $a_i(u,\xi,\eta)$ are called weight functions. If the matrix satisfies the equation (2.1) with $a_i(u,\xi,\eta) \neq 0$ it will be called a seven-vertex solution of CYBE. Throughout his paper, we denote

$$u_i = a_i(u,\xi,\eta) \quad v_i = a_i(v,\eta,\lambda) \quad w_i = a_i(u+v,\xi,\lambda) \quad i = 1,2,\ldots,8.$$

For seven-vertex-type solutions, the matrix equation (2.1) is equivalent to the following 19 equations:

$$u_2 w_3 v_2 - u_3 w_2 v_3 = 0 \tag{2.3a}$$

$$\left. \begin{array}{l} u_1 w_5 v_2 - u_5 w_1 v_2 - u_3 w_2 v_5 = 0 \\ u_2 w_6 v_1 - u_2 w_1 v_6 - u_6 w_2 v_3 = 0 \\ u_1 w_2 v_1 - u_2 w_1 v_2 - u_6 w_2 v_5 = 0 \end{array} \right\} \tag{2.3b}$$

$$\left.\begin{array}{c} u_4 w_6 v_2 - u_6 w_4 v_2 - u_3 w_2 v_6 = 0 \\ u_2 w_5 v_4 - u_2 w_4 v_5 - u_5 w_2 v_3 = 0 \\ u_4 w_2 v_4 - u_2 w_4 v_2 - u_5 w_2 v_6 = 0 \end{array}\right\} \qquad (2.3c)$$

$$\left.\begin{array}{c} u_1 w_5 v_3 - u_5 w_1 v_3 - u_2 w_3 v_5 = 0 \\ u_3 w_6 v_1 - u_3 w_1 v_6 - u_6 w_3 v_2 = 0 \\ u_1 w_3 v_1 - u_3 w_1 v_3 - u_6 w_3 v_5 = 0 \\ u_1 w_1 v_7 + u_7 w_2 v_4 - u_5 w_5 v_7 - u_2 w_7 v_1 = 0 \\ u_1 w_7 v_5 + u_7 w_6 v_2 - u_3 w_5 v_7 - u_6 w_7 v_1 = 0 \\ u_1 w_7 v_3 + u_7 w_6 v_6 - u_7 w_1 v_1 - u_4 w_3 v_7 = 0 \end{array}\right\} \qquad (2.3d)$$

$$\left.\begin{array}{c} u_4 w_6 v_3 - u_6 w_4 v_3 - u_2 w_3 v_6 = 0 \\ u_3 w_5 v_4 - u_3 w_4 v_5 - u_5 w_3 v_2 = 0 \\ u_4 w_3 v_4 - u_3 w_4 v_3 - u_5 w_3 v_6 = 0 \\ u_4 w_4 v_7 + u_7 w_2 v_1 - u_6 w_6 v_7 - u_2 w_7 v_4 = 0 \\ u_4 w_7 v_6 + u_7 w_5 v_2 - u_3 w_6 v_7 - u_5 w_7 v_4 = 0 \\ u_4 w_7 v_3 + u_7 w_5 v_5 - u_7 w_4 v_4 - u_1 w_3 v_7 = 0 \end{array}\right\} \qquad (2.3e)$$

Assuming $\check{R}(u, \xi, \eta)$ is a solution of equation (2.1). We can find there are five solution transformations in the system of equations (2.3):

(A) **Symmetry of interchanging indices.** The system of equations (2.3) is invariant if we interchange the two sub-indices 1 and 4 as well as the two sub-indices 5 and 6.

(B) **The scaling symmetry.** Multiplication of the solution $\check{R}(u, \xi, \eta)$ by an arbitrary function $\mathcal{F}(u, \xi, \eta)$ is still a solution of the equation (2.1).

(C) **Symmetry of the weight functions.** If the weight functions $a_2(u, \xi, \eta)$, $a_3(u, \xi, \eta)$, $a_7(u, \xi, \eta)$ are replaced by the new weight functions:

$$\tilde{a}_2(u, \xi, \eta) = \frac{N(\xi)}{N(\eta)} a_2(u, \xi, \eta) \qquad \tilde{a}_3(u, \xi, \eta) = \frac{N(\eta)}{N(\xi)} a_3(u, \xi, \eta)$$

$$\tilde{a}_7(u, \xi, \eta) = \frac{1}{sN(\xi)N(\eta)} a_7(u, \xi, \eta)$$

respectively, or $a_7(u, \xi, \eta)$, $a_5(u, \xi, \eta)$ and $a_6(u, \xi, \eta)$ are replaced by $-a_7(u, \xi, \eta)$ , $-a_5(u, \xi, \eta)$ and $-a_6(u, \xi, \eta)$, where $N(\xi)$ is an arbitrary function of $\xi$ and $s$ is a complex constant, the new matrix $\check{R}(u, \xi, \eta)$ is still a solution of (2.1).

(D) **Symmetry of spectral parameters.** If we take the new spectral parameter $\tilde{u} = \mu u$ where $\mu$ is a complex constant, the new matrix $\check{R}(\tilde{u}, \xi, \eta)$ is still a solution of (2.1).

(E) **Symmetry of the colored parameters.** If we take the new colored parameters $\zeta = f(\xi), \theta = f(\eta)$, where $f(\xi)$ is an arbitrary function, then the new matrix $\check{R}(u, \zeta, \theta)$ is also a solution of the equation (2.1).

Up to the solution transformation $\mathbf{B}$ and $\mathbf{C}$, we can assume

$$a_2(u,\xi,\eta) = 1 \quad a_3(u,\xi,\eta) = exp(ku)$$

without losing generality, the details and proof of which is omitted. Then the system of equations (2.3) can be simplified to the following 12 equations:

$$
\begin{aligned}
&u_1 w_5 - u_5 w_1 - v_5 exp(ku) = 0 \\
&w_6 v_1 - w_1 v_6 - u_6 exp(kv) = 0 \\
&u_1 v_1 - w_1 - u_6 v_5 = 0 \\
&u_1 w_1 v_7 + u_7 v_4 - u_5 w_5 v_7 - w_7 v_1 = 0 \\
&u_1 w_7 v_5 + u_7 w_6 - u_6 w_7 v_1 - w_5 v_7 exp(ku) = 0 \\
&u_7 w_6 v_6 - u_7 w_1 v_1 + u_1 w_7 exp(kv) - u_4 v_7 exp(ku + kv) = 0
\end{aligned}
\tag{2.7}
$$

plus six equations, that are called the counterparts of (2.7), obtained by interchanging the sub-indices 1 and 4 as well as 5 and 6 in each equation of (2.7).

Now we solve the equations obtained by setting u=0 and $\eta = \xi$ in (2.7). It is easy to get:

$$
\begin{aligned}
&a_1(0,\xi,\xi) = a_4(0,\xi,\xi) = 1 \\
&a_5(0,\xi,\xi) = a_6(0,\xi,\xi) = a_7(0,\xi,\xi) = 0
\end{aligned}
\tag{2.8}
$$

which are called the initial conditions of (2.1). Substituting (2.8) into (2.7) and letting $v = -u, \lambda = \xi$, we obtain

$$
\begin{aligned}
&a_5(u,\xi,\eta) = -a_5(-u,\eta,\xi) exp(ku) \\
&a_6(u,\xi,\eta) = -a_6(-u,\eta,\xi) exp(ku) \\
&a_4(u,\xi,\eta) a_4(-u,\eta,\xi) = a_1(u,\xi,\eta) a_1(-u,\eta,\xi) \\
&a_7(u,\xi,\eta) a_1(-u,\eta,\xi) = -a_7(-u,\eta,\xi) a_4(u,\xi,\eta)
\end{aligned}
\tag{2.9}
$$

Differentiating both sides of all equations in (2.7) and their counterparts with respect to the variable $v$ and letting $v = 0, \lambda = \eta$, by the virtue of initial conditions (2.8) we can get:

$$
\begin{aligned}
&u_1 u_5' - u_5 u_1' - m_5(\eta) exp(ku) = 0 \\
&u_6' + u_6 m_1(\eta) - u_1 m_6(\eta) - k u_6 = 0 \\
&u_1 m_1(\eta) - u_1' - u_6 m_5(\eta) = 0 \\
&u_1 u_7 m_5(\eta) + u_7 u_6' - u_5 m_7(\eta) exp(ku) - u_6 u_7 m_1(\eta) - u_6 u_7' = 0 \\
&k u_1 u_7 + u_1 u_7' + u_6 u_7 m_6(\eta) - u_1 u_7 m_1(\eta) - u_7 u_1' - u_4 m_7(\eta) exp(ku) = 0 \\
&(u_1^2 - u_5^2) m_7(\eta) + (m_4(\eta) - m_1(\eta)) u_7 - u_7' = 0
\end{aligned}
\tag{2.10a}
$$

and their counterparts, where and throughout this paper we denote

$$a_i'(u,\xi,\eta) = \frac{\partial a_i(u,\xi,\eta)}{\partial u} \quad m_i(\xi) = a_i'(u,\xi,\eta)\big|_{(u=0,\eta=\xi)}$$

$$i = 1, 2, \ldots, 7$$

We call $m_i(\xi)$ Hamiltonian coefficients of weight functions with respect to the spectral parameter or simply coefficients. Sometimes we write $m_i$ instead of $m_i(\xi)$ for brevity. If we differentiate (2.7) with respect to $u$ and let $u = 0, \eta = \xi$ and then replace the variables $v$ and $\lambda$ by $u$ and $\eta$, we can get:

$$u_5 m_1(\xi) + u_5' - u_1 m_5(\xi) - k u_5 = 0$$
$$u_1 u_6' - u_1' u_6 - m_6(\xi) exp(ku) = 0$$
$$u_1 m_1(\xi) - u_1' - u_5 m_6(\xi) = 0$$
$$u_1 u_7 m_1(\xi) - u_1 u_7' + u_1' u_7 + u_4 m_7(\xi) - u_5 u_7 m_5(\xi) = 0$$
$$u_5 u_7 m_1(\xi) + u_5 u_7' - u_5' u_7 + u_6 m_8(\xi) - k u_5 u_7 - u_1 u_7 m_6(\xi) = 0$$
$$(m_1(\xi) - m_4(\xi)) u_7 exp(ku) + (u_6^2 - u_1^2) m_7(\xi) + u_7' exp(ku) - k u_7 exp(ku) = 0$$

$$(2.10b)$$

and their counterparts.

From the last equation of (2.10a) and its counterpart we have:

$$2 u_7' = m_7(\eta)(u_1^2 + u_4^2 - u_5^2 - u_6^2) \qquad (2.11a)$$

We also get the following equation, from the last equation of (2.10b) and its counterpart,

$$2 u_7' = m_7(\xi)(u_1^2 + u_4^2 - u_5^2 - u_6^2) exp(-ku) + 2k u_7 \qquad (2.11b)$$

In what follows we will apply a computer algebraic method, given by the ref. 24, to calculate (2.7) and (2.10)s to find the most simple system of equations, including differential equations, satisfied by weight functions.

Now we eliminate the five weight functions $\{w_1, w_4, w_5, w_6, w_7\}$ in the equation (2.7) and their counterparts. Then we get seven polynomial equations without weight functions $w_i (i = 1, 4, \ldots, 7)$ If we differentiate the obtained seven equations with respect to the spectral parameter $v$, and let $v = 0, \lambda = \eta$ and substitute the initial condition (2.8) into them, we obtain the following seven polynomial equations:

$$m_6(\eta) exp(ku) + (m_1(\eta) + m_4(\eta)) u_4 u_6 - m_6(\eta)(u_1 u_4 + u_5 u_6) - k u_4 u_6 = 0$$
$$m_5(\eta) exp(ku) + (m_4(\eta) + m_1(\eta)) u_1 u_5 - m_6(\eta)(u_1 u_4 + u_5 u_6) - k u_1 u_6 = 0$$
$$m_7(\eta)(u_1^2 - u_4^2 - u_5^2 + u_6^2) + 2 u_7(m_4(\eta) - m_1(\eta)) = 0$$
$$m_7(\eta)(u_1^2 u_6 - u_5^2 u_6 + u_5 exp(ku)) - (m_5(\eta) + m_6(\eta)) u_1 u_7 - (k - m_1(\eta)$$
$$\qquad - m_4(\eta)) u_6 u_7 = 0$$
$$m_7(\eta)(u_1 u_5^3 - u_1^3 u_5 - u_1 u_6 exp(ku)) + m_5(\eta) u_7 exp(ku) + 2(m_1(\eta) - m_4(\eta)) u_5 u_7$$
$$\qquad + m_6(\eta) u_1 u_4 u_7 - m_5(\eta) u_5 u_6 u_7 = 0$$
$$m_7(\eta)(u_1^3 - u_1 u_5^2 - u_4 exp(ku)) + u_1 u_7(k + m_4(\eta) - 3 m_1(\eta)) + (m_5(\eta)$$
$$\qquad + m_6(\eta)) u_6 u_7 = 0$$
$$m_7(\eta) u_4(u_1^2 - u_5^2) + u_4 u_7(k - m_1(\eta) - m_4(\eta)) + (m_5(\eta) + m_6(\eta)) u_5 u_7 = 0$$

$$(2.12)$$

It is calculated that the system of equations (2.12) is equivalent to two groups of equations, applying computer algebraic method given by the ref.24. The first one is

$$m_5 exp(ku) + (m_1 + m_4)u_1 u_5 - m_5(u_1 u_4 + u_5 u_6) - ku_1 u_5 = 0$$
$$m_7 u_1 u_5(u_4^2 - u_1^2 + u_5^2 - u_6^2) + 2m_5 u_7(u_1 u_4 + u_5 u_6 - exp(ku))$$
$$+(2k - 4m_4)u_1 u_5 u_7 = 0$$
$$m_7 u_1(u_4^2 u_5 - u_5^2 u_6 + u_6 exp(ku)) - u_7(m_5 exp(ku) + m_6 u_1 u_4 - m_5 u_5 u_6) = 0$$
$$(m_7 u_5 u_6(u_4 u_6 - u_1 u_5) + m_7 u_4 u_5(u_5^2 - u_4^2) + m_7 exp(ku)(u_1 u_5 - u_4 u_6)$$
$$+m_6 u_7(u_4^2 - u_5^2)) = 0$$
$$u_1^2 u_5 - 2u_1 u_4 u_6 + u_4^2 u_5 - u_5^3 + u_5 u_6^2 = 0$$
$$-u_1 u_4^2 u_6 + u_1 u_5^2 u_6 + u_4^3 u_5 - u_4 u_5^3 - u_1 u_5 exp(ku) + u_4 u_6 exp(ku) = 0$$
$$-u_1^2 u_4 + 2u_1 u_5 u_6 + u_4^3 - u_4 u_5^2 - u_4 u_6^2 = 0.$$

$$(2.13a)$$

The second one is

$$m_5 exp(ku) + (m_1 + m_4)u_1 u_5 - m_5(u_1 u_4 + u_5 u_6) - ku_1 u_5 = 0$$
$$m_7 u_1 u_5(u_4^2 - u_1^2 + u_5^2 - u_6^2) + 2m_5 u_7(u_1 u_4 + u_5 u_6 - exp(ku))$$
$$+(2k - 4m_4)u_1 u_5 u_7 = 0$$
$$m_7 u_1(u_4^2 u_5 - u_5^2 u_6 + u_6 exp(ku)) - u_7(m_5 exp(ku) + m_6 u_1 u_4 - m_5 u_5 u_6) = 0$$
$$u_1 u_4 + u_5 u_6 = exp(ku).$$

$$(2.13b)$$

In the formulas (2.13a) and (2.13b) we write $m_i$ instead of $m_{(\eta)}$.

*Remark 1: When we perform the operation of eliminating indeterminates in a system of equations,*
according to the theorem of zero structure of algebraic varieties [16],the coefficient of the term with the highest degree of the indeterminate in the main polynomial equation(to be eliminated in the other polynomials)should not be identified with zero. In the event it is identified with zero,we should add the coefficient into the equations to produce a new system of equations. Otherwise,it is possible to lose some solutions. In our cases, we can discover that the coefficients of the terms which are eliminated are not identified with zero thanks to the initial conditions and the nondegenerate conditions we have set.

## 3. THE SOLUTIONS OF CYBE

From the two systems of equations (2.13a), (2.13b) and the differential equations (2.10a),(2.10b) we can obtain solutions of CYBE.

**3.1 THE FIRST CASE:** $k \neq 0$    In the case it is proved that the first system of equations has no solution. However, the second one has the following solution.

$$a_1(u,\xi,\eta) = a_4(u,\xi,\eta) = cosh(F(\xi) - F(\eta))exp(ku/2)$$

$$a_2(u,\xi,\eta) = 1$$

$$a_3(u,\xi,\eta) = exp(ku) \qquad\qquad\qquad (3.1)$$

$$a_5(u,\xi,\eta) = -a_6(u,\xi,\eta) = sinh(F(\xi) - F(\eta))exp(ku/2)$$

$$a_7(u,\xi,\eta) = F_7(\eta)exp(ku) - F_7(\xi)$$

where $k$ is a non-zero complex constant and $F(\xi), F_7(\xi)$ are two arbitrary function of colored parameter.

**3.2 THE SECOND CASE:** $k = 0$   in this case equations (2.11a) and (2.11b) change to:

$$2u_7' = m_7(\eta)(u_1^2 + u_4^2 - u_5^2 - u_6^2) = m_7(\xi)(u_1^2 + u_4^2 - u_5^2 - u_6^2)$$

which implies $m_7(\xi) = m_7(\eta) = \alpha$ is a complex constant independent of colored parameters. We can also prove

*Proposition:* In the case of $k = 0$, there is at least one between $m_5(\xi)$ (or $m_6(\xi)$) and $\alpha$ which is not zero identically. Otherwise, the solution will be independent of spectral parameter.

*Remark 2: We omit the proof of the proposition. The role of the proposition is to guarantee that the solution of CYBE do not lost in the operation to simplify the two groups of equations (2.13a) and (2.13b).*

Now let us simplify the two systems of equations (2.13a) and (2.13b) and find their solutions.

**3.2b Baxter type solutions:**   When $k = 0$ the first system of equations (2.13a) became

$$m_5 + (m_1 + m_4)u_1u_5 - m_5(u_1u_4 + u_5u_6) = 0$$
$$m_7u_1u_5(u_4^2 - u_1^2 + u_5^2 - u_6^2) + 2m_5u_7(u_1u_4 + u_5u_6 - 1)) + (-4m_4)u_1u_5u_7 = 0$$
$$m_7u_1(u_4^2u_5 - u_5^2u_6 + u_6 - u_7(m_5 + m_6u_1u_4 - m_5u_5u_6) = 0$$
$$m_7u_5u_6(u_4u_6 - u_1u_5) + m_7u_4u_5(u_5^2 - u_4^2) + m_7(u_1u_5 - u_4u_6)$$
$$\qquad\qquad +m_6u_7(u_4^2 - u_5^2) = 0$$
$$u_1^2u_5 - 2u_1u_4u_6 + u_4^2u_5 - u_5^3 + u_5u_6^2 = 0$$
$$-u_1u_4^2u_6 + u_1u_5^2u_6 + u_4^3u_5 - u_4u_5^3 - u_1u_5 + u_4u_6 = 0$$
$$-u_1^2u_4 + 2u_1u_5u_6 + u_4^3 - u_4u_5^2 - u_4u_6^2 = 0.$$

$$(3.2a)$$

It is proved from (3.2a) and initial condition (2.8) that then

$$m_1(\xi) = m_1(\eta) = \gamma, \quad m_5(\xi) = m_6(\eta) = \beta$$

which imply that $\gamma$ and $\beta$ are constants independent of colored parameter. Moreover, combining the system of equation with (2.10a) and applying the initial condition (2.8) and proposition, we have, by symbolic computation,

$$
\begin{aligned}
&a_1(u,\xi,\eta) = a_4(u,\xi,\eta), \\
&a_5(u,\xi,\eta) = a_6(u,\xi,\eta), \\
&a_7(u,\xi,\eta) = \frac{\alpha}{\beta} a_1(u,\xi,\eta) a_5(u,\xi,\eta), \\
&(u_5')^2 = \beta^2 - (\beta^2 - \gamma^2) u_5^2, \\
&(u_1')^2 = \beta^2 - (\beta^2 - \gamma^2) u_1^2.
\end{aligned}
\tag{3.3}
$$

So we immediately obtain the solutions of CYBE.

- **Subcase of $\beta^2 \neq \gamma^2$.**

$$
a_1(u,\xi,\eta) = a_4(u,\xi,\eta) = \frac{\cos(\lambda u + F(\xi) - F(\eta) - \theta)}{\cos\theta}
$$

$$
a_2(u,\xi,\eta) = a_3(u,\xi,\eta) = 1
$$

$$
a_5(u,\xi,\eta) = a_6(u,\xi,\eta) = \frac{\sin(\lambda u + F(\xi) - F(\eta))}{\cos\theta}
\tag{3.4a}
$$

$$
a_7(u,\xi,\eta) = \frac{\alpha\beta}{\lambda}\cos(\lambda u + F(\xi) - F(\eta) - \theta)\sin(\lambda u + F(\xi) - F(\eta))
$$

where

$$
\sin\theta = \frac{\gamma}{\beta}, \quad \cos\theta = \frac{\sqrt{\beta^2 - \gamma^2}}{\beta}, \quad \lambda = \sqrt{\beta^2 - \gamma^2}.
$$

- **Subcase of $\beta^2 = \gamma^2$** (in fact this case is he degenerate case)

$$
\begin{aligned}
&a_1(u,\xi,\eta) = a_4(u,\xi,\eta) = \beta u + F(\xi) - f(\eta) + 1 \\
&a_2(u,\xi,\eta) = a_3(u,\xi,\eta) = 1 \\
&a_5(u,\xi,\eta) = a_6(u,\xi,\eta) = \beta u + F(\xi) - f(\eta) \\
&a_7(u,\xi,\eta) = \frac{\alpha}{\beta}(\beta u + F(\xi) - f(\eta) + 1)(\beta u + F(\xi) - f(\eta))
\end{aligned}
\tag{3.4b}
$$

**3.2f.** **Free-fermion type solutions:** When $k = 0$ the second system of equations (2.13b) became

$$
\begin{aligned}
&m_5 + (m_1 + m_4)u_1 u_5 - m_5(u_1 u_4 + u_5 u_6) = 0 \\
&m_7 u_1 u_5(u_4^2 - u_1^2 + u_5^2 - u_6^2) + 2m_5 u_7(u_1 u_4 + u_5 u_6 - 1)) + (-4m_4)u_1 u_5 u_7 = 0 \\
&m_7 u_1(u_4^2 u_5 - u_5^2 u_6 + u_6 - u_7(m_5 + m_6 u_1 u_4 - m_5 u_5 u_6) = 0 \\
&u_1 u_4 + u_5 u_6 = 1.
\end{aligned}
\tag{3.2b}
$$

The fourth formula in (3.2b) is just free-fermion condition. As have done for the case of Baxter type, combining (3.2b) and (2.10a), (2.10b) and applying the initial condition (2.8) and the proposition, we have, by symbolic computation,

$$
\begin{aligned}
&u_1 u_4 + u_5 u_6 = 1, \\
&\alpha(u_1^2 + u_6^2 - u_4^2 - u_5^2) = 2(m_1(\eta) - m_4(\eta))u_7, \\
&\alpha(u_1 u_6 + u_4 u_5) = (m_6(\eta) + m_5(\eta))u_7, \\
&(u_7')^2 = \alpha^2 - ((m_5(\eta) + m_6(\eta))^2 - 4m_1(\eta)^2)u_7^2
\end{aligned}
\tag{3.5a}
$$

and

$$
\begin{aligned}
&u_1 u_4 + u_5 u_6 = 1, \\
&\alpha(u_1^2 + u_6^2 - u_4^2 - u_5^2) = 2(m_1(\xi) - m_4(\xi))u_7, \\
&\alpha(u_1 u_6 + u_4 u_5) = (m_6(\xi) + m_5(\xi))u_7, \\
&(u_7')^2 = \alpha^2 - ((m_5(\xi) + m_6(\xi))^2 - 4m_1(\xi)^2)u_7^2.
\end{aligned}
\tag{3.5b}
$$

So, we can obtain, from the fourth formulas in (3.5a) and (3.5b),

$$
\delta^2 = (m_5(\xi) + m_6(\xi))^2 - 4m_1(\xi)^2
\tag{4 - 2.6}
$$

is a complex constant independent of colored parameters. So we can obtain the solutions of CYBE, from (3.5a) and (3.5b)

**Subcase:** $m_5(\xi) = -m_6(\xi)$

$$
\begin{aligned}
&a_1(u, \xi, \eta) = a_4(u, \xi, \eta) = cosh(\beta u + F(\xi) - F(\eta)) \\
\\
&a_2(u, \xi, \eta) = a_3(u, \xi, \eta) = 1 \\
\\
&a_5(u, \xi, \eta) = -a_6(u, \xi, \eta) = sinh(\beta u + F(\xi) - F(\eta)) \\
\\
&a_7(u, \xi, \eta) = \alpha u + F(\xi) - F(\eta)
\end{aligned}
\tag{3.6}
$$

**Subcase:** $m_5(\xi) = m_6(\xi)$

*(A). $\delta \neq 0$ subcase*

$$
\begin{aligned}
&a_1(u, \xi, \eta) = X(\delta G(\eta) + \delta G(\xi)cos(\Omega) + 2G(\xi)H(\eta)sin(\Omega)) \\
\\
&a_2(u, \xi, \eta) = a_3(u, \xi, \eta) = 1 \\
\\
&a_4(u, \xi, \eta) = X(\delta G(\xi) + \delta G(\eta)cos(\Omega) - 2G(\eta)H(\eta)sin(\Omega)) \\
\\
&a_5(u, \xi, \eta) = -Y(\delta G(\xi) + \delta G(\eta)cos(\Omega) - 2G(\eta)H(\xi)sin(\Omega) \\
\\
&a_6(u, \xi, \eta) = -\delta G(\eta) + \delta G(\xi)cos(\Omega) - 2G(\xi)H(\eta)sin(\Omega) \\
\\
&a_7(u, \xi, \eta) = \frac{\alpha}{\delta}sin(\Omega)
\end{aligned}
\tag{3.7a}
$$

where

$$\Omega = \delta u + F(\xi) - F(\eta)$$

$$X = \sqrt{\frac{2(m_1(\xi) + m_1(\eta))sin(\delta u + F(\xi) - F(\eta))}{\delta(H_1^2 - H_4^2)}}$$

$$Y = \sqrt{\frac{2(m_1(\xi) - m_1(\eta))sin(\delta u + F(\xi) - F(\eta))}{\delta(H_5^2 - H_6^2)}}$$

and $F(\xi), G(\xi)$ and $H(\xi)$ are arbitrary functions of colored parameters.

*(B). For the subcase: $\delta = 0$*

$$a_1(u, \xi, \eta) = \frac{1}{2\alpha\sqrt{G(\xi)G(\eta)}}(\alpha(G(\xi) + G(\eta)) + 2G(\xi)G(\eta)(\alpha u + F(\xi) - F(\eta)))$$

$$a_2(u, \xi, \eta) = a_3(u, \xi, \eta) = 1$$

$$a_4(u, \xi, \eta) = \frac{1}{2\alpha\sqrt{G(\xi)G(\eta)}}(\alpha(G(\xi) + G(\eta)) - 2G(\xi)G(\eta)(\alpha u + F(\xi) - F(\eta)))$$

$$a_5(u, \xi, \eta) = \frac{1}{2\alpha\sqrt{G(\xi)G(\eta)}}(\alpha(G(\xi) - G(\eta)) + 2G(\xi)G(\eta)(\alpha u + F(\xi) - F(\eta)))$$

$$a_6(u, \xi, \eta) = \frac{1}{2\alpha\sqrt{G(\xi)G(\eta)}}(-\alpha(G(\xi) - G(\eta)) + 2G(\xi)G(\eta)(\alpha u + F(\xi) - F(\eta)))$$

$$(3.7b)$$

where $F(\xi)$ and $G(\xi)$ are arbitrary functions of colored parameters.

## 4. General solutions

In this paper, we have given six basic solutions of CYBE (2.1), which are (3.3), (3.4a), (3.4b),(3.6),(3.7a) and (3.7b). The six basic solutions can be classified into two types. one is of $m_1 = m_4$. We call them Baxter type solution. Another type is of $m_1 + m_4 = k$. The basic solution with $k = 0$, which satisfies the free-fermion condition, is called free-fermion solution. A theorem in ref. 24 tell us that the method that we apply to find solutions do not lost any solution of CYBE. So the six basic solutions together with the five solution transformations **A-E** will give all seven-vertex-type solutions of colored Yang-Baxter equation (2.1) and the general solutions can also be classified into two types. The first are Baxter-type solutions which can be get from the basic Baxter-type solution via some solution transformations. The second are free-fermion-type which can be obtained via the basic free-fermion type solutions via some solution transformations.

According to the standard method given by Baxter, for a given R matrix the spin-chain Hamiltonian is generally of the following form,

$$H = \sum_{j=1}^{N}(J_x\sigma_j^x\sigma_{j+1}^x + J_y\sigma_j^y\sigma_{j+1}^y + J_z\sigma_j^z\sigma_{j+1}^z + \frac{1}{2}(\sigma_j^z + \sigma j + 1^z))$$

where $\sigma^x, \sigma^y$ and $\sigma^z$ are Pauli matrices and the coupling constant are

$$J_x = \tfrac{1}{4}(m_5 + m_6 + m_7) \qquad J_y = \tfrac{1}{4}(m_5 + m_6 - m_7)$$

$$J_z = \tfrac{1}{4}(m_1 - m_3 + m_4 - m_2) \quad h = \tfrac{1}{4}(m_1 - m_3 - m_4 + m_2)$$

We can prove that in the six basic solutions the Hamiltonian coefficients obey, when $k = 0$,

$$m_1^2 = m_4^2 \quad m_5^2 = m_6^2$$

It follows from the solution transformations **B** and **D** that

$$(m_1 - m_3)^2 = (m_4 - m_2)^2 \quad m_5^2 = m_6^2$$

for general solutions. We can use the relations to give the couplings in Hamiltonian.

## ACKNOWLEDGMENTS

This paper is dedicated to Prof. Hiroshi Ezawa on the occasion for his 65th birthday. We are grateful for Prof. Ezawa's friendship and Prof. Christophen C. Bernido's invitation.

This work was supported by Scientific Project of Department of Science and technology in China, Natural Scientific Foundation of Chinese Academy of Sciences and Foundation of NSF.

## REFERENCES

1. Onsager, L. (1944) *Phys. Rev.*, **65**, 117.

2. Yang, C. N. (1967) *Phys. Rev. Lett.*, **19**, 1312-14.

3. Baxter, R, J. (1972) *Ann. Phys. Lpz.*, **70**, 193-288.

4. Zamolodchikov, A. B. (1979) *Ann. Phys. Lpz.*, **120**, 253-91.

5. Faddeev, L. D., Reshetikhin, N. Yu., Takhtajian, L. A. (1989) quantization of Lie groups and Lie algebras, *Yang-Baxter equation in integrable Systems, Advance Series in Mathematical Physics*, **10**, pp 299-309, Singapore: World Scientific.

6. Baxter, R. J. (1982) *Exactly Solved Models in statistical Mechanics*, London: Academic.

7. Jimbo, M. (1989) *Yang-Baxter equation in integrable systems.* Singapore: World Scientific.

8. Avan, J., Babelon, O., Billey, B. (1996) *Commu. Math. Phys.*, **178**, 281-299.

9. Akutsu, Y., Wadati, M. (1987) *J. Phys. Soc. Japan*, **56**, 839-42.

10. Belavin, A. A., Drinfel'd, V.G. (1982) *Funkt. Anal. Appl.*, **16**, 1-29; **16**, 159-180.

11. Cheng, Y., Ge, M.L., Xue, K. (1991) *Commu. Math. Phys*, **136**, 195-206.

12. Fei, S. M., Guo,H. Y., Shi, He. (1992) *J. Phys.*, **A 25**, 2711-2720.

13. Hou, B. Y., Ma, Z. Q. (1991) *J. Phys.*, **A 24**, 1363-1377.

14. Ge, M. Y., Xue, K. (1993) *J. Phys.*, **A 26**, 281.

15. Fan, C., Wu, F. Y. (1970) *Phys. Rev.*, **B2**, 723.

16. Cuerno, R., Gómez, C., López, E., Sierra, G. (1993) *Phys. Lett.*, **B307**, 56-60.

17. Murakami, J. (1992) *Int. J. Mod. Phys.*, **A7**, 765.

18. Ruiz-Altaba, M. (1992) *Phys. Lett.*, **277**, 326.

19. Bazhanov, V. V., Stroganov, Y. G. (1985) *Theor. Math. Fiz*, **62**, 253-60.

20. Delius, G. M., Gould, M. D., Zhang, Y. Z. (1994) *Nucl Phys.*, **B432**, 377.

21. Brachen, A. J., Gould, M. D., Zhang, Y. Z., Delius, G. M. (1994) *J.Math.Phys*, **A 27**, 6551.

22. Wang, S. K. (1996) *J.Phys.* **A 29**, 2259-2277.

23. Sun, X. D., Wang, S. K., Wu, K. (1996) *J.Math.Phys.*, **36**, 6043-6063.

24. Wu Wen-tsun 1978 *Sci. sinica*, **21**, 157-79.

# THE SUPER RS METHOD FOR QUANTUM AFFINE SUPERALGEBRA $U_q(\widehat{gl(1|1)})$ AND SUPER YANGIAN DOUBLE $DY(\widehat{gl(1|1)})$

## JIN-FANG CAI, KE WU, CHI XIONG

Institute of Theoretical Physics, Academia Sinica,
Beijing, 100080, P. R. China

## SHI-KUN WANG

Institute of Applied Mathematics, Academia Sinica,
Beijing, 100080, P. R. China

**Abstract:** We extend the Reshetikhin–Semenov-Tian-Shansky method to super case to define quantum affine superalgebra $U_q(\widehat{gl(1|1)})$ and super Yangian double $DY(\widehat{gl(1|1)})$; Use the Ding-Frenkel map to get their Drinfel'd realizations. The universal $R$-matrix of $DY(\widehat{gl(1|1)})$ is given.

## 1. INTRODUCTION

Quantum algebras, including quantum enveloping algebras, quantum affine algebras and Yangians, were introduced by Drinfel'd etc. [1–3] The Faddeev-Reshetikhin-Takhtajan (FRT) [4] and Reshetikhin–Semenov-Tian-Shansky (RS) [5] methods are convenient ways to define and construct these quantum algebras. Quantum affine algebras and Yangians [7–9] are related respectively with trigonometric and rational solutions of Yang-Baxter equation(YBE), The explicit isomorphism between RS realization and Drinfel'd new realization [2] of quantum affine algebras and Yangian double can be established through Gauss decomposition or Ding-Frenkel map [6]. However, for a non-standard solution (without spectrum parameter) of YBE, if we employ the usual FRT method, we would obtain a peculiar quantum algebra whose classical limit is neither a Lie algebra nor a Lie superalgebra. It was pointed out by Liao and Song [10]

73

that, to obtain a quantum Lie superalgebra from this nonstandard solution of
YBE, one must use the super extension of FRT method. Similar case happens
in constructing quantum affine superalgebras [11]. In the begining of this paper
(section 2 and 3), we review our works on the super extension of RS method
to define quantum affine superalgebra $U_q(\widehat{gl(1|1)})$ [13,14] and super Yangian
double $DY(\widehat{gl(1|1)})$ [15]. Drinfel'd realization of $U_q(\widehat{gl(1|1)})$ and $DY(\widehat{gl(1|1)})$
were obtained through Gauss decomposition. Similar works were done inde-
pendently by Zhang [16,17]. In section 4 and 5, the universal $\mathcal{R}$-matrix of
$DY(\widehat{gl(1|1)})$ was calculated out. By taking evaluation representation, the RS
realization of $DY(\widehat{gl(1|1)})$ was gotten from the properties of $\mathcal{R}$-matrix. The
result for super Yangian double without center extension has been published
in [18].

## 2.   DRINFEL'D REALIZATION OF QUANTUM AFFINE SU-PERALGEBRA $U_q(\widehat{gl(1|1)})$

It can be verified that following $R$-matrix:

$$R_{12}(z) = \begin{pmatrix} 1 & 0 & 0 & 0 \\ 0 & \frac{z-1}{zq-q^{-1}} & \frac{z(q-q^{-1})}{zq-q^{-1}} & 0 \\ 0 & \frac{(q-q^{-1})}{zq-q^{-1}} & \frac{z-1}{zq-q^{-1}} & 0 \\ 0 & 0 & 0 & -\frac{q-zq^{-1}}{zq-q^{-1}} \end{pmatrix} \tag{1}$$

satisfies the graded Yang-Baxter equation:

$$\eta_{12}R_{12}(z/w)\eta_{13}R_{13}(z)\eta_{23}R_{23}(w) = \eta_{23}R_{23}(w)\eta_{13}R_{13}(z)\eta_{12}R_{12}(z/w), \tag{2}$$

where $\eta_{ik,jl} = (-1)^{P(i)P(k)}\delta_{ij}\delta_{lk}$ and $P(1) = 0, P(2) = 1$. This solution is
free-fermion type and also satisfies the unitary condition: $R_{12}(z)R_{21}(z^{-1}) = 1$.
When $z = 0$ and $q$ replaced by $q^{-1}$, the $\eta R(z)$ degenerates to the non-standard
solution of YBE which was used in studying quantum superalgebra $U_q(gl(1|1))$
[10]. This solution $\eta R(z)$ can also be obtained from the non-standard solution
through Baxterization procedure [12].

From the above solution of graded YBE, we can define the quantum affine
superalgebra $U_q(\widehat{gl(1|1)})$ with a central extension employing super RS method
[5]. $U_q(\widehat{gl(1|1)})$ is an associative algebra with generators $\{l_{ij}^k|1 \le i,j \le 2, k \in \mathbb{Z}\}$ and center $c$, which subject to the following multiplication relations:

$$R_{12}(\tfrac{z}{w})L_1^\pm(z)\eta L_2^\pm(w)\eta = \eta L_2^\pm(w)\eta L_1^\pm(z)R_{12}(\tfrac{z}{w}) \tag{3}$$
$$R_{12}(\tfrac{z}{w_+})L_1^+(z)\eta L_2^-(w)\eta = \eta L_2^-(w)\eta L_1^+(z)R_{12}(\tfrac{z+}{w_-}) \tag{4}$$

here $z_{\pm} = zq^{\pm\frac{c}{2}}$ . We have used standard notation: $L_1^{\pm}(z) = L^{\pm}(z) \otimes 1$, $L_2^{\pm}(z) = 1 \otimes L^{\pm}(z)$ and $L^{\pm}(z) = \left(l_{ij}^{\pm}(z)\right)_{i,j=1,2}$, $l_{ij}^{\pm}(z)$ are generating functions (or currents) of $l_{ij}^k$: $l_{ij}^{\pm}(z) = \sum_{k=0}^{\infty} l_{ij}^{\pm k} z^{\pm k}$. This algebra admits the following coproduct, counit and antipode structure

$$\Delta\left(l_{ij}^{\pm}(z)\right) = \sum_{k=1}^{2} l_{kj}^{\pm}(zq^{\pm\frac{c_2}{4}}) \otimes l_{ik}^{\pm}(zq^{\mp\frac{c_1}{4}})(-1)^{(k+i)(k+j)}, \tag{5}$$

$$\epsilon\left(l_{ij}^{\pm}(z)\right) = \delta_{ij}, \qquad S\left({}^{st}L^{\pm}(z)\right) = \left[{}^{st}L^{\pm}(z)\right]^{-1}, \tag{6}$$

$$\Delta(c) = c \otimes 1 + 1 \otimes c, \tag{7}$$

$$\epsilon(c) = 0, \qquad S(c) = -c \tag{8}$$

where $c_1 = c \otimes 1$ , $c_2 = 1 \otimes c$ and $[{}^{st}L^{\pm}(z)]_{ij} = (-1)^{i+j} l_{ji}^{\pm}(z)$. It's easy to verify that the above coproduct, counit and antipode structure are compatible with the associative multiplication defined by eqs.(3) and (4), i.e. all $l_{ij}^k$ and $c$ satisfy

$$\Delta(ab) = \Delta(a)\Delta(b), \tag{9}$$

$$m(S \otimes id)\Delta(a) = m(id \otimes S)\Delta(a) = \epsilon(a) \cdot 1, \tag{10}$$

$$(\epsilon \otimes id)\Delta = (id \otimes \epsilon)\Delta = id, \tag{11}$$

$$S(ab) = S(b)S(a), \qquad \epsilon(ab) = \epsilon(a)\epsilon(b) \tag{12}$$

As what was done in Ding and Frenkel [6], $L^{\pm}(z)$ have the following unique Gauss decompositions :

$$L^{\pm}(z) = \begin{pmatrix} 1 & 0 \\ f^{\pm}(z) & 1 \end{pmatrix} \begin{pmatrix} k_1^{\pm}(z) & 0 \\ 0 & k_2^{\pm}(z) \end{pmatrix} \begin{pmatrix} 1 & e^{\pm}(z) \\ 0 & 1 \end{pmatrix} \tag{13}$$

where $e^{\pm}(z), f^{\pm}(z)$ and $k_i^{\pm}(z)$ $(i=1,2)$ is generating functions of $U_q(\widehat{gl(1|1)})$ and $k_i^{\pm}(z)$ $(i=1,2)$ are invertible. We set

$$X^{+}(z) = e^{+}(z_{-}) - e^{-}(z_{+}), \qquad X^{-}(z) = f^{+}(z_{+}) - f^{-}(z_{-}). \tag{14}$$

and introduce a transformation for the generating functions $X^{\pm}(z)$ and $k_i^{\pm}(z)$ $(i=1,2)$ to get the currents corresponding to generators of $gl(1|1)$:

$$E(z) = X^{+}(zq) \qquad\qquad F(z) = X^{-}(zq) \tag{15}$$

$$K^{\pm}(z) = k_1^{\pm}(zq)^{-1}k_2^{\pm}(zq) \qquad H^{\pm}(z) = k_2^{\pm}(zq)k_1^{\pm}(zq^{-1}) \tag{16}$$

then the (anti-)commutation relations among $E(z), F(z), K^{\pm}(z)$ and $H^{\pm}(z)$ are

$$[K^{\pm}(z) , K^{\pm}(w)] = [H^{\pm}(z) , H^{\pm}(w)] = 0 \tag{17}$$

$$[K^+(z)\,,\ K^-(w)] = [K^\pm(z)\,,\ H^\pm(w)] = 0 \tag{18}$$

$$H^+(z)H^-(w) = \left(\frac{(z_- q - w_+ q^{-1})(z_+ q^{-1} - w_- q)}{(z_+ q - w_- q^{-1})(z_- q^{-1} - w_+ q)}\right)^2 H^-(w)H^+(z) \tag{19}$$

$$K^\pm(z)H^\mp(w) = \frac{(w_\mp q - z_\pm q^{-1})(z_\mp q - w_\pm q^{-1})}{(w_\pm q - z_\mp q^{-1})(z_\pm q - w_\mp q^{-1})} H^\mp(w)K^\pm(z) \tag{20}$$

$$[K^\pm(z)\,,\ E(w)] = [K^\pm(z)\,,\ F(w)] = 0 \tag{21}$$

$$E(w)H^\pm(z) = \frac{z_\pm q - wq^{-1}}{z_\pm q^{-1} - wq} H^\pm(z)E(w) \tag{22}$$

$$F(w)H^\pm(z) = \frac{z_\mp q^{-1} - wq}{z_\mp q - wq^{-1}} H^\pm(z)F(w) \tag{23}$$

$$\{E(z)\,,\ E(w)\} = \{F(z)\,,\ F(w)\} = 0 \tag{24}$$

$$\{E(z)\,,\ F(w)\} = (q - q^{-1})\left[\delta\left(\frac{w_-}{z_+}\right)K^-(z_+) - \delta\left(\frac{z_-}{w_+}\right)K^+(w_+)\right] \tag{25}$$

The above relations are Drinfel'd's currents realization of quantum affine superalgebra $U_q(\widehat{gl(1|1)})$. Set $e^\pm(z_\mp q) = E^\pm(z)$ and $f^\pm(z_\pm q) = F^\pm(z)$, then $E(z) = E^+(z) - E^-(z)$ and $F(z) = F^+(z) - F^-(z)$. The coproduct structure of the generating functions $E^\pm(z), F^\pm(z), K^\pm(z)$ and $H^\pm(z)$ can be calculated directly from (5) :

$$\triangle\left(E^\pm(z)\right) = E^\pm(z) \otimes 1 + K^\pm(zq^{\mp\frac{c_1}{2}}) \otimes E^\pm(zq^{\mp c_1}) \tag{26}$$

$$\triangle\left(F^\pm(z)\right) = 1 \otimes F^\pm(z) + F^\pm(zq^{\pm c_2}) \otimes K^\pm(zq^{\pm\frac{c_2}{2}}) \tag{27}$$

$$\triangle\left(K^\pm(z)\right) = K^\pm(zq^{\pm\frac{c_2}{2}}) \otimes K^\pm(zq^{\mp\frac{c_1}{2}}) \tag{28}$$

$$\triangle\left(H^\pm(z)\right) = H^\pm(zq^{\pm\frac{c_2}{2}}) \otimes H^\pm(zq^{\mp\frac{c_1}{2}}) - (q + q^{-1})$$
$$F^\pm(zq^{-2\mp\frac{c_1}{2}\pm\frac{c_2}{2}})H^\pm(zq^{\pm\frac{c_2}{2}}) \otimes H^\pm(zq^{\mp\frac{c_1}{2}})E^\pm(zq^{-2\mp\frac{c_1}{2}\pm\frac{c_2}{2}}) \tag{29}$$

The antipode and counit structure for these currents is

$$S\left(K^\pm(z)\right) = K^\pm(z)^{-1} \tag{30}$$

$$S\left(E^\pm(z)\right) = -K^\pm(zq^{\pm\frac{c}{2}})^{-1}E^\pm(zq^{\pm c}) \tag{31}$$

$$S\left(F^\pm(z)\right) = -F^\pm(zq^{\mp c})K^\pm(zq^{\mp\frac{c}{2}})^{-1} \tag{32}$$

$$S\left(H^\pm(z)\right) = H^\pm(z)^{-1}$$
$$-(q + q^{-1})F^\pm(zq^{2\mp\frac{c}{2}})H^\pm(z)^{-1}K^\pm(zq^2)^{-1}E^\pm(zq^{2\pm\frac{c}{2}}) \tag{33}$$

$$\epsilon\left(K^\pm(z)\right) = \epsilon\left(H^\pm(z)\right) = 1 \tag{34}$$

$$\epsilon\left(E^\pm(z)\right) = \epsilon\left(F^\pm(z)\right) = 0 \tag{35}$$

The compatibility condition (10) of antipode and coproduct for $E^\pm(z)$, $F^\pm(z)$, $K^\pm(z)$ and $H^\pm(z)$ is easy to be checked.

## 3. SUPER YANGIAN DOUBLE $DY(\widehat{gl(1|1)})$

The definition of super Yangian double is similar to the quantum affine superalgebra. Super Yangians double with center extension $DY(\widehat{gl(1|1)})$ is a Hopf algebra generated by currents $L^\pm_{ij}(u)$ and center $c$ which obey the following multiplication relations:

$$R^\pm(u - v)L^\pm_1(u)\eta L^\pm_2(v)\eta = \eta L^\pm_2(v)\eta L^\pm_1(u)R^\pm(u - v) \tag{36}$$
$$R^-(u - v + c)L^-_1(u)\eta L^+_2(v)\eta = \eta L^+_2(v)\eta L^-_1(u)R^-(u - v) \tag{37}$$

where

$$R^\pm(u - v) = \rho^\pm(u - v)R(u - v)$$
$$= \rho^\pm(u - v)\begin{pmatrix} 1 & 0 & 0 & 0 \\ 0 & \frac{u-v}{u-v+1} & \frac{1}{u-v+1} & 0 \\ 0 & \frac{1}{u-v+1} & \frac{u-v}{u-v+1} & 0 \\ 0 & 0 & 0 & \frac{u-v-1}{u-v+1} \end{pmatrix} \tag{38}$$

and

$$\rho^+(x) = \prod_{n\geq 0} \frac{(x - 2n - 3)(x - 2n - 1)^2(x - 2n + 1)}{(x - 2n - 2)^2(x - 2n)^2} \tag{39}$$

$$\rho^-(x) = \prod_{n\geq 0} \frac{(x + 2n)^2(x + 2n + 2)^2}{(x + 2n - 1)(x + 2n + 1)^2(x + 2n + 3)} \tag{40}$$

$R(x)$ is Fermionic type rational solution of graded Yang-Baxter equation:

$$\eta_{12}R_{12}(x - y)\eta_{13}R_{13}(x - z)\eta_{23}R_{23}(y - z)$$
$$= \eta_{23}R_{23}(y - z)\eta_{13}R_{13}(x - z)\eta_{12}R_{12}(x - y) \tag{41}$$

The coproduct structures of $L^\pm_{ij}(u)$ are:

$$\Delta\left(L^\epsilon_{ij}(u)\right) = \sum_{k=1,2} (-1)^{(i+k)(k+j)} L^\epsilon_{kj}(u) \otimes L^\epsilon_{ik}(u - \delta_{\epsilon,+}c_1) \tag{42}$$

where $\epsilon = \pm, \delta_{+,+} = 1, \delta_{-,+} = 0$ and $c_1, c_2$ means $c \otimes 1, 1 \otimes c$ respectively.

As Ding-Frenkel map, $L^\pm(x)$ have the following decomposition

$$L^+(x) = \begin{pmatrix} 1 & 0 \\ f^+(x - c) & 1 \end{pmatrix}\begin{pmatrix} k^+_1(x) & 0 \\ 0 & k^+_2(x) \end{pmatrix}\begin{pmatrix} 1 & e^+(x) \\ 0 & 1 \end{pmatrix} \tag{43}$$

$$L^-(x) = \begin{pmatrix} 1 & 0 \\ f^-(x) & 1 \end{pmatrix}\begin{pmatrix} k^-_1(x) & 0 \\ 0 & k^-_2(x) \end{pmatrix}\begin{pmatrix} 1 & e^-(x) \\ 0 & 1 \end{pmatrix} \tag{44}$$

Let

$$k^{\pm}(u) = k_1^{\pm}(u)^{-1}k_2^{\pm}(u), \qquad h^{\pm}(u) = k_1^{\pm}(u-1)k_2^{\pm}(u) \qquad (45)$$

and

$$e(u) = e^+(u) - e^-(u), \qquad f(u) = f^+(u) - f^-(u) \qquad (46)$$

we can get the following commutation relations among $e(u), f(u), h^{\pm}(u)k^{\pm}(u)$ from (36) (37):

$$[k^{\pm}(u), k^{\pm}(v)] = [k^+(u), k^-(v)] = 0 \qquad (47)$$

$$[k^{\pm}(u), h^{\pm}(v)] = [h^{\pm}(u), h^{\pm}(v)] = 0 \qquad (48)$$

$$h^{\pm}(u)k^{\mp}(v) = k^{\mp}(v)h^{\pm}(u)\frac{(u-v-1)(u-v\mp c+1)}{(u-v+1)(u-v\pm c-1)} \qquad (49)$$

$$\frac{(u-v-2)(u-v-c+2)}{(u-v+2)(u-v-c-2)}h^+(u)h^-(v)$$

$$= h^-(v)h^+(u)\left[\frac{(u-v-1)(u-v-c+1)}{(u-v+1)(u-v-c-1)}\right]^2 \qquad (50)$$

$$[k^{\pm}(u), e(v)] = [k^{\pm}(u), f(v)] = 0 \qquad (51)$$

$$h^{\pm}(u)e(v) = \frac{u-v-1}{u-v+1}e(v)h^{\pm}(u) \qquad (52)$$

$$h^-(u)f(v) = \frac{u-v+1}{u-v-1}f(v)h^-(u) \qquad (53)$$

$$h^+(u)f(v) = \frac{u-v-c+1}{u-v-c-1}f(v)h^+(u) \qquad (54)$$

$$\{e(u), e(v)\} = \{f(u), f(v)\} = 0 \qquad (55)$$

$$\{e(u), f(v)\} = \delta(u-v)k^-(v) - \delta(u-v-c)k^+(u) \qquad (56)$$

where $\delta(u-v) = \sum_{k\in Z} u^k v^{-k-1}$. The coproduct structure of $e^{\pm}(u), f^{\pm}(u), k^{\pm}(u), h^{\pm}(u)$ can be derived directly from (42):

$$\Delta(e^{\epsilon}(u)) = e^{\epsilon}(u) \otimes 1 + k^{\epsilon}(u) \otimes e^{\epsilon}(u - \delta_{\epsilon,+}c_1) \qquad (57)$$

$$\Delta(f^{\epsilon}(u)) = 1 \otimes f^{\epsilon}(u) + f^{\epsilon}(u + \delta_{\epsilon,+}c_2) \otimes k^{\epsilon}(u + \delta_{\epsilon,+}c_2) \qquad (58)$$

$$\Delta(k^{\epsilon}(u)) = k^{\epsilon}(u) \otimes k^{\epsilon}(u - \delta_{\epsilon,+}c_1) \qquad (59)$$

$$\Delta(h^{\epsilon}(u)) = h^{\epsilon}(u) \otimes h^{\epsilon}(u - \delta_{\epsilon,+}c_1) -$$

$$2f^{\epsilon}(u - 1 - \delta_{\epsilon,+}c_1)h^{\epsilon}(u) \otimes h^{\epsilon}(u - \delta_{\epsilon,+}c_1)e^{\epsilon}(u - 1 - \delta_{\epsilon,+}c_1) \quad (60)$$

The Drinfel'd currents can be expanded into Drinfel'd generators:

$$e^{\pm}(u) = \pm\sum_{\substack{n\geq 0 \\ n<0}} e_n u^{-n-1} \qquad f^{\pm}(u) = \pm\sum_{\substack{n\geq 0 \\ n<0}} f_n u^{-n-1} \qquad (61)$$

$$h^\pm(u) = 1 \pm \sum_{\substack{n\geq 0 \\ n<0}} h_n u^{-n-1} \qquad k^\pm(u) = 1 \pm \sum_{\substack{n\geq 0 \\ n<0}} k_n u^{-n-1} \qquad (62)$$

$$e(u) = e^+(u) - e^-(u) \qquad f(u) = f^+(u) - f^-(u) \qquad (63)$$

If $c = 0$, these generators $\{e_n, f_n, h_n, k_n, n \in \mathbf{Z}\}$ satisfy the following multiplication relations

$$[h_m , h_n] = [h_m , k_n] = [k_m , k_n] = 0 \qquad (64)$$

$$[k_m , e_n] = [k_m , f_n] = 0 \qquad (65)$$

$$[h_0 , e_n] = -2e_n , \quad [k_0 , f_n] = 2f_n \qquad (66)$$

$$[h_{m+1} , e_n] - [h_m , e_{n+1}] + \{h_m , e_n\} = 0 \qquad (67)$$

$$[h_{m+1} , f_n] - [h_m , f_{n+1}] - \{h_m , f_n\} = 0 \qquad (68)$$

$$\{e_m , e_n\} = \{f_m , f_n\} = 0 \qquad (69)$$

$$\{e_m , f_n\} = -k_{m+n} \qquad (70)$$

## 4. UNIVERSAL $R$-MATRIX OF $DY(\widehat{gl(1|1)})$

The procedure of constructing of universal $R$-matrix of $DY(\widehat{gl(1|1)})$ is similar to the case of Yangian double $DY(\widehat{sl_2})$ [7,8]. As a quantum double, $DY(gl(1|1))$ consists of the super Yangian $Y(gl(1|1))$ and its dual $Y^*(gl(1|1))$ with opposite comultiplication. The super Yangian $Y(gl(1|1))$ is generated by $e^+(u), f^+(u), h^+(u), k^+(u)$ and $Y^*(gl(1|1))$ is generated by $e^-(u), f^-(u), h^-(u), k^-(u)$. There exists a Hopf pairing relation between $Y(gl(1|1))$ and $Y^*(gl(1|1)) : < , >$ which satisfies the conditions

$$< ab , c^* d^* > = < \Delta(ab) , c^* \otimes d^* > = < b \otimes a , \Delta(c^* d^*) > \qquad (71)$$

for any $a, b \in Y(gl(1|1))$ and $c^*, d^* \in Y^*(gl(1|1))$. We find that this pairing relations can be written as

$$< e^+(u) , f^-(v) > = \frac{1}{u-v}, \qquad < f^+(u) , e^-(v) > = \frac{1}{u-v}, \qquad (72)$$

$$< h^+(u) , k^-(v) > = \frac{u-v-1}{u-v+1}, \qquad < k^+(u) , h^-(v) > = \frac{u-v-1}{u-v+1} \qquad (73)$$

and the center element $c$ is dual to $d$: $< c , d > = 1$, where $[d , g(u)] = \frac{d}{du}g(u)$. The pairing relations between $e^\pm(u)$ and $f^\pm(v)$ have been canonical pairings:

$$< e_k , f_{-l-1} > = < f_k , e_{-l-1} > = -\delta_{kl} \qquad (74)$$

We have to find the canonical pairing of (73). From (73) and the coproduct structure of $k^{\pm}(u), h^{\pm}(u)$ (59)(60), we can get

$$< \frac{\mathrm{d}}{\mathrm{d}u} \ln h^+(u) \ , \ \ln k^-(v) >= \frac{1}{u-v-1} - \frac{1}{u-v+1}, \qquad (75)$$

$$< \frac{\mathrm{d}}{\mathrm{d}u} \ln k^+(u) \ , \ \ln h^-(v) >= \frac{1}{u-v-1} - \frac{1}{u-v+1}. \qquad (76)$$

If we suppose the pairing between $\frac{\mathrm{d}}{\mathrm{d}u} \ln h^+(u)$ and a certain $\phi(v)$ is canonical:

$$< \frac{\mathrm{d}}{\mathrm{d}u} \ln h^+(u) \ , \ \phi(v) >= \frac{1}{u-v}$$

then

$$< \frac{\mathrm{d}}{\mathrm{d}u} \ln h^+(u) \ , \ (T^{-1} - T)\phi(v) >= \frac{1}{u-v-1} - \frac{1}{u-v+1} \qquad (77)$$

where $T$ is a shift operator: $T\phi(v) = \phi(v-1)$. Comparing (75) with (77), we have

$$\phi(v) = (T^{-1} - T)^{-1} \ln k^-(v) = - \sum_{n \geq 0} \ln k^-(v + 2n + 1)$$

So

$$< \frac{\mathrm{d}}{\mathrm{d}u} \ln h^+(u) \ , \ - \sum_{n \geq 0} \ln k^-(v + 2n + 1) >= \frac{1}{u-v} \qquad (78)$$

Similarly, we have

$$< \frac{\mathrm{d}}{\mathrm{d}u} \ln k^+(u) \ , \ - \sum_{n \geq 0} \ln h^-(v + 2n + 1) >= \frac{1}{u-v} \qquad (79)$$

As the same discussion for $DY(sl_2)$ [7,8], the universal $\mathcal{R}$-matrix for $DY(\widehat{gl}(1|1))$ has the following form

$$\mathcal{R} = \mathcal{R}_+ \mathcal{R}_1 \mathcal{R}_2 \exp(c \otimes d) \mathcal{R}_- \qquad (80)$$

where

$$\mathcal{R}_+ = \overrightarrow{\prod_{n \geq 0}} \exp(-e_n \otimes f_{-n-1}) \qquad (81)$$

$$\mathcal{R}_- = \overleftarrow{\prod_{n \geq 0}} \exp(-f_n \otimes e_{-n-1}) \qquad (82)$$

$$\mathcal{R}_1 = \prod_{n \geq 0} \exp \left\{ \mathrm{Res}_{u=v} \left[ (-1)\frac{\mathrm{d}}{\mathrm{d}u}(\ln h^+(u)) \otimes \ln k^-(v + 2n + 1) \right] \right\} \qquad (83)$$

$$\mathcal{R}_2 = \prod_{n \geq 0} \exp \left\{ \mathrm{Res}_{u=v} \left[ (-1)\frac{\mathrm{d}}{\mathrm{d}u}(\ln k^+(u)) \otimes \ln h^-(v + 2n + 1) \right] \right\} \qquad (84)$$

here we have used the notations

$$\text{Res}_{u=v}\left(A(u) \otimes B(v)\right) = \sum_k a_k \otimes b_{-k-1}$$

for $A(u) = \sum_k a_k u^{-k-1}$ and $B(u) = \sum_k b_k u^{-k-1}$.

From the quasi-triangular property of the double, the universal $\mathcal{R}$-matrix satisfies

$$\mathcal{R}_{12} \cdot \mathcal{R}_{13} \cdot \mathcal{R}_{23} = \mathcal{R}_{23} \cdot \mathcal{R}_{13} \cdot \mathcal{R}_{12} \qquad (85)$$

$$(\Delta \otimes id)\mathcal{R} = \mathcal{R}_{13} \cdot \mathcal{R}_{23}, \qquad (id \otimes \Delta)\mathcal{R} = \mathcal{R}_{13} \cdot \mathcal{R}_{12} \qquad (86)$$

In dealing with the tensor product in the graded case, we must use the form $(A \otimes B) \cdot (C \otimes D) = (-1)^{P(B)P(c)} AC \otimes BD$, $P(B) = 0, 1$ for $B$ is bosonic and fermionic respectively.

## 5. EVALUATION REPRESENTATION OF $DY(\widehat{gl}(1|1))$

Let $\rho_x$ be taking two-dimensional evaluation representation for $DY(\widehat{gl}(1|1))$:

$$\rho_x(e_n) = \begin{pmatrix} 0 & 0 \\ x^n & 0 \end{pmatrix} \qquad \rho_x(f_n) = \begin{pmatrix} 0 & x^n \\ 0 & 0 \end{pmatrix} \qquad (87)$$

$$\rho_x(h_n) = \begin{pmatrix} x^n & 0 \\ 0 & -x^n \end{pmatrix} \qquad \rho_x(k_n) = \begin{pmatrix} -x^n & 0 \\ 0 & -x^n \end{pmatrix} \qquad (88)$$

and

$$L^+(x) = (\rho_x \otimes id)\exp(d \otimes c)(\mathcal{R}^{21})^{-1} \qquad L^-(x) = (\rho_x \otimes id)\mathcal{R} \qquad (89)$$

$$R^+(x-y) = (\rho_x \otimes \rho_y)(\mathcal{R}^{21})^{-1} \qquad R^-(x-y) = (\rho_x \otimes \rho_y)\mathcal{R}. \qquad (90)$$

We can find

$$(\rho_x \otimes id)\mathcal{R}_+ = \begin{pmatrix} 1 & 0 \\ -\sum_{n\geq 0} x^n f_{-n-1} & 1 \end{pmatrix} = \begin{pmatrix} 1 & 0 \\ f^-(x) & 1 \end{pmatrix}$$

$$(\rho_x \otimes id)\mathcal{R}_- = \begin{pmatrix} 1 & e^-(x) \\ 0 & 1 \end{pmatrix}$$

$$(\rho_x \otimes id)\mathcal{R}_1 = \begin{pmatrix} \frac{k^-(x+2n+1)}{k^-(x+2n)} & 0 \\ 0 & \frac{k^-(x+2n+1)}{k^-(x+2n+2)} \end{pmatrix}$$

$$(\rho_x \otimes id)\mathcal{R}_2 = \begin{pmatrix} \frac{h^-(x+2n+1)}{h^-(x+2n+2)} & 0 \\ 0 & \frac{h^-(x+2n+1)}{h^-(x+2n+2)} \end{pmatrix}$$

so

$$L^-(x) = \begin{pmatrix} 1 & 0 \\ f^-(x) & 1 \end{pmatrix} \begin{pmatrix} k_1^-(x) & 0 \\ 0 & k_2^-(x) \end{pmatrix} \begin{pmatrix} 1 & e^-(x) \\ 0 & 1 \end{pmatrix} \tag{91}$$

where

$$k_1^-(x) = \prod_{n \geq 0} \frac{k^-(x+2n+1)}{k^-(x+2n)} \frac{h^-(x+2n+1)}{h^-(x+2n+2)} \tag{92}$$

$$k_2^-(x) = \prod_{n \geq 0} \frac{k^-(x+2n+1)}{k^-(x+2n+2)} \frac{h^-(x+2n+1)}{h^-(x+2n+2)} \tag{93}$$

Similarly, we have

$$L^+(x) = \begin{pmatrix} 1 & 0 \\ f^+(x-c) & 1 \end{pmatrix} \begin{pmatrix} k_1^+(x) & 0 \\ 0 & k_2^+(x) \end{pmatrix} \begin{pmatrix} 1 & e^+(x) \\ 0 & 1 \end{pmatrix} \tag{94}$$

where

$$k_1^+(x) = \prod_{n \geq 0} \frac{k^+(x-2n-2)}{k^+(x-2n-1)} \frac{h^+(x-2n)}{h^+(x-2n-1)} \tag{95}$$

$$k_2^+(x) = \prod_{n \geq 0} \frac{k^+(x-2n)}{k^+(x-2n-1)} \frac{h^+(x-2n)}{h^+(x-2n-1)} \tag{96}$$

The above results of the decomposition of $L^\pm(x)$ is the same as (43) and (44).
We can also find

$$\rho_y(f^-(x)) = \begin{pmatrix} 0 & \frac{1}{x-y} \\ 0 & 0 \end{pmatrix} \qquad \rho_y(e^-(x)) = \begin{pmatrix} 0 & 0 \\ \frac{1}{x-y} & 0 \end{pmatrix}$$

$$\rho_y(h^-(x)) = \begin{pmatrix} \frac{x-y+1}{x-y} & 0 \\ 0 & \frac{x-y-1}{x-y} \end{pmatrix} \qquad \rho_y(k^-(x)) = \begin{pmatrix} \frac{x-y-1}{x-y} & 0 \\ 0 & \frac{x-y-1}{x-y} \end{pmatrix}$$

$$\rho_y(k_1^-(x)) = \begin{pmatrix} \rho^-(x-y) & 0 \\ 0 & \rho^-(x-y)\frac{x-y}{x-y+1} \end{pmatrix}$$

$$\rho_y(k_2^-(x)) = \begin{pmatrix} \rho^-(x-y)\frac{x-y-1}{x-y} & 0 \\ 0 & \rho^-(x-y)\frac{x-y-1}{x-y+1} \end{pmatrix}$$

then

$$R^-(x-y) = (\rho_x \otimes \rho_y)\mathcal{R} = \rho_y(L^-(x))$$

$$= \rho^-(x-y) \begin{pmatrix} 1 & 0 & 0 & 0 \\ 0 & 1 & 0 & 0 \\ 0 & \frac{1}{x-y} & 1 & 0 \\ 0 & 0 & 0 & 1 \end{pmatrix} \begin{pmatrix} 1 & 0 & 0 & 0 \\ 0 & \frac{x-y}{x-y+1} & 0 & 0 \\ 0 & 0 & \frac{x-y-1}{x-y} & 0 \\ 0 & 0 & 0 & \frac{x-y-1}{x-y+1} \end{pmatrix}$$

$$\begin{pmatrix} 1 & 0 & 0 & 0 \\ 0 & 1 & \frac{1}{x-y} & 0 \\ 0 & 0 & 1 & 0 \\ 0 & 0 & 0 & 1 \end{pmatrix}$$
$$= \rho^-(x-y)R(x-y) \tag{97}$$

Similarly, $R^+(x-y) = \rho^+(x-y)R(x-y)$. The scale factors $\rho^\pm(x-y)$ have been defined in (39) and (40). This result is the same as (38).

From above results, the multiplication and coproduct structure (36), (37) and (42) can be easily obtained from (85) and (86) through taking evaluation representation.

## 6. CONCLUSION AND REMARK

In this short article, we have reviewed our previous works on Drinfel'd realizations of quantum affine superalgebra and super Yangian double, then we studied the universal $R$-matrix and evaluation representation of $DY(\widehat{gl(1|1)})$.

These methods may be applied in studying quantum superalgebra corresponding to the solution of Yang-Baxter equation with colored parameter instead of spectrum parameter, as well as with dynamical parameter. Some of these works are under investigation.

## ACKNOWLEDGMENTS

This paper is dedicated to Prof. Hiroshi Ezawa on the occasion for his 65th birthday. We are grateful for Prof. Ezawa's friendship and Prof. Christophen C. Bernido's invitation.

This work was supported by Scientific Project of Department of Science and technology in China, Natural Scientific Foundation of Chinese Academy of Sciences and Foundation of NSF.

## REFERENCES

1. V.G. Drinfel'd, "Quantum Groups", *Proc. ICM-86(Berkeley)*. Vol. 1 (New York Academic Press 1986), p.789

2. V.G. Drinfel'd, *Soviet Math. Dokl.*, **36** (1988) 212

3. M. Jimbo, *Commun. Math. Phys.*, **102** (1986) 537

4. L.D. Faddeev, N.Yu. Reshetikhin and L.A. Takhtajan, *Algebraic analysis,* **1** (1988) 129

5. N.Yu. Reshetikhin and M.A. Semenov-Tian-Shansky, *Lett. Math. Phys.,* **19** (1990) 133

6. J. Ding and I.B. Frenkel, *Commun. Math. Phys.,* **156** (1993) 277

7. S.M. Khoroshkin and V.N. Tolstoy, *Lett. Math. Phys.,* **36** (1996) 373

8. S.M. Khoroshkin, preprint, q-alg/**9602031**

9. K. Iohara, *J. Phys. A: Math. Gen.* , **29** (1996) 4593

10. L. Liao and X.C. Song, *Mod. Phys. Lett.* , **A6** (1991) 959-968

11. H. Fan, B.Y. Hou and K.J. Shi, *J. Math. Phys.* , **38** (1997) 411

12. H. Yan, Y. Zhou and T.H. Zhu, *J. Phys. A: Math. Gen.* , **26** (1993) 935

13. J.F. Cai, S.K. Wang, K. Wu and W.Z. Zhao, *Commun. Theor. Phys.* **28** (1997) 1

14. J.F. Cai, S.K. Wang, K. Wu and W.Z. Zhao, *J. Phys. A: Math. Gen* , **31** (1998) 1989

15. J.F. Cai, G.X. Ju, K. Wu and S.K. Wang, *J. Phys. A: Math. Gen.* , **30** (1997) L347

16. Y.Z. Zhang, *J. Phys. A: Math. Gen.* , **30** (1997) 8325

17. Y.Z. Zhang, *Phys. Lett.* , **A234** (1997) 20

18. J.F. Cai, S.K. Wang, K. Wu and C. Xiong, *Commun. Theor. Phys.* **29** (1998) 173

# GREEN'S FUNCTION METHOD AND THEORY OF REPULSIVE BOSE GAS

FLORDIVINO BASCO[1,2] and HIDETOSHI FUKUYAMA[1]

[1]Department of Physics, University of Tokyo
Tokyo 113, Japan
[2]National Institute of Physics, University of the Philippines
Quezon City 1101, Philippines

**Abstract**    A unified theory of repulsive bose gas is constructed. The theory is based on the universal feature of the excitation spectra - linear and quadratic for small and large values of momenta, respectively. The ensemble of bosons is treated as a mixture of condensates and noncondensates. The measurable amount of condensates is determined by imposing that they coexist at equilibrium. This scenario of coexistence gives a natural explanation why the excitation spectra are gapless. Calculations made, based on this theory, yield the established results for one and three dimensions. For the first time, a new result is obtained in two dimensions indicating a strong dimensional crossover. The theory is consistent with vanishing of the *anomalous* self-energy for zero momentum and energy for dimensions less than or equal to three.

## 1  INTRODUCTION

It has been agreed that the procedure for treating interacting (repulsive) bose gas depends on the dimensions of the system. In one dimension, one uses the Bethe ansatz procedure[1] relying on the freedom that the momentum of each particle can be ordered and the result yields the absence of condensates (even with an infinitesimally small interaction). In three dimensions, one assumes that the amount of condensates is macroscopic a la Bogoliubov[2] prescription. Based on this prescription, different methods have been introduced, some of which are: canonical[2], binary-collision[3], pseudo-potential[4] and Green's function[5, 6, 7] methods. Among these methods, the Green's function method appears to be the most promising in giving the exact description of the system, including the ground-state energy ($GS$ energy), the amount of condensate and the thermodynamical properties. For the two-dimensional case[8], however, to our knowledge no established procedure has so far been reported.

In this paper, we will give one mathematical construct to describe the interacting bose gas in any number of dimensions. We will show that our procedure recovers the result in one[1] and three[2, 3, 4, 5, 6, 7] dimensions, and hence, can be applied in two dimensions. The only assumption in this theory is that the potential is repulsive. The theory is based on the crossover nature of the excitation spectra: linear and quadratic in momenta for low and high energy regions, respectively. These properties of the excitation spectra for low and high energies are universal features of an ensemble of interacting bosons, i.e., independent of dimension, $d$, of the system. Utilizing these universal features of bosons an "ad hoc" form of the Green's function that interpolates from small to large momenta and energy limits has been proposed, where the ground-state energy ($GS$ energy) and amount of condensates are calculated. The result of the calculations reproduces the well known results in one[1] and three[2, 3, 4, 5, 6, 7, 9] dimensions, and for the first time, obtains a new result in two dimensions.

The previous theoretical works on the Green's function method (limited to three dimensions) assumed a finite value of the *anomalous* self-energy for zero momentum and energy, $\Sigma_{12}(0)$[5, 6, 7, 9], however it was shown to vanish for $d \leq 3$[10, 11]. Hence, theory accounting for the vanishing of $\Sigma_{12}(0) = 0$ ($d \leq 3$) is essential to establish.

## 2   GENERAL FORMALISM

### 2.1   Difficulties With The Zero-Momentum State

At zero temperature, an ideal gas undergoes bose condensation and all the particles are in their lowest energy states. If $\hat{N}_k = a_k^* a_k$ is the occupation number operator of the one particle state with momentum $\mathbf{k}$, the distribution function is given by $\hat{N}_k = N$ for $k = 0$ and zero otherwise. When the interaction is switched on, the condensed state is only partially removed, and $N_0$ ($< N$) remains macroscopic. Moreover, neither the creation operator $a_0^*$ nor the annihilation operator $a_0$ annihilates the ground state wavefunction. Hence, a perturbation calculation treating $a_0$ or $a_0^*$ as operators is not feasible since it renders the Wick theorem inapplicable and, thus makes application of the standard perturbation method impossible. For a very large value of $N$ (or $N_0$), the distinction between $N_0^{1/2}$ or $(N_0 + 1)^{1/2}$ is immaterial and $a_0$ and $a_0^*$ can be treated as $c$ numbers. The usual Bogoliubov[2] prescription can be applied: $a_0^* \simeq a_0 = N_0^{1/2}$. This procedure obviously neglects fluctuations in the occupation number of the condensate. Let us introduce the intensive quantities by the operators, $\psi_k = \Omega^{-1/2} a_k$ for $\mathbf{k} \neq 0$ and $\psi_0^* = (N_0/\Omega)^{1/2} \equiv n_0^{1/2}$ for $\mathbf{k} = 0$. The nonzero component, $\mathbf{k} \neq 0$, has the following commutation relations: $[\psi_k, \psi_l] = [\psi_k^*, \psi_l^*] = 0$ and $[\psi_k, \psi_l^*] = \Omega^{-1} \delta_{Kr}^d (\mathbf{k} - \mathbf{l})$ where the Kronecker delta $\delta_{Kr}^d(\mathbf{k})$ is equal to one if the argument is zero, and zero otherwise. Nevertheless, the Bogoliubov procedure of replacement of $a_0$ and $a_0^*$ (or $\psi_0$ and $\psi_0^*$) by $c$ numbers describes the correct ground state in the thermodynamic limit, i.e., $N \to \infty$, $\Omega \to \infty$, $n \equiv N/\Omega \to$ constant, $n_0 \equiv N_0/\Omega \to$ constant.

We are, therefore, led to express the field operator as $\hat{\psi}(\mathbf{x}) = n_0^{1/2} + \sum' e^{i\mathbf{k}\cdot\mathbf{x}} \psi_k$, where $\sum'$ implies that $\mathbf{k} = 0$ is not included in the summation. After such a separation of $\hat{\psi}(\mathbf{x})$ into two parts, we can now consider $\psi_0$ and $\psi_0^*$ as $c$ numbers, and the only remaining problem is how to determine $n_0$. To determine $n_0$, we will let the condensates and noncondensates coexist at equilibrium, i.e., their chemical potentials are the same.

We are interested in the Hamiltonian $\hat{H} = \hat{H}_0 + \hat{V}$ where the kinetic energy $\hat{H}_0$ and the interaction energy $\hat{V}$ are given as:

$$\hat{H}_0 = \sum' \frac{k^2}{2m} a_k^* a_k = \Omega \sum' \frac{k^2}{2m} \psi_k^* \psi_k, \qquad \text{and}$$
$$\hat{V} = \tfrac{1}{2}\Omega \sum_{k_1,k_2,k_3,k_4=0} \psi_{k_4}^* \psi_{k_3}^* V_{k_1-k_3} \psi_{k_1} \psi_{k_2}, \tag{1}$$

where $m$ is the bare mass of bosons and $V_k$ is the Fourier component of the pair interaction $V(\mathbf{x}) = \sum_{\mathbf{k}} e^{i\mathbf{k}\cdot\mathbf{x}} V_k$. The pair interaction $V(\mathbf{x})$ is assumed to depend only on the relative distance between the two bosons. Thus, $V_k$ is expected to be an even function of momentum $\mathbf{k}$.

The substitution $\psi_0^* \simeq \psi_0 = n_0^{1/2}$ does not change the kinetic part, but the interaction part changes in a fundamental way and can be decomposed into eight distinct parts according to the number of factors of $n_0^{1/2}$:

$$E_0 = \tfrac{1}{2} n_0^2 V_0 \delta(0)$$
$$\hat{V}_1 = \tfrac{1}{2} n_0 \sum' V_{k_1} \psi_{k_1} \psi_{k_2} \delta(k_1 + k_2)$$
$$\hat{V}_2 = \tfrac{1}{2} n_0 \sum' \psi_{k_4}^* \psi_{k_3}^* V_{-k_3} \delta(-k_3 - k_4)$$
$$\hat{V}_3 = 2(\tfrac{1}{2} n_0) \sum' \psi_{k_3}^* V_{-k_3} \psi_{k_2} \delta(k_2 - k_3)$$
$$\hat{V}_4 = 2(\tfrac{1}{2} n_0) \sum' \psi_{k_3}^* V_{k_1-k_3} \psi_{k_1} (\mathbf{x}) \delta(k_1 - k_3)$$
$$\hat{V}_5 = 2(\tfrac{1}{2} n_0^{1/2}) \sum' \psi_{k_4}^* \psi_{k_3}^* V_{k_1-k_3} \psi_{k_1} \delta(k_1 - k_3 - k_4)$$
$$\hat{V}_6 = 2(\tfrac{1}{2} n_0^{1/2}) \sum' \psi_{k_3}^* V_{k_1-k_3} \psi_{k_1} \psi_{k_2} \delta(k_1 + k_2 - k_3)$$
$$\hat{V}_7 = \tfrac{1}{2} \sum' \psi_{k_4}^* \psi_{k_3}^* V_{k_1-k_3} \psi_{k_1} \psi_{k_2} \delta(k_1 + k_2 - k_3 - k_4), \tag{2}$$

where $\delta(\mathbf{k}) \equiv \Omega^{-1} \delta_{Kr}^d(\mathbf{k})$. The term with the single $\psi_k$ or $\psi_k^*$ alone inside the summation symbol is not allowed by the conservation of momentum and thus, does not appear in the interaction, eq.(2).

## 2.2  Two Species Picture: Condensates And Nonconden-sates

The replacement of $\psi_0$ and $\psi_0^*$ by $n_0^{1/2}$, their average value, simplies the problem and the Hamiltonian becomes $\hat{H}(N_0) = \hat{H}_0 + \hat{V}(N_0)$. As a consequence, the fluctuations of $\psi_0$ and $\psi_0^*$ are neglected and the total number of particles is no longer conserved because $\hat{V}(N_0)$ does not commute anymore with the operator $\hat{N}' = \sum' a_k^* a_k$. Thus, we must impose the subsidiary condition $\langle \hat{N} \rangle = N$.

This subsidiary condition cannot be imposed on the ground state of $\hat{H}$ since $\hat{N}$ does not commute with $\hat{H}$. The only possible way to satisfy $\langle \hat{N} \rangle = N$ is to resolve with the method of the undetermined multiplier. For this purpose, the relevant Hamiltonian is

$$\hat{\tilde{H}}(\mu_0, \mu) = \hat{H}(N_0) - \mu_0 N_0 - \mu \hat{N}' \tag{3}$$

where $\mu_0$ are $\mu$ are the Lagrange multipliers that can be identified as the chemical potentials of condensates and noncondensates, respectively, with

$$\hat{N}' = \hat{N} - N_0 \qquad \text{and} \qquad N = \langle \hat{N}' \rangle + N_0, \tag{4}$$

where the expectation value in $\langle \hat{N}' \rangle$ is with respect to the ground-state of $\hat{\tilde{H}}$.

With the understanding that every average is taken with respect to the ground-state of $\hat{\tilde{H}}$, let us introduce $\tilde{E}$ and $E$ as the expectation values of $\hat{\tilde{H}}$ and $\hat{H}$, respectively. The two $GS$ energies are related to each other by

$$\tilde{E}(\mu_0, \mu) = E(N_0, N') - \mu_0 N_0 - \mu N'. \tag{5}$$

In the thermodynamic level, eq.(5) is the usual Legendre transformation that changes the original independent variables $N_0$ and $N$ (or $N' = N - N_0$) into a new set of independent variables, $\mu_0$ and $\mu$, and does not imply that the number of independent variables increased from two to four. For example, in the context of canonical ensemble, the chemical potentials are given in terms of $N_0$ and $N'$: $\mu_0 = \frac{\partial E}{\partial N_0} \big)_{N'}$ and $\mu = \frac{\partial E}{\partial N'} \big)_{N_0}$. Similarly, in the context of grand canonical ensemble, the average number of particles are given in terms of $\mu_0$ and $\mu$: $N_0 = -\frac{\partial \tilde{E}}{\partial \mu_0} \big)_{\mu}$, and $N' = -\frac{\partial \tilde{E}}{\partial \mu} \big)_{\mu_0}$.

It is, however, more convenient to work with $\mu$ and $N_0$ as a new set of independent variables. To proceed, let us define

$$E^*(N_0, N) \equiv E(N_0, N' = N - N_0) \tag{6}$$

The corresponding Legendre transformation is

$$E'(N_0, \mu) \equiv \tilde{E}(\mu_0, \mu) + \mu_0 N_0 = E^*(N_0, N) - \mu N'. \tag{7}$$

In this set of independent variables the set of "equations of state" is

$$\frac{\partial E'}{\partial N_0} \bigg)_{\mu} = \mu_0 = \frac{\partial E^*}{\partial N_0} \bigg)_{N} + \mu \qquad \text{and} \qquad \frac{\partial E'}{\partial \mu} \bigg)_{N_0} = -N'. \tag{8}$$

Since the order of differentiation is immaterial, one has

$$\frac{\partial \mu_0}{\partial \mu} \bigg)_{N_0} = -\frac{\partial N'}{\partial N_0} \bigg)_{\mu}. \tag{9}$$

## 2.3 Coexistence Of Condensates And Noncondensates

In the following, we will shift to intensive variables $(n = \frac{N}{\Omega}, n_0 = \frac{N_0}{\Omega}, n' = \frac{N'}{\Omega})$ instead of working on extensive variables $(N, N_0, N')$.

So far, we have two degrees of freedom; either $n_0$ and $n$ or $n_0$ and $\mu$. To trim down the degree of freedom into one, one should impose an equilibrium condition. The natural condition for multi-species system is to let every species coexist with each other at equilibrium. In our case, this condition is given by

$$\mu_0(n_0(\mu), \mu) = \mu, \tag{10}$$

i.e., condensates and noncondensates coexist. This treatment of coexistence offers an interesting point of view; one can think the condensates as particles' "reservoir" of the noncondensates. It implies that the noncondensates can be produced and also absorbed out of the condensates. The processes of production or absorption always occur in pairs so that the total momentum of a pair is zero.

Application of this condition, the Hugenholtz-Pines[7] condition (as followed from eq.(8)),

$$\frac{\partial}{\partial n_0}\left(\frac{E^*(n_0, n(n_0(\mu), \mu))}{\Omega}\right)_n = 0, \tag{11}$$

became meaningful. This implies that the amount of condensates at the equilibrium naturally minimize the $GS$ energy. Consequently, we have

$$\frac{d}{dn}\left(\frac{E}{\Omega}\right) = \left(\frac{dn_0}{dn}\frac{\partial}{\partial n_0} + \frac{\partial}{\partial n}\right)\frac{E^*(n_0, n)}{\Omega} = \mu \tag{12}$$

where $\frac{dn_0}{dn} \equiv \frac{dn_0}{d\mu}\left(\frac{dn}{d\mu}\right)^{-1}$.

This result can be used to determine the physical solution. To do this, one simply calculates $\frac{E}{\Omega}$ and eliminates $n_0$ using eq.(16), and one is left with the first order differential equation with $\mu = \frac{d}{dn}\left(\frac{E}{\Omega}\right) \equiv \frac{d}{dn}\left(\frac{E(n)}{\Omega}\right)$. However, the implicit dependence of $E$ and $\mu$ on the "microscopic" sound velocity (the sound velocity derived from the slope of the linear part of the energy spectrum) yields a term in the form $\frac{d\mu}{dn} \equiv \frac{d^2}{dn^2}\left(\frac{E(n)}{\Omega}\right)$ which is of second order differential equation. This is another new aspect in our formalism — solving a second order differential equation, instead of solving either eq.(10) or eq.(11), to find the physical solution.

## 3 NONCONDENSATES, $GS$ ENERGY AND GREEN'S FUNCTIONS

### 3.1 Noncondensates And Ground-State Energy

Let us proceed by constructing the Green's function. It is easy to see that the corresponding Hamiltonian $\hat{H}'$ of $E'$ is

$$\hat{H}' = \hat{H} - \mu\hat{N}' = E_0(n_0) + \hat{K} \tag{13}$$

where $\hat{K} = \Omega \sum_{k \neq 0} \psi_k^*(\frac{k^2}{2m} - \mu)\psi_k + \sum_{i=1}^{7} \hat{V}_i$. The single-particle Green's function of the noncondensates is given by

$$iG(\mathbf{x_1} - \mathbf{x_2}, t_1 - t_2) = \langle \psi_0 | T\psi(\mathbf{x_1}t_1)\psi^*(\mathbf{x_2}t_2) | \psi_0 \rangle \tag{14}$$

where $|\psi_0\rangle$ is the ground state wave function of the interacting system and $\psi(\mathbf{x}t)$ are the field operators in Heisenberg representation. The signature factor of all time ordering is assumed to be +1. With the usual definition of the Fourier transform, $iG(\mathbf{x}, t) = i(2\pi)^{-d} \int d^d k e^{i\mathbf{k}\cdot\mathbf{x}} G(\mathbf{k}, t)$, one finds

$$iG(\mathbf{k}, t - t')\delta^d(\mathbf{k} - \mathbf{k}') = \langle \psi_0 | T\psi_k(t)\psi_{k'}^*(t') | \psi_0 \rangle. \tag{15}$$

and the $n'$ is given by

$$n'(n, n_0) \equiv n - n_0 = i(2\pi)^{-d-1} \int d^d k \int_C d\epsilon G(\mathbf{k}, \epsilon) e^{i\epsilon\delta} \tag{16}$$

where we impose an implicit limit of $\delta \to 0^+$ so that $t > t'$ always. The contour C of $\epsilon$-integration is the real axis from $-\infty$ to $+\infty$. The convergence factor $e^{i\epsilon\delta}$ makes it possible for contour C to extend and close to a semicircle of infinite radius in the upper half plane of the complex plane.

The $GS$ energy can be expressed also in terms of the Green's function by calculating the equation of motion of $\psi_k(t)$ and $\psi_{k'}^*(t')$. (The details are discussed in Refs. Basco, Abri and Fetter).

$$\frac{E}{\Omega} = \tfrac{1}{2}n\mu + (2\pi)^{-d-1} \int d^d k \int d\epsilon \, \tfrac{1}{2} \left(\epsilon + \frac{k^2}{2m}\right) iG(\mathbf{k}, \epsilon) e^{i\epsilon\delta}. \tag{17}$$

The first contribution is basically a "vacuum" expectation value, $\frac{1}{2}n_0\mu$, which renormalizes to $\frac{1}{2}n\mu$ due to the presence of noncondensates. The second contribution comes from two different excitations; collective, $\epsilon(\mathbf{k})$, and individual, $\frac{k^2}{2m}$. The factor $\frac{1}{2}$ reflects that these contributions should be averaged. The collective excitations is the low energy contribution, favors the formation of quasiparticles and is always negative. On the other hand, the individual excitations is the high energy contribution, free particle-like and always positive. From here, we can see the presence of a trade-off of energies between the collective ($< 0$) and the individual ($> 0$) excitations, and the dominance of the former to the latter favors the presence of bose condensation, and vice versa. This competition determines the degree of formation of condensates.

## 3.2  Green's Functions

The absence of the conservation of particles arising from bose condensation provides sources and sinks of the noncondensates out of the condensates. Treating the condensates as "reservoir" of pair particles; the noncondensates are emitted and absorbed in pairs with momenta $\mathbf{k}$ and $-\mathbf{k}$. This leads us to introduce two "new" Green's functions, namely, $G_{12}$ for the process representing the "emission" of pairs from the condensates and $G_{21}$ for the process representing the "absorption" of pairs by the condensates. Let us refer to them as *anomalous* Green's functions. Corresponding to the Green's functions $G_{12}$ and $G_{21}$ we have also two *anomalous* self-energies, $\Sigma_{12}$ and $\Sigma_{21}$, respectively, other than the *normal* self-energy $\Sigma_{11}$ which is related to $G$. The Dyson's equation for this system is

$$
\begin{aligned}
G(p) &= G^0(p) + G^0(p)\Sigma_{11}(p)G(p) + G^0(p)\Sigma_{12}(p)G_{21}(p) \\
G_{12}(p) &= G^0(p)\Sigma_{12}(p)G(-p) + G^0(p)\Sigma_{11}(p)G_{12}(p) \\
G_{21}(p) &= G^0(-p)\Sigma_{21}(p)G(p) + G^0(-p)\Sigma_{11}(-p)G_{21}(p)
\end{aligned}
\tag{18}
$$

which is first derived by Beliaev[5] where $p = (\mathbf{k}, \epsilon)$. The solution of these system of equations is

$$
G(p) = \frac{\epsilon + \frac{k^2}{2m} - \mu + S(p) - A(p)}{D(p)} \quad \text{and} \quad G_{12}(p) = G_{21}(p) = -\frac{\Sigma_{12}(p)}{D(p)}
\tag{19}
$$

where

$$
D(p) = [\epsilon - A(p)]^2 - \left[\frac{k^2}{2m} - \mu + S(p)\right]^2 + \Sigma_{12}^2(p),
\tag{20}
$$

$S(p) = \frac{1}{2}[\Sigma_{11}(p) + \Sigma_{11}(-p)]$ and $A(p) = \frac{1}{2}[\Sigma_{11}(p) - \Sigma_{11}(-p)]$. Due to absence of gap of the excitation spectra[7, 8], all the Green's functions have the same limiting form of

$$
G(p) = -G_{12}(p) = \frac{\frac{n_0}{n} mc^2}{\epsilon^2 - c^2 k^2 + i\delta} \quad (p \to 0).
\tag{21}
$$

## 4  THE REPULSIVE BOSE GAS

### 4.1  Formalism

Let us start with the form of the Green's function. For finite density of bosons, the most singular contribution of the Green's function is given by its limiting form for small $\epsilon$ and $\mathbf{k}$, eq.(21), and hence, is a good approximation for $G(\mathbf{k}, \epsilon)$. However, we can not use this alone for all range of energies and momenta or else we will overestimate the degree of interactions, especially in the case of very small density $n$ of bosons where the effect of interaction is not so important. This follows from the fact that the magnitude of the sound velocity $c$, which reflects the strength of interaction and measures the stiffening of excitation spectra from $k^2$ (for $c = 0$) to $k$ (for $c \neq 0$), is proportional to the density of bosons $n$,

$mc^2 = n \left(\frac{dn}{d\mu}\right)^{-1}$. This implies that the form of the Green's function that we are seeking should have a correct limiting form for small values of $\mathbf{k}$ and $\epsilon$, and should be reduced to free-particle Green's function for large values of $\mathbf{k}$ and $\epsilon$.

A simple form of the Green's function that will satisfy the above requirement of crossover of momenta and energy is

$$G(\mathbf{k}, \epsilon) = \frac{\epsilon + \frac{k^2}{2m} + \frac{n_0}{n}mc^2}{\epsilon^2 - \left(\frac{k^2}{2m}\right)^2 - c^2k^2 + i\delta}. \tag{22}$$

It is plain to see that this proposed or "ad hoc" Green's function satisfies the necessary requirements of crossover of momenta and energy:

$$G(\mathbf{k}, \epsilon) \sim \begin{cases} \frac{\frac{n_0}{n}mc^2}{\epsilon^2 - c^2k^2 + i\delta}, & \text{for small } \epsilon \text{ and } k, \\ \frac{1}{\epsilon - \frac{k^2}{2m} + i\delta}, & \text{for large } \epsilon \text{ and } k. \end{cases} \tag{23}$$

With the correct form of the Green's function, eq.(22), we can now proceed to calculate the amount of condensates and the $GE$ energy. The integration over the energy variable can be done straightforwardly and gives the amount of noncondensates, $n - n_0$, and the $GS$ energy, $E$, as

$$n - n_0 = \int \frac{d^d\mathbf{k}}{(2\pi)^d} n_k \qquad \text{and} \tag{24}$$

$$\frac{E}{\Omega} = \frac{1}{2}n\mu + \int \frac{d^d\mathbf{k}}{(2\pi)^d} \frac{1}{2}\left(-\epsilon_k + \frac{k^2}{2m}\right) n_k, \tag{25}$$

respectively, where $\epsilon_k \equiv \sqrt{\left(\frac{k^2}{2m}\right)^2 + c^2k^2}$ is the energy dispersion and $n_k \equiv \frac{-\epsilon_k + \frac{k^2}{2m} + \frac{n_0}{n}mc^2}{2\,\epsilon_k}$ is the occupation of the state $k$. For small values of $k$, this $n_{bfk}$ to

$$n_k = \frac{1}{2}\frac{n_0}{n}\frac{mc}{k} \qquad (k \to 0) \tag{26}$$

where it is easy to see that the finiteness of $n_0$ is intimately related to macroscopic (or infinite) occupation of single-particle quantum, i.e., $k = 0$.

The apparent divergence for large values of $k$ is remedied by introducing a cut-off $\Lambda$ such that $n_k = 0$ for $|\mathbf{k}| = k < \Lambda$. This is related to the strength of interaction $\gamma$ and, hence, scales with $c$ and can be written as $\Lambda = 2m\kappa c$ with $\kappa$ appears as effective coupling of interaction. We note that it is possible that $n_k \neq 0$ for $|\mathbf{k}| = k > \Lambda$, however, we expect that $n_k$ is very small and its contribution to the result of integrations, eq.(4.1a) and eq.(4.1b), is negligible.

The physical solution, with the help of eq.(4.1), is determined using the self-consistent equations

$$\mu = \left(\frac{d}{dn}\right)\frac{E}{\Omega} \quad \text{and} \quad mc^2 = n\left(\frac{d\mu}{dn}\right). \tag{27}$$

This set of self-consistent equations consists of two equations as opposed to the treatment of Hugenholtz-Pines (in three dimensions) who used only the first equation and assumed $\Sigma_{12}(0) \neq 0$ ($d \leq 3$) which is showed to be incorrect[10, 11]. We note that our procedure do not contradict with the case $\Sigma_{12}(0) = 0$ ($d \leq 3$) as been discussed in Ref. Basco. The second equation guarantees the equality of "microscopic" and "macroscopic" sound velocity which was not taken into account in their work.

## 4.2   One-dimensional Bose Gas.

Let us now apply our formalism to the one-dimensional bose gas and proceed by considering the expression for the amount of noncondensates, eq.(24). The indicated integration diverges infraredly, unless $n_0$ is zero. This result is independent of the magnitude of $\Lambda$ and hence, the strength of interaction, $\gamma$. This is due only on the linearity of excitation for small momenta. Thus, we can conclude that the condensates do not exist in one dimension even if the interaction is very weak. This does not invalidate our starting point concerning the form of our excitation spectra. In fact, the exact analysis (based on Bethe ansatz method) by Lieb[14] showed that our excitation spectra coincided with his exact calculation for small values of momenta. Indeed, our conclusion that the condensates do not exist in one dimension is consistent with the Bethe ansatz calculation of Lieb and Liniger[1] that a finite amount of interaction leads to a finite value of $n_{k=0} < \infty$ (or $n_0$ is not macroscopic) and hence, no condensate.

## 4.3   Three-dimensional Bose Gas.

Let us proceed with the situation for $2d$ and $3d$. First we will consider the situation in three dimensions before going directly to the case of two-dimensional bose gas to enable us to see if there is any possible dimensional crossover of $2d$ from $1d$ and $3d$.

Let us begin by carrying the momentum integration of eq.(4.1) both for the amount of noncondensates, $n - n_0$, and the ground state energy, $\frac{E}{\Omega}$. The result is

$$n - n_0 = \frac{\frac{1}{3}\pi^{-2}(mc)^3 a_1(\kappa)}{1 + \pi^{-2}n^{-1}(mc)^3 a_2(\kappa)} \quad \text{and} \tag{28}$$

$$\frac{E}{\Omega} = \frac{1}{2}n\mu + \frac{1}{15}\pi^{-2}m^{-1}(mc)^5\left[a_3(\kappa) - 5(1 - \frac{n_0}{n})a_4(\kappa)\right] \tag{29}$$

where $a_1(\kappa) \equiv 2(\omega^3 - \kappa^3) - 3w + 1$, $a_2(\kappa) \equiv \omega - 1$, $a_3(\kappa) \equiv 12(\omega^5 - \kappa^5) - 5(\omega^3 - 5\kappa^3) + 15\omega - 2$, $a_4(\kappa) \equiv (\omega^3 - \kappa^3) - 3w + 2$, and $\omega \equiv (\kappa^2 + 1)^{1/2}$. Then, let us scale the variables by introducing dimensionless variables as follows:

$$\begin{pmatrix} n \\ n_0 \end{pmatrix} = k_0^3 \pi^{-2} \begin{pmatrix} x \\ y \end{pmatrix} \qquad \text{and} \qquad \begin{pmatrix} \pi^2 k_0^{-3} \frac{E}{\Omega} \\ \mu \end{pmatrix} = \frac{k_0^2}{2m} \begin{pmatrix} e \\ u \end{pmatrix} \qquad (30)$$

where $k_0$ is a unit momentum that accounts for the correct dimension. With this, the fraction of noncondensates, $1 - \frac{n_0}{n} = 1 - \frac{y}{x}$, and the dimensionless $GS$ energy, $e$, are given as

$$1 - \frac{n_0}{n} = 1 - \frac{y}{x} = \frac{\frac{1}{3} c_0^3 a_1(\kappa)}{x + c_0^3 a_2(\kappa)} \qquad \text{and}$$

$$e = \frac{1}{2} xu + \frac{2}{15} c_0^5 \left[ a_3(\kappa) - 5 \left( 1 - \frac{y}{x} \right) a_4(\kappa) \right], \qquad (31)$$

respectively, where $c_0$ is the dimensionless sound velocity defined by $c_0^2 \equiv \frac{1}{2} x\dot{u} = \frac{1}{2} x\ddot{e}$ and the dot means differentiation with respect to $x$. Immediately, we arrive at the self-consistent equation as

$$e - \frac{1}{2} x\dot{e} = \frac{1}{15} 2^{-3/2} x^{5/2} \ddot{e}^{5/2} \frac{A(\kappa) + 2^{-3/2} B(\kappa) x^{1/2} \ddot{e}^{3/2}}{1 + 2^{-3/2} x^{1/2} \ddot{e}^{3/2}} \qquad (32)$$

where $A(\kappa) \equiv a_3(\kappa)$ and $B(\kappa) \equiv a_2(\kappa)a_3(\kappa) - \frac{5}{3}a_1(\kappa)a_4(\kappa)$. This differential equation seems very difficult to solve. However, for the physical value of interest, $x < 1$, it permits a series solution in $z \equiv x^{1/2}$. The first few terms in the expansion are

$$e(x) = b_1(\kappa) x^2 + b_2(\kappa) x^{5/2} + b_3(\kappa) x^3 \qquad (33)$$

where (by setting the mass $m$ as unity) $b_1(\kappa) = \frac{k_0 f_0}{\pi^2}$, $b_2(\kappa) = -\frac{8}{15} a_3 b_1^{3/2}$, $b_3(\kappa) = \frac{2}{3}(a_3^2 + \frac{2}{3} a_1 a_4) b_1^3$, and $f_0 \equiv -\frac{\pi^2}{2k_0} a_3(\kappa)$ is the "effective" scattering amplitude. We note that $-2 \le a_3 \le 0$ and $f_0$ acquires a maximum value of $f_0^{max} = 4\pi a$, where $a$ is the "effective" scattering length given as $a \equiv (\frac{4k_0}{\pi})^{-1}$. The corresponding expressions of the dimensionless chemical potential, $u(x)$, and the square of the sound velocity, $c_0^2$, are

$$u(x) = 2 b_1(\kappa) x + \frac{5}{2} b_2(\kappa) x^{3/2} + 3 b_3(\kappa) x^2 \qquad \text{and} \qquad (34)$$

$$c_0^2 = b_1(\kappa) x + \frac{15}{8} b_2(\kappa) x^{3/2} + 3 b_3(\kappa) x^2, \qquad (35)$$

respectively. By using eq.(31), the fraction of the noncondensates as well as the fraction of the condensates follow with a leading expression of $1 - \frac{n_0}{n} = \frac{1}{3\pi} (n f_0^3)^{1/2} a_1(\kappa)$. At once, in the limit $\kappa \to \infty$, we recover the results of Lee and Yang[3] and Lee, Huang and Yang[4] (who both used different methods) for dilute hard core bosons.

## 4.4    Two-dimensional Bose Gas.

In $1d$, we found out the absence of condensates even if the strength of the interaction is very small. This is consistent with the result of the Bethe ansatz calculation [1]. In $3d$, however, an opposite situation occurs. Even if the interaction is infinitely strong as in the case of hard core bosons, a macroscopic amount of condensates exists. Hence, we will expect some peculiarity about the physics of $2d$ bose gas.

Following what we did in $3d$ we integrate eq.(4.1) and arrive at expressions for noncondensates, $n - n_0$, and $GE$ energy, $\frac{E}{\Omega}$, to be

$$n - n_0 = \frac{(2\pi)^{-1} (mc)^2 c_1(\kappa)}{1 + (2\pi)^{-1} n^{-1}(mc)^2 c_2(\kappa)} \qquad \text{and}$$

$$\frac{E}{\Omega} = \tfrac{1}{2} n\mu - (4\pi)^{-1} m^{-1} (mc)^4 \left[ c_1^2 + (1 - \tfrac{n_0}{n})(c_1 - c_2) \right], \qquad (36)$$

respectively, where $c_1(\kappa) \equiv \kappa \left( \sqrt{\kappa^2 + 1} - \kappa \right)$ and $c_2(\kappa) \equiv ln \left( \sqrt{\kappa^2 + 1} + \kappa \right)$. Once again we introduce dimensionless parameters as follows:

$$\begin{pmatrix} n \\ n_0 \end{pmatrix} = \frac{k_0^2}{4\pi} \begin{pmatrix} x \\ y \end{pmatrix} \qquad \text{and} \qquad \left( \frac{4\pi k_0^{-2} \frac{E}{\Omega}}{\mu} \right) = \frac{k_0^2}{2m} \begin{pmatrix} e \\ u \end{pmatrix} \qquad (37)$$

to rewrite eq.(36) as

$$1 - \frac{x}{y} = \frac{\ddot{e}c_1}{1 + \ddot{e}c_2} \qquad \text{and}$$

$$e = \frac{1}{2}xu - 2c_0^4 \left[ c_1^2 + \left( 1 - \frac{y}{x} \right)(c_1 - c_2) \right]. \qquad (38)$$

The self-consistent equation is

$$e - \tfrac{1}{2}x\dot{e} = \tfrac{1}{2}c_1 x^2 \ddot{e}^2 \, \frac{c_1 + \ddot{e}(c_1 - c_2 + c_1 c_2)}{1 + \ddot{e}c_2}. \qquad (39)$$

The solution for eq.(39) can be written as

$$e(x) = x^2 f \left( ln \, x^{-1} \right) \qquad (40)$$

where $f(z)$ is an analytic function of $z (\equiv ln \, x^{-1})$ and satisfies

$$\dot{f} = c_1 (2f - 3\dot{f} + \ddot{f})^2 \left[ \frac{c_1 + (2f - 3\dot{f} + \ddot{f}) (c_1 - c_2 + c_1 c_2)}{1 + c_2 (2f - 3\dot{f} + \ddot{f})} \right] \qquad (41)$$

Figure 1: The fraction of noncondensates, $1 - \frac{y}{x}$, (relative to the density of bosons) expressed in terms of dimensionless bosons' density, $x$. The fraction of condensates is given by $\frac{y}{x}$. For $2d$ and $3d$, the functional forms of their leading term are explicitly written above and the proportionality constants depend on the strength of interaction $\gamma(= \gamma(\Lambda))$.

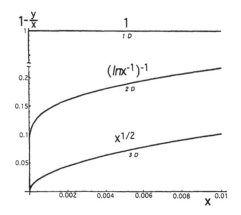

Again, the dot implies differentiation with respect to argument. For small $x$, the energy, $e(x)$, chemical potential, $u(x)$, and noncondensates, $1 - \frac{y}{x}$, have the following leading contributions:

$$e(x) \sim \frac{x^2}{4c_1^2} \left( \ln x^{-1} \right)^{-1}$$

$$u(x) \sim \frac{x}{2c_1^2} \left( \ln x^{-1} \right)^{-1}$$

$$1 - \frac{y}{x} \sim \frac{1}{2c_1} \left( \ln x^{-1} \right)^{-1} \tag{42}$$

The results of this section can be summarized in Fig.1 where the fraction of excitations, $1 - \frac{y}{x}$, in every dimension is plotted as a function of the dimensionless density of bosons, $x$. The strong singular behavior of $2d$, for small $x$, can be attributed to strong dimensional crossover. For example, the $2d$ system tries to copy the behaviors of $1d$ and $3d$ as absence of condensate and presence of macroscopic amount of condensates, respectively.

## 5   SUMMARY AND DISCUSSION

We developed a theory based on the crossover from linear (small $k$) to quadratic (large $k$) in momenta of the excitation spectra and gave a unified description of interacting bose gas in one, two and three dimensions.

The building block of the theory is the coexistence of condensates and non-condensates in equilibrium. This condition defines the amount of condensates and naturally accounts for the gaplessness of the excitation spectra[15].

Based on the proposed form of the Green's function the amount of condensates and noncondensates are calculated and the results agree with expected understanding of absence of condensates in one dimension[1] and macroscopic amount of condensates in three dimensions[2, 3, 4, 7, 9]. The result for two dimensions indicate a very strong dimensional crossover. We note, in two dimensions, that the functional dependence of the fraction of noncondensates, $1 - \frac{n_0}{n} \propto (lnn^{-1})^{-1}$, which has nontrivial form, is the first analytical result in the field of two-dimensional bose gas.

Lastly, we reiterate that our proposed theory is the first that can deal with the $2d$ bose system and offer a new avenue for reinvestigation of many peculiar behaviors of two-dimensional superfluidity[16]. This theory can also be applied in high-$T_c$ where the physics is governed by the $CuO$-plane. For example, the local constraint of one particle per lattice which always relaxed into global average may now be treated by ascribing a hard core to every particle. As applied to the $t$-$J$ model, the $g$-on mean field theory[17] predicts a wider region of intermediate statistics when the hard core is taken into account[18].

## ACKNOWLEDGEMENTS

F.B. is grateful to the Hitachi Scholarship Foundation for financial support during the period of his doctoral study at the University of Tokyo. The comments and suggestions of Dr. H. Kohno were valuable for the completion of this work. The dicussions of F.B. with Profs. P. Stamp, M. Ogata, D. Yoshioka, N. Nagaosa, H. Namaizawa, and T. Ando are highly appreciated. This work is financially supported by a Grant-in-Aid for Scientific Research on Priority Area "Anomalous Metallic State near the Mott Transition"(07237102) from the Ministry of Education, Science, Sports and Culture.

## References

[1] E. H. Lieb and W. Liniger, *Phys. Rev.* **130**, 1605 (1963).

[2] N. N. Bogoliubov, *J. Phys. U.S.S.R* **9**, 23 (1947).

[3] T. D. Lee and C. N. Yang, *Phys. Rev.* **105**, 1119 (1957).

[4] T. D. Lee, K. Huang and C. N. Yang, *Phys. Rev.* **105**, 1136 (1957).

[5] S. T. Beliaev, *J. Exptl. Theoret. Phys. U.S.S.R* **34**, 417 (1958) [translation: *Soviet Phys. JETP* **7**, 289 (1958)].

[6] S. T. Beliaev, *J. Exptl. Theoret. Phys. U.S.S.R* **34**, 433 (1958) [translation: *Soviet Phys. JETP* **7**, 299 (1958)].

[7]  N. M. Hugenholtz and D. Pines, *Phys. Rev.* **116**, 489 (1959).

[8]  F. Basco (December 1997) $g$-on Mean Field Theory of the $t$-$J$ Model. ScD *Dissertation*, University of Tokyo, Japan.

[9]  T. Gavoret and P. Nozieres, *Ann. Phys.* **28**, 349 (1964).

[10]  A. A. Nepomnyashchĭiand$Yu.A.Nepomnyashchĭi, J.Exptl.Theoret.Phys. U.S.S.R$**21**, 3(1975)[*translation : SovietPhys.JETP***21**, 1(1975)].

[11]  A.A. Nepomnyashchĭiand$Yu.A.Nepomnyashchĭi, J.Exptl.Theoret.Phys. U.S.S.R.$**75**, 976(1978)[*translation : SovietPhys.JETP***48**, 493(1978)].

[12]  A.A. Abrikosov, L. P. Gorkov and I. E. Dzyaloshinski, *Methods of Quantum Field Theory in Statistical Physics* (1963).

[13]  A. L. Fetter and J. D. Walecka, *Quantum Theory of Many-Particle Systems* (1971).

[14]  E. H. Lieb, *Phys. Rev.* **130**, 1616 (1963).

[15]  F. Basco and H. Fukuyama, to be published.

[16]  C.E. Campbell, J.G. Dash and M. Schick: *Phys. Rev. Lett.* **26**, 966 (1971).

[17]  F. Basco, H. Kohno, H. Fukuyama and G. Baskaran: *J. Phys. Soc. Jpn.* **65**, 687 (1996); *Physica C* **282-287**, 907 (1997); to be published.

[18]  F. Basco, H. Kohno and H. Fukuyama, in preparation.

# ON AN EXPRESSION OF THE GROUND STATE ENERGY OF THE SPIN-BOSON HAMILTONIAN

MASAO HIROKAWA[1]

Department of Mathematics, Tokyo Gakugei University
Koganei 184-8501, Japan

**Abstract**    An expression of the ground state energy $E_{SB}$ of the spin-boson Hamiltonian $H_{SB}$ is considered. The expression in the cases of both massive and massless bosons is given by a nonperturbative method.

The spin-boson model, which describes a two-level system coupled to a quantized Bose field, has been investigated as a simplified model for atomic systems interacting with a quantized radiation or phonon field ( [LCDFGW, Am, Ar1, D, FaNV, Gér, HüSp1, HüSp2, Sp1, ArHi1, JP3] and references therein). Several properties of the ground states of the model are of interest. Especially, we are interested in expressions of the ground state energy of the model, because for each Hamiltonian we can actually observe its energies only at every state, neither the Hamiltonian nor the state according to the standard quantum theory. For the spin-boson Hamiltonian $H_{SB}$, recently attention has been paid to the ground states as the eigenvectors of $H_{SB}$ with eigenvalue equal to the infimum of its spectrum to develop nonperturbative method ( [Sp3, (ii) on p.5], [ArHi1, ArHi2]) and analyze spectral properties and the process of radiative decay ( [HüSp1, HüSp2]). Talking of the ground states of this type model, we here note that in  [T1, T2, T3, T4] Tomonaga argued the ground state of the model which has relation to the spin-boson model in order to get rid of physical difficulties caused by applying the perturbation theory to the model. Moreover, recently Bach, Fröhlich and Sigal argued the ground state, spectrum and resonance for a model of nonrelativistic quantum electrodynamics ( [BaFrSig1, BaFrSig2, BaFrSig3, BaFrSig4]). Especially they established the method of renormalization group to investigate resonances in quantum electrodynamics, which is of great value for many who deal with problems on the resonances in the case of massless bosons. For the generalized spin-boson model, Arai and the author showed that, under certain conditions, there exists a ground state of the generalized model in  [ArHi2] by a nonperturbative method, and we gave a formula for

[1]Research is supported by the Grant-In-Aid No.09740092 for Encouragement of Young Scientists from the Ministry of Education, Japan.

the asymptotic behavior of the ground state energy of the generalized model in the strong coupling region ( [ArHi2, Proposition 1.4]). In this paper we focus our attention on the expression of the ground state energy of the (standard) spin-boson Hamiltonian $H_{SB}$ in the cases of both massive and massless bosons. Especially, it is important that we clarify the expression in the case of massless bosons, because we cannot apply the regular perturbation theory to $H_{SB}$ in the case. Thus we try nonperturbative approach to our problem in this paper.

For physical reality, we consider the situation where bosons move in the 3-dimensional Euclidean space $\mathbf{R}^3$. We take a Hilbert space of bosons to be

$$\mathcal{F}_b = \mathcal{F}\left(L^2\left(\mathbf{R}^3\right)\right) = \bigoplus_{n=0}^{\infty} \left[\otimes_s^n L^2\left(\mathbf{R}^3\right)\right], \tag{1}$$

the symmetric Fock space over $L^2(\mathbf{R}^3)$ ($\otimes_s^n \mathcal{K}$ denotes the $n$-fold symmetric tensor product of a Hilbert space $\mathcal{K}$, $\otimes_s^0 \mathcal{K} \equiv \mathbf{C}$). Let $\Omega_0$ be the Fock vacuum in $\mathcal{F}_b$.

In this paper, we set both of $\hbar$ and $c$ one, i.e., $\hbar = c = 1$, where $\hbar$ is the Planck constant divided by $2\pi$, and $c$ the velocity of the light. A function $\omega_r$ is given by

$$\omega_r(k) := \sqrt{|k|^2 + m^2}, \qquad m \geq 0, \, k \in \mathbf{R}^3,$$

which is the energy of the relativistic bosons with mass $m$ and momentum $k$.

We denote by $d\Gamma(\omega_r)$ the second quantization of the multiplication operator $\omega_r$ on $L^2(\mathbf{R}^3)$ and set

$$H_b = d\Gamma(\omega_r) = \int_{\mathbf{R}^3} dk \, \omega_r(k) a^+(k) a(k),$$

where $a(k)$ and $a^+(k)$ are the operator-valued distribution kernels of the smeared annihilation and creation operators respectively:

$$a(f) = \int_{\mathbf{R}^3} dk \, a(k) f(k), \tag{2}$$

$$a^+(f) = \int_{\mathbf{R}^3} dk \, a^+(k) f(k) \tag{3}$$

for every $f \in L^2(\mathbf{R}^3)$ on $\mathcal{F}_b$.

*Remark 1.* In [ArHi1, ArHi2], we used the definition,

$$a(f) = \int_{\mathbf{R}^3} dk \, a(k) f(k)^*, \qquad f \in L^2(\mathbf{R}^3),$$

as the annihilation operator $a(f)$ according to the custom for mathematics, where $f(k)^*$ denotes the complex conjugate of $f(k)$ ($k \in \mathbf{R}^3$), but we here employ (2) as the definition of $a(f)$ according to the way of physics.

The Segal field operator $\phi_s(f)$ ($f \in L^2(\mathbf{R}^3)$) is given by

$$\phi_s(f) := \frac{1}{\sqrt{2}} \left(a^+(f) + a(f)\right). \tag{4}$$

Let $\lambda$ be a real-valued continuous function on $\mathbf{R}^3$ satisfying the following conditions:

(A)       $\lambda(k) = \lambda(-k)$ $(k \in \mathbf{R}^3)$, and $\lambda, \lambda/\omega_r \in L^2(\mathbf{R}^3)$.

*Remark 2.* Since $\lambda, \lambda/\omega_r \in L^2(\mathbf{R}^3)$, we have $\lambda/\sqrt{\omega_r} \in L^2(\mathbf{R}^3)$.

The Hamiltonian of the spin-boson model is defined by

$$H_{SB} := \frac{\mu}{2}\sigma_3 \otimes I + I \otimes H_b + \sqrt{2}\alpha\sigma_1 \otimes \phi_s(\lambda)$$

acting in the Hilbert space

$$\mathcal{F} := \mathbf{C}^2 \otimes \mathcal{F}_b = \mathcal{F}_b \oplus \mathcal{F}_b, \tag{5}$$

where $\sigma_1, \sigma_3$ are the standard Pauli matrices,

$$\sigma_1 = \begin{pmatrix} 0 & 1 \\ 1 & 0 \end{pmatrix}, \qquad \sigma_3 = \begin{pmatrix} 1 & 0 \\ 0 & -1 \end{pmatrix}.$$

In (5), we identified $\mathbf{C}^2 \otimes \mathcal{F}_b$ with $\mathcal{F}_b \oplus \mathcal{F}_b$. So, $H_{SB}$ has the following representation on $\mathcal{F}_b \oplus \mathcal{F}_b$ and we employ it in this paper:

$$H_{SB} = \begin{pmatrix} H_b + \frac{\mu}{2} & \sqrt{2}\alpha\phi_s(\lambda) \\ \sqrt{2}\alpha\phi_s(\lambda) & H_b - \frac{\mu}{2} \end{pmatrix}.$$

For a linear operator $T$ on a Hilbert space, we denote its domain by $D(T)$. It is well-known that $H_{SB}$ is self-adjoint with $D(H_{SB}) = D(I \otimes H_b)$ and

$$-\frac{|\mu|}{2} - \alpha^2 \left\| \frac{\lambda}{\sqrt{\omega_r}} \right\|_{L^2}^2 \leq H_{SB},$$

where $\| \cdot \|_{L^2}$ denotes the norm of $L^2(\mathbf{R}^3)$.

For a self-adjoint operator $T$ bounded from below, we denote by $E(T)$ the infimum of the spectrum $\sigma(T)$ of $T$:

$$E(T) = \inf \sigma(T).$$

In this paper, an eigenvector of $T$ with eigenvalue $E(T)$ is called a *ground state* of $T$ (if it exists). We say that $T$ has a (resp. unique) ground state if $\dim \ker(T - E(T)) \geq 1$ (resp. $\dim \ker(T - E(T)) = 1$). We call $E(T)$ the *ground state energy* of $T$ if $T$ is a Hamiltonian.

For $H_{SB}$ we set

$$E_{SB}(\mu, \alpha) := E(H_{SB}).$$

By the variational principle ( [Ar1, Theorem 2.4] and [D, p.161]), we have

$$E_{SB}(\mu, \alpha) \leq -\frac{|\mu|}{2}e^{-2\alpha^2\|\lambda/\omega_r\|_{L^2}^2} - \alpha^2 \left\| \frac{\lambda}{\sqrt{\omega_r}} \right\|_{L^2}^2. \tag{6}$$

Under certain assumptions, we know that $H_{SB}$ has a ground state ( [HüSp1, ArHi1, Sp3] and see Remark 4(1) in this paper).

DEFINITION 1. We say *a vector* $\Psi \in \mathcal{F} \equiv \mathcal{F}_b \oplus \mathcal{F}_b$ *overlaps with a ground state* $\Omega_{SB}(\mu, \alpha)$ if and only if there exists the ground state $\Omega_{SB}(\mu, \alpha)$ of $H_{SB}$ such that

$$\langle \Psi , \Omega_{SB}(\mu, \alpha) \rangle_{\mathcal{F}} \neq 0,$$

where $\langle \ , \ \rangle_{\mathcal{F}}$ is the standard inner product of $\mathcal{F} \equiv \mathcal{F}_b \oplus \mathcal{F}_b$.

From now on, according to the custom for the physicists, all the inner products of the Hilbert spaces appearing in this paper have the linearity on the right hand side.

If a ground state $\Omega_{SB}(\mu, \alpha)$ of $H_{SB}$ exists, for $\Omega_{SB}(\mu, \alpha)$ we set

$$\Omega_{SB}(\mu, \alpha) = \begin{pmatrix} \Omega_1 \\ \Omega_2 \end{pmatrix} \in \mathcal{F} \equiv \mathcal{F}_b \oplus \mathcal{F}_b.$$

It is well known that, if $f \in L^2(\mathbf{R}^3)$, we can define a self-adjoint operator $P(f)$ by

$$P(f) := i \left\{ a^+ (f) - a (f) \right\},$$

thus, if $\lambda/\omega_r \in L^2(\mathbf{R}^3)$, we have two unitary operators $U_\pm$ defined by

$$U_\pm := \exp \left[ \pm i \alpha P(\lambda/\omega_r) \right]. \tag{7}$$

We define two unit vectors $\Omega_\pm \in \mathcal{F} \equiv \mathcal{F}_b \oplus \mathcal{F}_b$ by

$$\Omega_\pm := \frac{1}{2} \begin{pmatrix} U_+\Omega_0 \mp U_-\Omega_0 \\ U_+\Omega_0 \pm U_-\Omega_0 \end{pmatrix}.$$

We have the following proposition on an upper bound of the ground state energy:

PROPOSITION 2. *Assume (A). Then,*

$$E_{SB}(\mu, \alpha)$$
$$\leq -\alpha^2 \int_{\mathbf{R}^3} dk \frac{\lambda(k)^2}{\omega_r(k)}$$
$$- \lim_{\beta \to \infty} \frac{1}{\beta} \ln \left\{ 1 + \beta \frac{|\mu|}{2} \exp \left[ -2\alpha^2 \int_{\mathbf{R}^3} dk \frac{\lambda(k)^2}{\omega_r(k)^2} \right] \right.$$
$$+ \sum_{\ell=1}^{\infty} \left( \frac{\mu}{2} \right)^{2\ell} \int_0^\beta d\beta_1 \int_0^{\beta_1} d\beta_2 \cdots \int_0^{\beta_{2\ell-1}} d\beta_{2\ell}$$
$$\exp \left[ -2\alpha^2 \int_{\mathbf{R}^3} dk \frac{\lambda(k)^2}{\omega_r(k)^2} \left( 2G_{\beta_1, \cdots, \beta_{2\ell}}(k) + 2\ell \right) \right]$$
$$+ \sum_{\ell=1}^{\infty} \left( \frac{|\mu|}{2} \right)^{2\ell+1} \int_0^\beta d\beta_1 \int_0^{\beta_1} d\beta_2 \cdots \int_0^{\beta_{2\ell}} d\beta_{2\ell+1}$$
$$\exp \left[ -2\alpha^2 \int_{\mathbf{R}^3} dk \frac{\lambda(k)^2}{\omega_r(k)^2} \left( 2G_{\beta_1, \cdots, \beta_{2\ell}}(k) \right. \right.$$
$$\left. \left. \left. + 2F_{\beta_1, \cdots, \beta_{2\ell+1}}(k) + (2\ell + 1) \right) \right] \right\},$$

*where*

$$G_{\beta_1,\cdots,\beta_{2\ell}}(k) = - \sum_{p=1}^{\ell} e^{-(\beta_{2p-1}-\beta_{2p})\omega_r(k)}$$

$$+ \sum_{p,q=1;p<q}^{\ell} \left( e^{-\beta_{2p-1}\omega_r(k)} - e^{-\beta_{2p}\omega_r(k)} \right) \left( e^{\beta_{2q-1}\omega_r(k)} - e^{\beta_{2q}\omega_r(k)} \right)$$

$$\leq 0, \tag{8}$$

$$F_{\beta_1,\cdots,\beta_{2\ell+1}}(k) = e^{\beta_{2\ell+1}\omega_r(k)} \sum_{p=1}^{\ell} \left( e^{-\beta_{2p-1}\omega_r(k)} - e^{-\beta_{2p}\omega_r(k)} \right) \leq 0. \tag{9}$$

*Remark 3.* The upper bound (6) of $E_{SB}(\mu,\alpha)$ by the variational principle ( [Ar1, Theorem 2.4], [D, p.161]) is given by estimating $G_{\beta_1,\cdots,\beta_{2\ell}}(k)$ and $F_{\beta_1,\cdots,\beta_{2\ell+1}}(k)$ at 0 from above in our Proposition, which is the most rough estimation in ours.

The statement of our main theorem is made as follows:

THEOREM 3. *Assume (A).*

(i) *Suppose that a ground state $\Omega_{SB}(\mu,\alpha)$ of $H_{SB}$ exists.*

(i)$_+$   *If $\Omega_+$ overlaps with the ground state $\Omega_{SB}(\mu,\alpha)$, then $E_{SB}(\mu,\alpha)$ is given by*

$$E_{SB}(\mu,\alpha)$$
$$= -\alpha^2 \int_{\mathbf{R}^3} dk \frac{\lambda(k)^2}{\omega_r(k)}$$
$$- \lim_{\beta\to\infty} \frac{1}{\beta} \ln\left\{ 1 + \beta\frac{\mu}{2} \exp\left[ -2\alpha^2 \int_{\mathbf{R}^3} dk \frac{\lambda(k)^2}{\omega_r(k)^2} \right] \right.$$
$$+ \sum_{\ell=1}^{\infty} \left(\frac{\mu}{2}\right)^{2\ell} \int_0^{\beta} d\beta_1 \int_0^{\beta_1} d\beta_2 \cdots \int_0^{\beta_{2\ell-1}} d\beta_{2\ell}$$
$$\exp\left[ -2\alpha^2 \int_{\mathbf{R}^3} dk \frac{\lambda(k)^2}{\omega_r(k)^2} \left( 2G_{\beta_1,\cdots,\beta_{2\ell}}(k) + 2\ell \right) \right]$$
$$+ \sum_{\ell=1}^{\infty} \left(\frac{\mu}{2}\right)^{2\ell+1} \int_0^{\beta} d\beta_1 \int_0^{\beta_1} d\beta_2 \cdots \int_0^{\beta_{2\ell}} d\beta_{2\ell+1}$$
$$\exp\left[ -2\alpha^2 \int_{\mathbf{R}^3} dk \frac{\lambda(k)^2}{\omega_r(k)^2} \left( 2G_{\beta_1,\cdots,\beta_{2\ell}}(k) \right.\right.$$
$$\left.\left.\left. + 2F_{\beta_1,\cdots,\beta_{2\ell+1}}(k) + (2\ell+1) \right) \right] \right\},$$

*where $G_{\beta_1,\cdots,\beta_{2\ell}}$ and $F_{\beta_1,\cdots,\beta_{2\ell+1}}$ are given in (8) and (9) respectively.*

(i)$_-$ *If $\Omega_-$ overlaps with the ground state $\Omega_{SB}(\mu, \alpha)$, then $E_{SB}(\mu, \alpha)$ is given by*

$$
\begin{aligned}
& E_{SB}(\mu, \alpha) \\
& = -\alpha^2 \int_{\mathbf{R}^3} dk \frac{\lambda(k)^2}{\omega_r(k)} \\
& \quad - \lim_{\beta \to \infty} \frac{1}{\beta} \ln \Bigg\{ 1 - \beta \frac{\mu}{2} \exp\left[ -2\alpha^2 \int_{\mathbf{R}^3} dk \frac{\lambda(k)^2}{\omega_r(k)^2} \right] \\
& \qquad + \sum_{\ell=1}^{\infty} \left(\frac{\mu}{2}\right)^{2\ell} \int_0^\beta d\beta_1 \int_0^{\beta_1} d\beta_2 \cdots \int_0^{\beta_{2\ell-1}} d\beta_{2\ell} \\
& \qquad\qquad \exp\left[ -2\alpha^2 \int_{\mathbf{R}^3} dk \frac{\lambda(k)^2}{\omega_r(k)^2} \left( 2G_{\beta_1, \cdots, \beta_{2\ell}}(k) + 2\ell \right) \right] \\
& \qquad - \sum_{\ell=1}^{\infty} \left(\frac{\mu}{2}\right)^{2\ell+1} \int_0^\beta d\beta_1 \int_0^{\beta_1} d\beta_2 \cdots \int_0^{\beta_{2\ell}} d\beta_{2\ell+1} \\
& \qquad\qquad \exp\Big[ -2\alpha^2 \int_{\mathbf{R}^3} dk \frac{\lambda(k)^2}{\omega_r(k)^2} \Big( 2G_{\beta_1, \cdots, \beta_{2\ell}}(k) \\
& \qquad\qquad\qquad\qquad + 2F_{\beta_1, \cdots, \beta_{2\ell+1}}(k) + (2\ell+1) \Big) \Big] \Bigg\},
\end{aligned}
$$

*where $G_{\beta_1, \cdots, \beta_{2\ell}}$ and $F_{\beta_1, \cdots, \beta_{2\ell+1}}$ are given in (8) and (9) respectively.*

(ii) *(for massive bosons) Let $m > 0$. Then there exists a ground state $\Omega_{SB}(\mu, \alpha)$ of $H_{SB}$. Moreover, if*

$$
\frac{1}{2}|\mu| \left( 1 - e^{-2\alpha^2 \|\lambda/\omega_r\|_{L^2}^2} \right) < m,
$$

*then either $\Omega_+$ or $\Omega_-$ overlaps with the ground state $\Omega_{SB}(\mu, \alpha)$ at least.*

(iii) *(for massive or massless bosons) Let $m \geq 0$, and $|\alpha| \|\lambda/\omega_r\|_{L^2}^2 < 1$. Then there exists a ground state $\Omega_{SB}(\mu, \alpha)$ of $H_{SB}$. Moreover, if $\lambda/\omega_r^2 \in L^2(\mathbf{R}^3)$ with*

$$
|\mu\alpha| \left\| \frac{\lambda}{\omega_r^2} \right\|_{L^2} < \frac{1}{2},
$$

*then either $\Omega_+$ or $\Omega_-$ overlaps with the ground state $\Omega_{SB}(\mu, \alpha)$ at least.*

*Remark 4.* (1) Spohn showed a necessary and sufficient condition for existence of a ground state of $H_{SB}$ (condition (A) implies the condition) whose statement appears in [Sp3, Comment (ii) after Theorem 1], and its proof in his unpublished note (1991) was clarified in [Sp4].

(2) In the case of massive bosons, by applying regular perturbation theory (e.g. [RSi3, Theorem XII.8 and Theorem XII.9]), we can easily obtain an expression of the ground state energy of the spin-boson model as the perturbation series. For instance, Davies had the expression in the case of massive bosons by regarding $I \otimes H_b + \sqrt{2}\alpha\sigma_1 \otimes \phi_s(\lambda)$ and $\frac{\mu}{2}\sigma_3 \otimes I$ as the free part and perturbation term respectively ( [D, Theorem 10]): Let $\mu < 0$ and $\alpha > 0$.

By inserting $|\mu|/2$, $\alpha^2 \|\lambda/\sqrt{\omega_r}\|_{L^2}^2$, and $\|\lambda/\sqrt{\omega_r}\|_{L^2}^{-1}\lambda$ into $\varepsilon$, $\Lambda$, and $f$ in [D, Theorem 10], we know that $F_1$, $F$, $N$, and the Hamiltonian $\mathbf{H}$ in [D, Theorem 10] is given by $\omega_r > 0$, $H_b$, $\alpha^2\|\lambda/\omega_r\|_{L^2}^2$, and

$$H'_{SB} := -\frac{\mu}{2}\sigma_1 \otimes I + I \otimes H_b + \sqrt{2}\alpha\sigma_3 \otimes \phi_s(\lambda)$$

respectively. We note here that $H_{SB}$ and $H'_{SB}$ are unitary equivalent (see Lemma 2(i) in this paper). So, by [D, Theorem 10], for sufficiently small $|\mu|$,

$$E_{SB}(\mu,\alpha) = -\alpha^2 \int_{\mathbf{R}^3} dk \frac{\lambda(k)^2}{\omega_r(k)} - \frac{|\mu|}{2}\exp\left[-2\alpha^2 \int_{\mathbf{R}^3} dk \frac{\lambda(k)^2}{\omega_r(k)^2}\right] + 0\left(\mu^2\right).$$

For arbitrary fixed $m > 0$ and $\mu \neq 0$ (resp. $\alpha \neq 0$), sufficiently small $|\alpha|$ (resp. $|\mu|$) satisfies the inequality in (ii). Thus Theorem 3(i)$_\pm$ with (ii) may be regarded as a result which improves the one obtained by regular perturbation theory. Note that (10) is a nonperturbative estimate in $\alpha$, since the left hand side of (10) is non-polynomial in $\alpha$.

(3) To author's best knowledge, Theorem 3(i)$_\pm$ with (iii) is the first which establishes a concrete expression of the ground state energy of the spin-boson model $H_{SB}$ in the case of *massless bosons*.

COROLLARY. *Assume (A). Fix $\alpha \in \mathbf{R}$.*
(i) *Suppose that, for $\mu,\mu' > 0$, $\Omega_{SB}(\mu,\alpha)$ and $\Omega_{SB}(\mu',\alpha)$ exist, and $\Omega_+$ overlaps with both of them. Then*

$$E_{SB}(\mu',\alpha) \leq E_{SB}(\mu,\alpha) \text{ if } \mu < \mu'.$$

(ii) *Suppose that, for $\mu,\mu' < 0$, $\Omega_{SB}(\mu,\alpha)$ and $\Omega_{SB}(\mu',\alpha)$ exist, and $\Omega_-$ overlaps with both of them. Then*

$$E_{SB}(\mu',\alpha) \geq E_{SB}(\mu,\alpha) \text{ if } \mu < \mu'.$$

The basic idea to prove our main theorem is as follows: If there exist a ground state $\Omega_{SB}(\mu,\alpha)$ of $H_{SB}$ and a vector $\Psi \in \mathcal{F} \equiv \mathcal{F}_b \oplus \mathcal{F}_b$ such that $\Psi$ overlaps with $\Omega_{SB}(\mu,\alpha)$, then by Bloch's formula ( [Blo, (12)] and see Lemma 2.4 in this paper), we have

$$E_{SB}(\mu,\alpha) = -\lim_{\beta\to\infty} \frac{1}{\beta}\ln\langle\Psi, e^{-\beta H_{SB}}\Psi\rangle_{\mathcal{F}}.$$

So, our problem is reduced to that of how to find such $\Psi$ that we now try to calculate $\langle\Psi, e^{-\beta H_{SB}}\Psi\rangle_{\mathcal{F}}$ in the concrete. We can use several unitary transformations and the Du Hammel formula so that we can apply the Feynman-Kac-Nelson formula for the free field, and find that either $\Omega_+$ or $\Omega_-$ is one of the answers for the problem above. Here, it is important that we employ the Feynman-Kac-Nelson formula for the *free field* because we can calculate actually and concretely the ground state energy of the spin-boson model.

## References

[Am]      A.Amann, Ground states of a spin-boson model, *Ann. Phys.* **208** (1991), 414-448.

[Ar1]     A.Arai, An asymptotic analysis and its application to the non-relativistic limit of the Pauli-Fierz and a spin-boson model, *J. Math. Phys.* **31** (1990), 2653-2663.

[Ar2]     A.Arai, Perturbation of embedded eigenvalues: a general class of exactly soluble models in Fock spaces, *Hokkaido Math. Jour.* **19** (1990), 1-34.

[ArHi1]   A.Arai and M.Hirokawa, On the Spin-Boson Model, in Proceedings of the symposium "Quantum Stochastic Analysis and related Fields" held at the RIMS, Kyoto University, November 27-29, 1995: *RIMS Kokyuroku* No. **957** (1996), 16-35.

[ArHi2]   A.Arai and M.Hirokawa, On the Existence and Uniqueness of Ground States of a Generalized Spin-Boson Model *J. Funct. Anal.* **151** (1997), 455-503.

[BaFrSig1] V.Bach, J.Fröhlich and I.M.Sigal, Mathematical theory of nonrelativistic matter and radiation, *Lett. Math. Phys.* **34** (1995), 183-201.

[BaFrSig2] V.Bach, J.Fröhlich and I.M.Sigal, Quantum electrodynamics of confined non-relativistic particles (to appear in *Adv. in Math.*), preprint, 1996; revised version 1997.

          ftp://ftp.ma.utexas.edu/pub/mp_arc/papers/97-414

[BaFrSig3] V.Bach, J.Fröhlich, I.M.Sigal and A.Soffer, Positive commutators and spectrum of nonrelativistic QED, preprint, 1997.

          ftp://ftp.ma.utexas.edu/pub/mp_arc/papers/97-268

[BaFrSig4] V.Bach, J.Fröhlich and I.M.Sigal, Renormalization group analysis, of spectral problem in quantum field theory (to appear in *Adv. in Math.*), preprint, 1997.

          ftp://ftp.ma.utexas.edu/pub/mp_arc/papers/97-415

[Blo]     C.Bloch, Sur la détermination de l'état fondamental d'un système de particules, *Nucl.Phys.* **7** (1958), 451-458.

[D]       E.B.Davies, Symmetry breaking for molecular open system, *Ann. Inst. H. Poincaré* **A 35** (1981), 149-171.

[EAr]     H. Ezawa and A. Arai, "Quantum fields theory and statistical mechanics," Nihon Hyōron Sha, Tokyo, 1988 (in Japanese).

[FaNV]    M.Fannes, B.Nachtergaele and A.Verbeure, The equilibrium state of the spin-boson model, *Comm. Math. Phys.* **114** (1988), 537-548.

[Gér]     C. Gérard, Asymptotic completeness for the spin-boson model with a particle number cutoff, *Rev. Math. Phys.* **8** (1996), 549-589.

[GliJaf]  J.Glimm and A.Jaffe, The $\lambda(\varphi^4)_2$ quantum field theory without cutoffs: II. The field operators and the approximate vacuum, *Ann. of Math.* **91** (1970), 362-401.

[Hi]      M.Hirokawa, An inverse problem for a quantum particle interacting with a Bose field and long-time behavior of canonical correlation functions in an infinite volume limit: an application of a solvable model called the rotating wave approximation (to appear in *J. Math. Soc. of Japan*), preprint, 1997.

[HüSp1]   M.Hübner and H.Spohn, Spectral properties of the spin-boson Hamiltonian, *Ann. Inst. H. Poincaré* **62** (1995), 289-323.

[HüSp2]  M.Hübner and H.Spohn, Radiative decay: nonperturbative approach, *Rev. Math. Phys.* **7** (1995), 363-387.

[JP1]    V.Jakšić and C.-A.Pillet, On a model for quantum friction. I. Fermi's golden rule and dynamics at zero temperature, *Ann. Inst. H. Poincaré* **62** (1995), 47-68.

[JP2]    V.Jakšić and C.-A.Pillet, On a model for quantum friction. II. Fermi's golden rule and dynamics at positive temperature, *Comm. Math. Phys.* **176** (1996), 619-644.

[JP3]    V.Jakšić and C.-A.Pillet, Spectral theory of thermal relaxation in the spin-boson model, IAMP XIIth ICMP 13-19 July 1997.

[LCDFGW] A.J.Leggett, S.Chakravarty, A.T.Dorsey, M.P.A.Fisher, A.Garg, and W.Zwerger, Dynamics of the dissipative two-state system, *Rev. Mod. Phys.* **59** (1987), 1-85.

[RSi1]   M.Reed and B.Simon, "Methods of Modern Mathematical Physics Vol.I," Academic Press, New York, 1972.

[RSi2]   M.Reed and B.Simon, "Methods of Modern Mathematical Physics Vol.II," Academic Press, New York, 1975.

[RSi3]   M.Reed and B.Simon, "Methods of Modern Mathematical Physics Vol.IV," Academic Press, New York, 1978.

[SiHoe]  B.Simon and R.Hoegh-Krohn, Hypercontractive semigroups and two dimensional self-adjoint Bose fields, *J. Funct. Anal.* **9** (1972), 121-180.

[Sp1]    H.Spohn, Ground state(s) of the spin-boson Hamiltonian, *Comm. Math. Phys.* **123** (1989), 277-304.

[Sp2]    H.Spohn, Asymptotic completeness for Rayleigh scattering, preprint, 1996.

[Sp3]    H.Spohn, Ground state of a quantum particle coupled to a scalar Boson field, preprint, 1997.
         ftp://ftp.ma.utexas.edu/pub/mp_arc/papers/97-174

[Sp4]    In private communication with Prof.Spohn at Hokkaido Univ. during the first week of Oct. '96.

[T1]     S.Tomonaga, On the effect of the field reactions on the interaction of mesotrones and nuclear particles I, *Prog. Theo. Phys.* **1** (1946), 83-101.

[T2]     S.Tomonaga, On the effect of the field reactions on the interaction of mesotrones and nuclear particles II, *Prog. Theo. Phys.* **1** (1946), 109-124.

[T3]     S.Tomonaga, On the effect of the field reactions on the interaction of mesotrones and nuclear particles III, *Prog. Theo. Phys.* **2** (1947), 6-24.

[T4]     S.Tomonaga, On the effect of the field reactions on the interaction of mesotrones and nuclear particles IV, *Prog. Theo. Phys.* **2** (1947), 63-70.

# SPIN-POLARONS IN HIGH-T$_C$ SUPERCONDUCTIVITY

## AUGUSTO A. MORALES, JR. AND DANILO M. YANGA

National Institute of Physics, College of Science
University of the Philippines, Diliman, Quezon City 1101, Philippines

**Abstract:** A spin polaron Hamiltonian is constructed in this paper based on the slave-particle formulation. In particular, this mechanism can be used to characterize the insulating state of high-T$_c$ superconductors. By introducing convenient approximations, we were able to obtain a Hamiltonian that agrees with that of Martinez and Horsch.

## I. Preliminaries

Considerable progress has been achieved in understanding the properties of strongly correlated systems[1-10]. However, the wealth of experimental data that is obtained has yet to be accurately explained by analytical methods. Several mechanisms have been proposed, one of the most prominent of these is the slave-particle formulation. We shall use this formulation to investigate spin polarons in high critical temperature superconductors.

In the paper of Martinez and Horsch on spin polarons[11], the motion of a single hole in a two-dimensional Heisenberg antiferromagnet was analyzed based on the t-J model, where holes were described as spinless fermions (holons) and spins were treated as normal bosons. Using the linear spin-wave theory, spin dynamics was discussed with the assumption of long-range antiferromagnetic order, thereby closely resembling the conventional polaron problem. Making use of the finite-cluster geometries for the numerical solution, the holon Green's function was calculated self-consistently within the Born approximation. The result revealed that even in the strong-coupling regime, the formulation of the dynamics of a holon in the t-J model was quite valuable.

In this paper, we shall devise a reformulated t-J Hamiltonian in terms of the slave-particle formulation. The reformulated Hamiltonian does not contain a kinetic part for either holons or spinons, rather a coupling term involving holon and spinon fields arises. The absence of a kinetic term, however, precisely emphasizes its similarity with

109

A. MORALES AND D. M. YANGA

the polaron problem. In particular, this work proposes a mechanism based on spin polarons in the insulating state of high-$T_c$ superconductors.

## II. Spin Polarons in High $T_c$ Superconductors in the Slave-Particle Formulation

In the large Hubbard potential limit, the t-J Hamiltonian near half-filling reads

$$H = -t \sum_{\langle ij \rangle, \sigma} \left( c_{i\sigma}^\dagger c_{j\sigma} + h.c. \right) + J \sum_{\langle ij \rangle} \left[ \left( S_{iz} S_{jz} - \tfrac{1}{4} \right) + \tfrac{\alpha}{2} \left( S_{i+} S_{j-} + S_{i-} S_{j+} \right) \right], \tag{1}$$

where, as usual the $c_{i\sigma}^\dagger$ ($c_{i\sigma}$) is the creation (annihilation) operator, the $S_i$, $S_j$ are spin angular momentum operators, $\sigma$ is the spin index, and all summations are done at the nearest neighbor sites.    Note that when the constant $\alpha = 1$, we obtain the Heisenberg limit, while for $\alpha = 0$, we obtain the Ising limit.

In the slave-particle formulation, the electron creation and annihilation operators[12-13] are re-expressed in terms of the holon, $e_i$, the doublon, $d_i$, and the spinon $s_{i\sigma}$, fields in accordance with

$$c_{i\sigma} = e_i^\dagger s_{i\sigma} + \sigma d_i s_{i,-\sigma}^\dagger \tag{2}$$

and subject to the constraint

$$e_i^\dagger e_i + d_i^\dagger d_i + \sum_\sigma s_{i\sigma}^\dagger s_{i\sigma} = 1. \tag{3}$$

The usual anticommuting properties of electron operators are recovered only if the above constraint is imposed.

Following Martinez-Horsch, we perform a canonical transformation by making a rotation of the spin operators about the $S_{jx}$ axis by 180°:

$$S_{j\pm} \rightarrow S_{j\mp}; \; S_{jz} \rightarrow -S_{jz}; \; c_{j\sigma} \rightarrow c_{j,-\sigma}. \tag{4}$$

In this manner, the Neel configuration $|\uparrow\downarrow\uparrow\downarrow...\rangle$ is changed into a ferromagnetic state $|\uparrow\uparrow\uparrow\uparrow...\rangle$ with all spins up due to the above canonical transformation. This scheme removes the distinction between sublattices. We then have

$$H = -t \sum_{\langle ij \rangle, \sigma} \left( c_{i\sigma}^\dagger c_{j-\sigma} + h.c. \right) + J \sum_{\langle ij \rangle} \left[ \frac{\alpha}{2} \left( S_{i+} S_{j+} + S_{i-} S_{j-} \right) - \left( S_{iz} S_{jz} + \frac{1}{4} \right) \right]. \tag{5}$$

Expressing (5) in terms of the slave particles, the t-J Hamiltonian takes the form

$$H = -t \sum_{\langle i,j \rangle \sigma} (e_i e_j^\dagger) \sum_\sigma s_{i\sigma}^\dagger s_{j-\sigma} + h.c.)$$

$$+ J \sum_{\langle i,j \rangle \sigma} \left[ \frac{\alpha}{2} e_i e_i^\dagger \left( s_{i\sigma}^\dagger s_{i,-\sigma}^\dagger s_{j,-\sigma}^\dagger s_{j\sigma} \right) e_j e_j^\dagger - \frac{1}{4} e_i e_i^\dagger \left( s_{i\sigma}^\dagger s_{i\sigma} s_{j\sigma}^\dagger s_{j\sigma} - s_{i\sigma}^\dagger s_{i\sigma} s_{j,-\sigma}^\dagger s_{j,-\sigma} \right) e_j e_j^\dagger \right] - \frac{JNZ}{4} \tag{6}$$

where N is the number of lattice sites and Z is the coordination number (the number of nearest neighbors that a lattice site possesses). Note that we have projected out the doublon operators perturbatively as in Rice transformation. This takes care of the no double occupancy constraint.

If we carry out the summation over the spin index $\sigma$ and introduce the operator relation $\sum_j e_j^\dagger s_{i\alpha}^\dagger s_{j\beta} \equiv \sum_{j(i)} e_j^\dagger z_{j(i)}$, then the above Hamiltonian becomes

$$H = -t \sum_{i,j(i)} (e_i e_j^\dagger z_{j(i)} + h.c.) + J \sum_{i,j(i)} \left[ \frac{\alpha}{2} e_i e_i^\dagger \left( z_i z_{j(i)} + z_i^\dagger z_{j(i)}^\dagger \right) e_j e_j^\dagger \right] + other\ terms, \tag{7}$$

where j(i) denotes the neighboring sites of i. This is the desired spin polaron Hamiltonian.

Finally, we consider the $H_t$ term in (6) and rewrite it as

$$H_t = -t \sum_{\langle ij \rangle} e_i e_j^\dagger \left( s_{i\alpha}^\dagger s_{j\beta} + s_{i\beta}^\dagger s_{j\alpha} \right).$$

(8)

If we use the approximation $s_{i\alpha}^\dagger \left( \sqrt{1 - s_{j\alpha}^\dagger s_{j\beta}} \right) \cong s_{i\alpha}^\dagger \equiv z_i^\dagger$, we can recast (8) as

$$H_t \cong -t \sum_{\langle ij \rangle} \left( e_i e_j^\dagger z_j + h.c. \right).$$

(9)

If we do exactly the same thing with the $H_J$ term in (6), we readily obtain a result that agrees with the Martinez-Horsch spin-polaron Hamiltonian.

Observe that the agreement becomes exact if in our slave-particle formulations, the holon fields anticommute, while the spinon fields commute. However, note that in our derivation of the spin-polaron Hamiltonian, we did not use the Holstein-Primakoff transformation anymore and discarded the unphysical assumption that when the electron's spin is down, it is a holon, while it is up, it becomes a composite particle consisting of a holon and a spin boson.

## III. Conclusions.

Through the use of a mathematical construct, the slave-fermion formulation, the effective spin polaron Hamiltonian was derived which we believe is the viable low-energy Hamiltonian in the insulating state (which is the main concern of this work) of high temperature superconductors. The interaction of a holon field with the spinon operator was obtained by performing a 180° spin rotation on the B sublattices of the original t-J Hamiltonian. Physically this corresponds to a change from a Neel configuration state to a configuration state with all spins parallel.

The reformulated spin polaron Hamiltonian makes use of the widely accepted idea of spin-charge separation, and in the end, re-derive the spin-polaron theory of Martinez and Horsch where the holon field corresponds to the spin-up electron operator while the spin-down electron operator is a composite particle made up of a holon and a normal spin boson. We have shown that using slave-fermion decomposition in the MH Hamiltonian can be derived better.

The prospective work involves the calculation of the onset of the Neel temperature and other physical quantities in the insulating state. We hope that this scheme can be extended to the conducting state, in particular, the superconducting aspect.

## REFERENCES

1.  G. D. Mahan, and Ji-Wei Wu, *Phys. Rev. B* **39**, 265 (1989).
2.  P. W. Anderson, *Science*, **235**, 1196 (1987).
3.  S. Kivelson, D. Rokhsar, and J. Sethna, *Phys. Rev. B* **35**, 8865 (1987).
4.  G. Baskaran, Z. Zou, and P. W. Anderson, *Solid State Comm.*, **63**, 973, (1987).
5.  G. Baskaran, *Int. Jour. Mod. Phys. B* **1**, 539, (1988).
6.  G. Baskaran and P. W. Anderson, *Phys. Rev. B* **37**, 580, (1988).
7.  P. Coleman, *Phys. Rev. B* **29**, 3035, (1984).
8.  Z. Zou and P. W. Anderson, *Phys. Rev. B* **37**, 627, (1988).
9.  G. Baskaran, *Physica Scripta*, **T 27**, 53, (1989).
10. R. B. Laughlin, *Springer Series in Materials Science*, Vol. 11, Ed. H. Kamimura and A. Ushiyama, Springer-Verlag, Berlin, Heidelberg, 1989.
11. G. Martinez and P. Horsch, *Phys. Rev. B* **44**, 317, (1991).
12. S. Schmitt-Rink, C. M. Varma, and A. E. Ruckenstein, *Phys. Rev. Lett.* **60**, 2791, (1988).
13. C. L. Kane, P. A. Lee, and N. Read, *Phys. Rev. D* **39**, 6880 (1989).

# FUNCTIONAL INTEGRALS
# AND THEIR APPLICATIONS

# AN INTRODUCTION TO COORDINATE FREE QUANTIZATION AND ITS APPLICATION TO CONSTRAINED SYSTEMS[1]

JOHN R. KLAUDER[a] and SERGEI V. SHABANOV[b,2]

[a]Departments of Physics and Mathematics,
 University of Florida, Gainesville FL-32611, USA
[b]Institute for Theoretical Physics, FU-Berlin,
 Arnimallee 14, D-14195, Berlin, Germany

**Abstract**     Canonical quantization entails using Cartesian coordinates, and Cartesian coordinates exist only in flat spaces. This situation can either be questioned or accepted. In this paper we offer a brief and introductory overview of how a flat phase space metric can be incorporated into a covariant, coordinate-free quantization procedure involving a continuous-time (Wiener measure) regularization of traditional phase space path integrals. Additionally we show how such procedures can be extended to incorporate systems with constraints and illustrate that extension for special systems.

## 1. INTRODUCTION

In order to quantize a system with constraints it is of course first necessary to have a quantization procedure for systems without constraints. Although the quantization of systems without constraints would seem to be well in hand due to the pioneering work of Heisenberg, Schrödinger, and Feynman, it is a less appreciated fact that all of the standard methods of quantization are consistent only in Cartesian coordinates [1]. As a consequence it follows that the usual

---

[1]To appear in the Proceedings of the 2nd Jagna Workshop, Jagna, Bohol, Philippines, January, 1998.
[2]On leave from Laboratory of Theoretical Physics, JINR, Dubna, Russia.

quantization procedures depend—or at least seem to depend—on choosing the right set of coordinates before promoting c-numbers to q-numbers. This circumstance gives rise to an apparently unwanted coordinate dependence on the very process of quantization. For systems without constraints this is generally not a major problem because an underlying Euclidean space expressed in terms of Cartesian coordinates can generally be assumed. However, for systems with constraints, the configuration space—let alone the frequently more complicated phase space—are generally incompatible with a flat Euclidean structure needed to carry Cartesian coordinates. Hence, before we can properly quantize systems with constraints it will be necessary for us to revisit the quantization of systems without constraints in order to present a coordinate-free procedure for such cases. Only then will we be able to undertake the program represented by the title to this contribution.

All individuals engaged in quantization have a natural inclination to seek a quantization procedure that is as coordinate independent as possible, and in particular, does not depend on using Cartesian coordinates inasmuch as the use of such coordinates seems to contradict the ultimate goal of a coordinate-free approach. When faced with the need for quantizing in Cartesian coordinates a number of workers currently seek alternative quantization schemes which avoid completely any reference to Cartesian coordinates. The schemes of geometric quantization and of deformation quantization, among possibly other approaches as well, fit into the category of efforts to eliminate the central role played by Cartesian coordinates, and indeed to construct a fully coordinate independent formulation. It is entirely natural to presume that quantization should be coordinate independent, and so this approach is most reasonable. While these disciplines can be considered and analyzed from a mathematical viewpoint without any conceptual difficulty, it is not a priori evident that just because these methods have the word "quantization" in their name that they have an automatic connection with physics. Indeed, it may be argued that this is not always the case, and this conclusion pertains to the fact that the result does not in general agree with the results of ordinary quantization in the physical sense of the term. Thus such contemporary methods are acceptable as mathematical exercises but should not be taken necessarily as leading to a coordinate-free formulation of quantization as it is needed and used in physics.

Such a circumstance naturally leads to the question: Can we find a coordinate-free form of quantization that does agree with physics, i.e., as a test case to quantize the anharmonic oscillator in accord with the usual quantum mechanical result as obtained, say, from the Schrödinger prescription? The answer in our opinion is yes, and the first part of this article is devoted to a brief review of that subject. The extension of that procedure to systems with constraints is currently in progress, and a preliminary account of part of that work forms the second part of this article.

## 2.  COORDINATE-FREE QUANTIZATION

In 1948 Feynman proposed the path integral in Lagrangian form, and in 1951 he extended the path integral to a phase space form. It is generally acknowledged that the phase space formulation is more widely applicable than the Lagrangian formulation, and it is the phase space version on which we shall focus. In particular, the expression for the usual propagator is formally given by

$$\mathcal{N} \int \exp\{i \int [p\dot{q} - h(p,q)]\, dt\}\, \mathcal{D}p\, \mathcal{D}q \ . \tag{1}$$

As it stands, however, this expression is formal and needs to be properly defined. There are several ways to do so. One way involves what may be referred to as a continuous-time regularization in the form [2]

$$K(p'',q'',T;p',q',0)$$
$$= \lim_{\nu \to \infty} \mathcal{N} \int \exp\{i \int [p\dot{q} - h(p,q)]\, dt\}\, \exp[-(1/2\nu) \int (\dot{p}^2 + \dot{q}^2)\, dt]\, \mathcal{D}p\, \mathcal{D}q. \tag{2}$$

In this expression we have introduced an additional term in the integrand which is formally set to unity in the limit $\nu \to \infty$. However, that extra factor in the integrand serves as a *regularizing factor*, and this fact can be seen in the alternative—and mathematically precise—expression given by

$$K(p'',q'',T;p',q',0) = \lim_{\nu \to \infty} 2\pi e^{\nu T/2} \int \exp\{i \int [p\, dq - h(p,q)\, dt]\}\, d\mu_W^\nu(p,q) \ , \tag{3}$$

where $\mu_W^\nu$ denotes a Wiener measure on a flat phase space expressed in Cartesian coordinates and in which $\nu$ denotes the diffusion constant. In addition, the nature of the regularization forces one to pin (i.e., fix) the values of both *p and q* at the *initial and final times*, namely, $p'' = p(T), q'' = q(T)$ and $p' = p(0), q' = q(0)$. This leads to a nontraditional representation of the propagator, and as we assert below the very regularization itself leads to a *canonical coherent state representation*. Observe, moreover, that (3) is mathematically well-defined and totally ambiguity free. As such, there can be no ambiguity in factor ordering within the quantization procedure at this point, and it is significant that the very regularization chosen has *selected*—or perhaps even better, *preselected*—a particular operator ordering, namely antinormal ordering.

More specifically, on the basis of being a positive-definite function, one may show that

$$K(p'',q'',T;p',q',0) \equiv \langle p'',q'' | e^{-i\mathcal{H}T} | p',q' \rangle \ , \tag{4}$$

$$|p, q\rangle \equiv e^{-iqP} e^{ipQ} |0\rangle , \qquad (Q + iP) |0\rangle = 0 , \qquad \langle 0|0\rangle = 1 , \quad (5)$$

$$[Q, P] = i\mathbf{1} , \tag{6}$$

$$\mathbf{1} = \int |p, q\rangle \langle p, q| \, dp \, dq / 2\pi , \tag{7}$$

$$\mathcal{H} = \int h(p, q) |p, q\rangle \langle p, q| \, dp \, dq / 2\pi , \tag{8}$$

the last formula being an alternative expression for antinormal ordering. We do not attempt to prove these remarks here; for that the reader may consult the literature [3]. Instead, we limit our discussion to an overview of the general scheme. In that line it is important to note the behavior of these expressions under a *canonical coordinate transformation*. In the classical theory, we often let

$$r\,ds = p\,dq + dF(s, q) \tag{9}$$

symbolize a canonical change of coordinates where the function $F$ serves as the "generator" of the coordinate change. In the quantum theory, as described here, the paths $p(t)$ and $q(t)$ represent sample paths of a Wiener process, i.e., Brownian motion paths. As such these paths are continuous but nowhere differentiable, and thus they are more singular than the classical path behavior (e.g., $C^2$) for which (9) normally holds. The integral $\int p\,dq$ appearing in (3) is initially undefined due to the distributional nature of the paths involved. There are two standard prescriptions to deal with such stochastic integrals, one due to Itô (I), the other due to Stratonovich (S) [4]. The two prescriptions may be characterized by continuum limits of two distinct discretization procedures. If $p_l \equiv p(l\epsilon)$ and $q_l \equiv q(l\epsilon)$ for $l \in \{0, 1, 2, 3, \ldots\}$, then

$$\int_I p\,dq = \lim_{\epsilon \to 0} \Sigma p_l (q_{l+1} - q_l) , \qquad \int_S p\,dq = \lim_{\epsilon \to 0} \Sigma \tfrac{1}{2}(p_{l+1} + p_l)(q_{l+1} - q_l) . \quad (10)$$

Generally, the results of these two approaches disagree, and it is a feature of the Itô prescription ("nonanticipating") that the rules of ordinary calculus are generally not obeyed. The Itô prescription has other virtues, but they are not of interest to us here. Instead, we adopt the Stratonovich ("midpoint") prescription because it possesses the important feature that the ordinary laws of the classical calculus do in fact hold for Brownian motion paths. In particular, therefore, (9) also holds for the paths that enter the regularized form of the phase space path integral, and as a consequence, we find that

$$\int p\,dq = \int r\,ds - \int dF(s, q)$$
$$\equiv \int r\,ds + \int dG(r, s) = \int r\,ds + G(r'', s'') - G(r', s') . \tag{11}$$

If we couple this relation with the fact that $h$ transforms as a *scalar*, i.e., namely, that

$$\bar{h}(r, s) \equiv h(p(r, s), q(r, s)) = h(p, q) , \tag{12}$$

then we learn that under a canonical change of coordinates, the coherent state propagator becomes

$$\overline{K}(r'', s'', T; r', s', 0)$$
$$\equiv e^{i[G(r'',s'')-G(r',s')]} K(p(r'', s''), q(r'', s''), T; p(r', s'), q(r', s'), 0)$$
$$= \lim_{\nu \to \infty} 2\pi e^{\nu T/2} \int \exp\{i\int[rds + dG(r, s) - \overline{h}(r, s)]\, dt\}\, d\overline{\mu}_W^\nu(r, s) \quad (13)$$

Here $\overline{\mu}_W^\nu(r, s)$ denotes the Wiener measure on a flat two-dimensional phase space no longer expressed, in general, in Cartesian coordinates but rather in curvilinear coordinates. Observe that the form of this expression is exactly that as given in the original coordinates apart from the presence of the total derivative $dG$, which leads to nothing more than a phase change for the coherent states. In particular, based on the positive-definite nature of the transformed function we may conclude that

$$\overline{K}(r'', s'', T; r', s', 0) = \langle r'', s'' | e^{-i\mathcal{H}T} | r', s' \rangle , \quad (14)$$
$$|r, s\rangle \equiv e^{-iG(r,s)}\, e^{-iq(r,s)P}\, e^{ip(r,s)Q} |0\rangle , \quad (Q + iP)|0\rangle = 0, \quad \langle 0|0\rangle = 1, (15)$$
$$\mathbf{1} = \int |r, s\rangle\langle r, s|\, dr\, ds/2\pi , \quad (16)$$
$$\mathcal{H} = \int \overline{h}(r, s)|r, s\rangle\langle r, s|\, dr\, ds/2\pi . \quad (17)$$

Observe carefully that the coherent states have *not* changed under the coordinate transformation, only their *names* have changed. In addition, the operator $\mathcal{H}$ has *not* changed, only the functional form of the (lower) symbol associated with it has changed. Thus we have achieved a completely covariant formulation of quantum theory!

As an example of such a coordinate change, we may cite the simple case of the harmonic oscillator for which $h(p, q) = (p^2 + q^2)/2$. If we introduce new canonical coordinates (action angle variables) according to $r = (p^2 + q^2)/2$ and $s = \tan^{-1}(q/p)$—namely, where $F(q, s) = -q^2 \cot(s)/2$ and $G(r, s) = r\cos(s)\sin(s)$—then it follows that

$$\tfrac{1}{2}(P^2 + Q^2 + 1) = \int \tfrac{1}{2}(p^2 + q^2)\, |p, q\rangle\langle p, q|\, dp\, dq/2\pi = \int r\, |r, s\rangle\langle r, s|\, dr\, ds/2\pi, (18)$$

which clearly illustrates the fact that although a classical coordinate change has been carried out, all quantum operators, such as $Q$, $P$, and particularly the Hamiltonian $\mathcal{H}$, have remained completely unchanged!

The foregoing scenario may be readily extended to deal with multiple degrees of freedom, say $N$, and—reverting to the original Cartesian coordinates— we find (using the summation convention) that

$$\langle p'', q'' | e^{-i\mathcal{H}T} | p', q' \rangle$$

$$\equiv \lim_{\nu \to \infty} (2\pi)^N e^{N\nu T/2} \int \exp\{i\int[p_j dq^j - h(p,q)dt]\} \, d\mu_W^\nu(p,q) \,, \qquad (19)$$

$$|p,q\rangle \equiv e^{-iq^j P_j} e^{ip_j Q^j} |0\rangle \,, \qquad (Q^j + iP_j)|0\rangle = 0 \,, \qquad \langle 0|0\rangle = 1 \,, \,(20)$$

$$[Q^j, P_k] = i\delta_k^j \mathbf{1} \,, \qquad (21)$$

$$\mathbf{1} = \int |p,q\rangle \langle p,q| \, d\mu_N(p,q) \,, \qquad (22)$$

$$\mathcal{H} = \int h(p,q) \, |p,q\rangle \langle p,q| \, d\mu_N(p,q) \,, \qquad (23)$$

$$d\mu_N(p,q) \equiv \Pi_{j=1}^N dp_j \, dq^j / 2\pi \,, \qquad (24)$$

and where we have used the notation $p = \{p_j\}_{j=1}^N$ and $q = \{q^j\}_{j=1}^N$. We next turn our attention to the inclusion of constraints.

## 3. CONSTRAINTS AND THE PROJECTION METHOD
### Classical Preliminaries

From the classical point of view some of the equations of motion that follow from an action principle are just exactly *not* equations of motion in that they do not involve time derivatives but rather conditions that must be satisfied among the canonical variables. Consider a classical phase space action principle of the form

$$I = \int [p_j \dot{q}^j - h(p,q) - \lambda^a \phi_a(p,q)] \, dt \,, \qquad (25)$$

where $\lambda^a = \lambda^a(t)$ denote Lagrange multipliers and $\phi_a(p,q)$ denote constraints, and $1 \leq a \leq K \leq 2N$. Stationary variation with respect to the dynamical variables $p_j$ and $q^j$ leads to the equations

$$\dot{q}^j = \partial h(p,q)/\partial p_j + \lambda^a \partial \phi_a(p,q)/\partial p_j$$
$$= \{q^j, h(p,q)\} + \lambda^a \{q^j, \phi_a(p,q)\} \,, \qquad (26)$$
$$\dot{p}_j = -\partial h(p,q)/\partial q^j - \lambda^a \partial \phi_a(p,q)/\partial q^j$$
$$= \{p_j, h(p,q)\} + \lambda^a \{p_j, \phi_a(p,q)\} \,, \qquad (27)$$

where the last version of each equation is written in terms of Poisson brackets. In turn, stationary variation with respect to the Lagrange multipliers leads to the constraint equations

$$\phi_a(p,q) = 0 \qquad (28)$$

the fullfilment of which defines the *constraint hypersurface*. All processes, dynamics included, takes place on the constraint hypersurface. It follows that

$$\dot{\phi}_a(p,q) = \{\phi_a(p,q), h(p,q)\} + \lambda^b \{\phi_a(p,q), \phi_b(p,q)\} = 0 \,. \qquad (29)$$

Assuming that the set $\{\phi_a\}$ is a complete set of the constraints, two possible scenarios may hold. In the first scenario

$$\{\phi_a(p,q), \phi_b(p,q)\} = c_{ab}{}^c \phi_c(p,q) , \qquad (30)$$

$$\{\phi_a(p,q), h(p,q)\} = h_a{}^b \phi_b(p,q) . \qquad (31)$$

This situation, termed *first class constraints*, implies that if the constraints are satisfied at one time then they will be satisfied for all time in the future as a consequence of the equations of motion. Observe in this case that the time dependence of the Lagrange multipliers $\{\lambda^a\}$ is not determined by these equations. To solve the equations for the variables $p_j(t)$ and $q^j(t)$ it is necessary to choose the Lagrange multipliers which then constitutes a "gauge choice". Nothing that is deemed physical can depend on just which gauge choice has been selected, and any observable, say $O(p,q)$, must satisfy the relation

$$\{\phi_a(p,q), O(p,q)\} = o_a{}^b \phi_b(p,q) . \qquad (32)$$

In the second situation, (30) fails, or (30) and (31) both fail, to hold, and as a consequence, the Lagrange multipliers must assume a special time dependence in order to satisfy (29). In short, in this case, consistency of the equations of motion determines the Lagrange multipliers. This case is termed *second class constraints*. Of course, one may also have a mixed case composed of some first and some second class constraints. In this case some of the Lagrange multipliers are determined while others are not.

The coefficients $c_{ab}{}^c$, $h_a{}^b$, and $o_a{}^b$ above may also depend on the phase space variables. However, for convenience, we shall restrict attention hereafter to those cases where these coefficients are simply constants.

## Quantization à la Dirac

We next take up the topic of quantization of these systems. For the purposes of the present paper we shall confine our attention to the case of first class constraints. (The case of second class constraints has been discussed elsewhere using the methods of the present paper [5, 6].)

According to Dirac [7], quantization of first class systems proceeds along the following line. First quantize the system as if there were no constraints, namely, introduce kinematical operators $\{P_j\}$ and $\{Q^j\}$, which fulfill (21), and a Hamiltonian operator $\mathcal{H} = \mathcal{H}(P,Q)$ (modulo some choice of ordering). For a general operator $W(P,Q)$ adopt the dynamical equation

$$i\dot{W}(P,Q) = [W(P,Q), \mathcal{H}] \qquad (33)$$

as usual. The constraint operators are assumed to fulfill commutation relations similar to (30) and (31), namely,

$$[\Phi_a(P,Q), \Phi_b(P,Q)] = ic_{ab}{}^c \Phi_c(P,Q) , \qquad (34)$$

$$[\Phi_a(P,Q), \mathcal{H}(P,Q)] = ih_a{}^b\Phi_b(P,Q) \ . \tag{35}$$

Next impose the constraints to select the physical Hilbert subspace in the form

$$\Phi_a(P,Q)\,|\psi\rangle_{\text{phy}} = 0 \ . \tag{36}$$

If zero lies in the discrete spectrum of the constraint operators this equation offers no difficulties, and we shall content ourselves with that case. On the other hand, if zero lies in the continuous spectrum, then some subtlies are involved, and one example of where such issues are discussed is [5]. Observe that (34) and (35) demonstrate the consistency of imposing the constraints and the fact that if they are imposed at one time then they will hold for all subsequent time. This imposition of the constraints at the initial time may be called an *initial value equation*, just as in the classical theory.

### The Projection Method

We note that the commutation relations among the constraint operators is that of a Lie algebra. (Indeed, including the Hamiltonian and noting (35), we observe that the constraints plus the Hamiltonian also form a Lie algebra.) For present purposes we assume that the group generated by this Lie algebra is compact, and we denote the group elements by

$$e^{i\xi^a\Phi_a(P,Q)} \ . \tag{37}$$

Let $\delta\xi$ denote the normalized group invariant measure, $\int \delta\xi = 1$, and consider the operator

$$\mathbf{E} \equiv \int e^{i\xi^a\Phi_a(P,Q)}\,\delta\xi \ . \tag{38}$$

It is a modest exercise to establish that $\mathbf{E}^2 = \mathbf{E}^\dagger = \mathbf{E}$, which are just the criteria that make $\mathbf{E}$ a projection operator. The fact that

$$e^{i\tau^a\Phi_a(P,Q)}\mathbf{E} = \int e^{i\tau^a\Phi_a(P,Q)}e^{i\xi^a\Phi_a(P,Q)}\delta\xi$$
$$\equiv \int e^{i(\tau\cdot\xi)^a\Phi_a(P,Q)}\delta\xi = \int e^{i\xi^a\Phi_a(P,Q)}\delta\xi = \mathbf{E} \ , \tag{39}$$

due simply to the invariance of the measure, establishes that $\mathbf{E}$ is a projection operator onto the subspace where $\Phi_a = 0$ for all $a$, i.e., a projection operator onto the physical Hilbert subspace. We note further, based on (35), that

$$e^{-i\mathcal{H}T}\,\mathbf{E} = \mathbf{E}\,e^{-i\mathcal{H}T}\mathbf{E} = \mathbf{E}\,e^{-i\mathbf{E}\mathcal{H}\mathbf{E}T}\,\mathbf{E} \ , \tag{40}$$

Suppose we consider a formal phase space path integral for a system with the classical action functional (25). The formal path integral reads

$$\mathcal{N}\int \exp\{i\textstyle\int[p_j\dot{q}^j - h(p,q) - \lambda^a\phi_a(p,q)]\,dt\}\,\mathcal{D}p\,\mathcal{D}q$$
$$= \langle p'',q''|\,\mathbf{T}\,e^{-i\int[\mathcal{H}+\lambda^a\Phi_a]\,dt}\,|p',q'\rangle \ , \tag{41}$$

which, as written, evidently depends on the choice of the Lagrange multipliers. Now let us impose the quantum version of the initial value equation, namely let us force the system at the initial time to lie in the physical Hilbert subspace. This we may do by considering the expression

$$\int \langle p'', q''| \, \mathbf{T} \, e^{-i\int [\mathcal{H}+\lambda^a \Phi_a]\, dt} \, |\overline{p}', \overline{q}'\rangle \langle \overline{p}', \overline{q}'|\mathbf{E}|p', q'\rangle \, d\mu_N(\overline{p}', \overline{q}') \,, \qquad (42)$$

which has the effect of projecting the propagator onto the physical subspace at time zero. Using the resolution of unity for the coherent states (22), it is straightforward to determine that

$$\langle p'', q''| \, \mathbf{T} \, e^{-i\int [\mathcal{H}+\lambda^a \Phi_a]\, dt} \, |\overline{p}', \overline{q}'\rangle \langle \overline{p}', \overline{q}'|\mathbf{E}|p', q'\rangle \, d\mu_N(\overline{p}', \overline{q}')$$
$$= \langle p'', q''| \, \mathbf{T} \, e^{-i\int [\mathcal{H}+\lambda^a \Phi_a]\, dt} \, \mathbf{E}\,|p', q'\rangle$$
$$= \langle p'', q''| \, e^{-i\mathcal{H}T} \, e^{i\tau^a \Phi_a} \, \mathbf{E}\,|p', q'\rangle$$
$$= \langle p'', q''| \, e^{-i\mathcal{H}T} \, \mathbf{E}\,|p', q'\rangle \,. \qquad (43)$$

Here we have also used the properties of the Lie group relations (34) and (35) to separate the operators in the exponent into two factors, where $\{\tau^a\}$ denote parameters made of the the the functions $\{\lambda^a(t)\}$ and the constants appearing in (34) and (35). However, whatever the choice of the Lagrange multipliers $\{\lambda^a\}$, i.e., whatever the choice of the parameters $\{\tau^a\}$, and as indicated in the last line, *the result is completely independent of the Lagrange multipliers.* That is to say, the propagator projected on the subspace spanned by $\mathbf{E}$ is already, and automatically, gauge invariant. Thus, for first class constraint systems, all that is necessary to achieve a gauge invariant propagator is to project onto the proper subspace at the initial time.

We may introduce the projection operator $\mathbf{E}$ also by integrating over the Lagrange multipliers $\{\lambda^a(t)\}$ with one or another suitable measure. Let $C(\lambda)$ denote a (possibly complex) normalized measure, $\int dC(\lambda) = 1$, with the property that it introduces into the integrand at least one copy of the projection operator $\mathbf{E}$; we say at least one since if two (or more) are introduced the result will be the same because $\mathbf{E}^2 = \mathbf{E}$, etc. Hence, the desired propagator on the physical subspace may also be written as

$$\langle p'', q''| \, e^{-i\mathcal{H}T} \, \mathbf{E}\,|p', q'\rangle$$
$$= \mathcal{N} \int \exp\{i\int [p_j \dot{q}^j - h(p,q) - \lambda^a \phi_a(p,q)]\, dt\} \, \mathcal{D}p\,\mathcal{D}q\, dC(\lambda) \,, \quad (44)$$

where, as already noted, the normalized measure $C$ is designed to introduce one (or more) projection operators $\mathbf{E}$ [5, 8].

Commentary

Readers familiar with the proposal of quantization of systems with first class constraints by Faddeev will note a significant difference in the measure for the Lagrange multipliers. In Faddeev's treatment [9] the measure for the Lagrange multipliers is taken to be $\mathcal{D}\lambda$, namely a formally flat measure designed to introduce $\delta$-functionals of the classical constraints. Such a choice generally leads to a divergence in some of the remaining integrals due to a nonappearance of the variables conjugate to the constraints. Auxiliary conditions in the form of dynamical gauge fixing are necessary, along with the attendant Faddeev-Popov (F-P) determinant needed to ensure formal canonical coordinate covariance. As is well known, a global choice of dynamical gauge fixing is generally impossible, which then leads to Gribov ambiguities, and their associated difficulties. All these issues arose from using a different measure for the Lagrange multipliers than that which is chosen in the present paper. In our choice the measure for the Lagrange multipliers is *normalized*, $\int dC(\lambda) = 1$, namely, an *average over Lagrange multipliers*, for which any divergence is manifestly impossible! It is of course true that the measure for the classical dynamical variables ($p$ and $q$) is fixed (formally as $\mathcal{D}p\,\mathcal{D}q$) by the requirements of consistency, but there is no requirement that the Lagrange multipliers must be integrated just as the classical variables here, namely, from $-\infty$ to $\infty$ with a flat weighting. Who says we must enforce the *classical* constraints when doing the *quantum* theory? No one of course, and in a general sense that is the only freedom that has been used to avoid dynamical gauge fixing, F-P determinants, Gribov ambiguities, ghosts, ghosts of ghosts, etc., and all the other machinery of the BRST and BFV formalisms [10, 11]. Our results involve only the original phase space variables as augmented by the Lagrange multipliers, do not in any way entail enlargement or even in many cases a reduction in the number of the original variables, lead to results that are entirely gauge invariant and satisfactory, and at the same time avoid altogether the whole galaxy of issues listed above which are instigated by using a flat measure for the Lagrange multipliers.

## 4. YANG-MILLS TYPE GAUGE MODELS

In this section we illustrate the imposition of constraints for a special class of models which we call Yang-Mills type [12]. In the classical theory, in which we assume we have chosen Cartesian coordinates in phase space, the constraints are taken as

$$\phi_a(p, q) = p_a + A_{ab}{}^c q^b p_c\,, \tag{45}$$

where the parameters $A_{ab}{}^c$ are antisymmetric in the indices $b, c$. This constraint induces a shift in the $a$th coordinate and a rotation in the $b - c$ plane. Such constraints are broad enough to cover the usual Yang-Mills theories (where $p$ plays the role of the electrical field strength $E$ and $q$ the role of the vector potential $A$). It is clear that such constraints commute among themselves to form a Lie algebra. In the quantum theory the constraint operators are taken to be

$$\Phi_a(P,Q) = P_a + A_{ab}{}^c Q^b P_c , \qquad (46)$$

and it is an attractive feature of such constraints that

$$e^{i\Omega^a \Phi_a} |p,q\rangle = |p^\Omega, q^\Omega\rangle , \qquad (47)$$

namely that the unitary transformation generated by the constraints takes one coherent state into another coherent state. Here

$$p^\Omega \equiv e^{-\Omega^a \, \mathrm{ad} \, \phi_a} p , \qquad q^\Omega \equiv e^{-\Omega^a \, \mathrm{ad} \, \phi_a} q , \qquad (48)$$

where $\mathrm{ad} \, \phi_a(\cdot) \equiv \{\phi_a, \cdot\}$ and the exponential is defined by its power series expansion.

The inclusion of such constraints into a dynamical system is quite straightforward. For that goal we first note that

$$\int \langle p'', q'' | e^{-i\mathcal{H}T} e^{i\Omega^a \Phi_a} |p', q'\rangle \, \delta\Omega = \int \langle p'', q'' | e^{-i\mathcal{H}T} |p'^\Omega, q'^\Omega\rangle \, \delta\Omega$$
$$= \langle p'', q'' | e^{-i\mathcal{H}T} \, \mathbf{E} \, |p', q'\rangle , \qquad (49)$$

which asserts that in order to insert the desired projection operator it is only necessary to average the initial coherent state labels over the gauge transformations they experience. In turn this means that

$$\langle p'', q'' | e^{-i\mathcal{H}T} \, \mathbf{E} \, |p', q'\rangle$$
$$= \lim_{\nu \to \infty} (2\pi)^N e^{N\nu T/2} \int \exp\{i\int [pdq - h(p,q) \, dt]\} \, d\mu_W^\nu(p,q) \, \delta\Omega , \quad (50)$$

where it is understood that the Wiener paths are pinned initially at $p^\Omega$ and $q^\Omega$. We can make this expression appear more familiar by making a change of variables within the well defined path integral. In particular, we let

$$p(t) \to e^{\int_t^T ds \, \omega^a(s) \, \mathrm{ad} \, \phi_a} \, p(t) , \qquad q(t) \to e^{\int_t^T ds \, \omega^a(s) \, \mathrm{ad} \, \phi_a} \, q(t) , \qquad (51)$$

where the functions $\omega^a(s)$ are arbitrary save for the condition that

$$\int_0^T \omega^a(s) \, ds = \Omega^a . \qquad (52)$$

Since $(p'^\Omega)^{-\Omega} \equiv p'$ and $(q'^\Omega)^{-\Omega} \equiv q'$, this transformation has the effect of removing any influence of the gauge transformation on the initial labels $p', q'$, and instead redistributing that influence throughout the time evolution of the path integral. As a proper coordinate transformation within a well defined path integral, we can readily determine the effect of such a change of variables. In particular, appealing to a formal notation for clarity, we find after such a variable change that (50) becomes

$$\langle p'', q'' | e^{-i\mathcal{H}T} \mathbf{E} | p', q' \rangle = \lim_{\nu \to \infty} \mathcal{N} \int \exp\{i\int[p_j \dot{q}^j - \omega^a \phi_a(p, q) - h(p, q)] \, dt\}$$
$$\times \exp\{-(1/2\nu)\int[(\dot{p} - \omega^a\{\phi_a, p\})^2 + (\dot{q} - \omega^a\{\phi_a, q\})^2] \, dt\} \, \mathcal{D}p \, \mathcal{D}q \, \delta\Omega. \quad (53)$$

In this expression a "new" term has appeared in the classical action that looks like a sum of Lagrange multipliers times the constraints, and drift terms have arisen in the Wiener measure regularization. Note also what this formula states: On the right side is a path integral which superficially depends on the functions $\{\omega^a(s)\}$, $0 \le s \le T$. On the left side, there is no such dependence. In other words, although the path integral *appears* to depend on $\{\omega^a\}$, in fact it does not. Therefore we are free to *average* the right-hand side of (53) over the functions $\{\omega\}$ and still obtain the desired answer. Let the measure $C(\omega)$ denote such a measure chosen so as to include the initial average over the variables $\Omega$ as well, and normalized so that $\int dC(\omega) = 1$. The only requirement we impose on this measure is that it introduce, as did the original measure over the variables $\Omega$, at least one projection operator $\mathbf{E}$. In this case we find the important phase space path integral representation given by

$$\langle p'', q'' | e^{-i\mathcal{H}T} \mathbf{E} | p', q' \rangle$$
$$= \lim_{\nu \to \infty} \mathcal{N} \int \exp\{i\int[p_j \dot{q}^j - \omega^a \phi_a(p, q) - h(p, q)] \, dt\}$$
$$\times \exp\{-(1/2\nu)\int[(\dot{p} - \omega^a\{\phi_a, p\})^2 + (\dot{q} - \omega^a\{\phi_a, q\})^2] \, dt\} \, \mathcal{D}p \, \mathcal{D}q \, dC(\omega). \quad (54)$$

Here we can really see the variability of the Lagrange multipliers in the path integration and how the proper choice of measure for them can lead to the desired gauge invariant result without any additional complications.

The only topic left to discuss is what should we take for the measure $C$. In fact there are many answers to that question, but, for brevity, we shall only indicate one of them. We suppose that our compact gauge group is semisimple and therefore admits a group-induced, positive definite metric $g_{ab}(\omega)$. Given that metric on the group manifold we introduce a Wiener measure formally given by

$$dC(\omega) = \mathcal{M} \exp[-\tfrac{1}{2}\int g_{ab} \, \dot{\omega}^a \dot{\omega}^b \, dt] \, \Pi_t \delta\omega(t) , \quad (55)$$

where this measure is *not* pinned either at $t = 0$ or at $t = T$. A normalized measure without pinning is made possible because the space of variables $\{\omega^a\}$

at any one time is compact. The formal constant $\mathcal{M}$ is chosen to ensure $\int dC(\omega) = 1$.

In summary, we have illustrated how the use of coherent states and a natural flat phase space metric can be used to develop a coordinate-free quantization procedure for systems without constraints as well as for systems with constraints. We hope this introductory paper may encourage the reader to delve further into this fascinating subject.

## DEDICATION

It is a pleasure to dedicate this paper to the 65th birthday of Hiroshi Ezawa, which was the main event celebrated at the 2nd Jagna Workshop. Over a number of years, one of the authors (J.R.K.) has enjoyed numerous interactions with the honoree including, but not limited to, two years of close collaboration at Bell Laboratories, and several visits to Tokyo to share in the scientific and social life of Japan. Such great personal interactions and experiences are truly what makes life worthwhile!

## ACKNOWLEDGEMENTS

It is a great pleasure to thank the organizers Chris and Victoria Bernido, and their extended families, for hosting such a pleasant meeting, and which additionally offered the participants a delightful glimpse of Philippine country life. It was an additional pleasure to meet old friends and to make new ones among the local participants. We hope the series of Jagna workshops will continue for many years to come!

# References

[1] P.A.M. Dirac, *The Principles of Quantum Mechanics*, (Oxford University Press, Oxford, Fourth Edition, 1958), p. 114.

[2] J.R. Klauder, *Annals of Physics*, **188**, 120 (1988).

[3] I. Daubechies and J.R. Klauder, *J. Math. Phys.*, **26**, 2239 (1985).

[4] See, e.g., H. Cramér and M.R. Leadbetter, *Stationary and Related Stochastic Processes* (John Wiley & Sons, Inc., 1968), Sec. 3.3.

[5] J.R. Klauder, *Annals of Physics*, **254**, 419 (1997).

[6] J.R. Klauder and S.V. Shabanov, *Nucl. Phys.*, **B511**, 713 (1998).

[7] P.A.M. Dirac, *Lectures on Quantum Mechanics* (Yeshiva University, N.Y., 1964).

[8] S.V. Shabanov, *Path integral in holomorphic representation without gauge fixation*, JINR preprint, E2-89-678, Dubna, 1989 (unpublished); see also S.V. Shabanov, in *Path Integrals, Dubna '96* (Publishing Department, Joint Institute for Nuclear Research, Dubna, Russia, 1996), p. 133.

[9] L.D. Faddeev, *Theor. Math. Phys.*, **1**, 1 (1970).

[10] M. Henneaux and C. Teitelboim, *Quantization of Gauge Systems* (Princeton University Press, Princeton, 1991).

[11] V.N. Gribov, *Nucl. Phys.*, **B139**, 1 (1978); I.M. Singer, *Commun. Math. Phys.*, **60**, 7 (1978); see also J. Govaerts, *Hamiltonian Quantization and Constrained Dynamics* (Leuven University Press, 1991).

[12] J.R. Klauder and S.V. Shabanov, *Phys. Lett. B*, **398**, 116 (1997).

# PHYSICS ON AND NEAR CAUSTICS
## – A Simpler Version –

**Pierre Cartier**
Ecole Normale Supérieure, F-75005 Paris

*and*

**Cécile DeWitt-Morette**
Department of Physics and Center for Relativity
The University of Texas, Austin, Texas 78712-1081

### Forward

The Workshop Directors selected "Physics on and near caustics" among the topics I had offered. I was both delighted and concerned. It is a lovely subject, easy to present in one lecture, but I could not contribute to the Proceedings because, together with Pierre Cartier, I had published an article with the same title in the Proceedings of a NATO-ASI held at the Institut des Etudes Scientifiques de Cargèse in September 1996.[1] However, while preparing my talk I simplified, *ipso facto* improved, the original presentation; and the Directors wishing that the RCTP Proceedings reflect the second Jagna Workshop felt it desirable to include the following text. It borrows heavily from the original text, and we are grateful to Plenum Press for their permission to use it. The original text contains proofs and techniques which are not reproduced here.

### Abstract

Physics *on* caustics is obtained by studying physics *near* caustics. Physics on caustics is at the cross-roads of calculus of variation and functional integration. More precisely consider a space $\mathcal{P}\mathbf{M}^d$ of paths $x : [t_a, t_b] \rightarrow \mathbf{M}^d$, on a $d$-dimensional manifold $\mathbf{M}^d$. Consider two of its subspaces:

i) The $2d$-dimensional space $\mathcal{U}$ of critical points of an action functional $S : \mathcal{P}\mathbf{M}^d \rightarrow \mathbf{R}; q \in \mathcal{U} \Leftrightarrow \frac{\delta S}{\delta q} = 0$.

ii) The space $\mathcal{P}_{\mu,\nu}\mathbf{M}^d$ of paths satisfying $d$ initial conditions $(\mu)$ and $d$ final conditions $(\nu)$.

The nature of the intersection $\mathcal{P}_{\mu,\nu}\mathbf{M}^d \cap \mathcal{U}$ is the key to identifying and analyzing caustics. Caustics occur when the roots of $S'(q) = 0$, for $S$ restricted to $\mathcal{P}_{\mu,\nu}\mathbf{M}^d$, are not isolated points in $\mathcal{U}$. Said in different terms, caustics occur when the hessian $S''(q)$ at a critical point $q$ is a degenerate quadratic form.

We consider two cases:

i) The roots define a subspace of non zero dimension $\ell$ of $\mathcal{U}$. For example, the classical system with action $S$ is constrained by conservation laws.

ii) There is a multiple root of $S'(q) = 0$. For example the classical flow has an envelope.

In both situations there is, at least, one nonzero Jacobi field along $q$ with $d$ initial vanishing boundary conditions $(\mu)$ and $d$ final vanishing boundary conditions $(\nu)$, i.e. $q(t_a)$ and $q(t_b)$ are conjugate points along $q$. In both cases we say "$q(t_b)$ is on the caustics".

The strict WKB approximations of functional integrals for the system $S$ "break down" on caustics, but the full semiclassical expansion, including contributions from $S'(q)$ and $S'''(q)$ (and possibly higher derivatives), as well as $S''(q)$, yield the physical properties of the system $S$ on and near the caustics. The study of caustics displays classical physics as a limit of quantum mechanics.

As an example of a semiclassical approximation where the strict WKB approximation breaks down, we give the glory scattering cross-sections; there, both situations i) and ii) occur simultaneously: the system is constrained by conservation laws, and the classical flow has an envelope.

# I  Introduction

Interesting phenomena occur when a flow of classical paths is caustic forming. Glory scattering, rainbows, orbiting, etc...are only a few examples of physics on and near caustics. Conservation laws are, as we shall see, another example.

Physics on caustics is at the cross-roads of calculus of variation and functional integration. More precisely consider a space $\mathcal{P}\mathbf{M}^d$ of $(L^{2,1})$ paths[2]

$$x : \mathbf{T} \rightarrow \mathbf{M}^d \quad , \quad \mathbf{T} = [t_a, t_b]$$

on a $d$-dimensional manifold $\mathbf{M}^d$. Consider two of its subspaces:

i) the $2d$-dimensional subspace $\mathcal{U}$ of critical points of an action functional

(I.1) $$S : \mathcal{P}\mathbf{M}^d \rightarrow \mathbf{R}$$

(I.2) $$q \in \mathcal{U} \Leftrightarrow \frac{\delta S}{\delta q} = 0 \, ;$$

ii) the space $\mathcal{P}_{\mu,\nu} \mathbf{M}^d$ of paths satisfying $d$ boundary conditions $(\mu)$ at $t = t_a$, and $d$ boundary conditions $(\nu)$ at $t = t_b$.

The nature of the intersection

(I.3) $$\mathcal{U}_{\mu,\nu} := \mathcal{P}_{\mu,\nu} \, \mathbf{M}^d \cap \mathcal{U}$$

is the key to identifying and analyzing caustics.

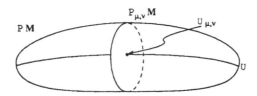

Figure 1: $\mathcal{U}$ is the 2d dimensional space of classical motions (critical points of the action functional $S$); $\mathcal{P}_{\mu,\nu} \, \mathbf{M}^d$ is the infinite dimensional space of paths satisfying $d$ initial conditions and $d$ final conditions; $\mathcal{U}_{\mu,\nu}$ is their intersection.

The intersection $\mathcal{U}_{\mu,\nu}$ is analyzed in terms of the tangent spaces at $q$ of $\mathcal{P}_{\mu,\nu} M^d$ and $\mathcal{U}^{2d}$.

1. *A basis for* $T_q \mathcal{P}_{\mu,\nu} M^d$, $q \in \mathcal{U}^{2d}$, $S : \mathcal{P}_{\mu,\nu} M^d \to \mathbf{R}$.

Consider the restriction of the action functional $S$ to $\mathcal{P}_{\mu,\nu} M$ and expand $S(x)$ around $S(q)$.[3]

(I.4)   $$S(x) = S(q) + S'(q) \cdot \xi + \frac{1}{2!} S''(q) \cdot \xi\xi + \frac{1}{3!} S'''(q) \cdot \xi\xi\xi + \cdots$$

The vector field $\xi \in T_q \mathcal{P}_{\mu,\nu} M^d$ has $d$ vanishing boundary conditions at $t_a$ and $d$ boundary conditions at $t_b$ dictated by the $(\mu)$ and $(\nu)$ conditions satisfied by all $x \in \mathcal{P}_{\mu,\nu} M^d$.

A good basis for $T_q \mathcal{P}_{\mu,\nu} M^d$ diagonalizes the hessian $S''(q) \cdot \xi\xi$.

(I.5)   $$S''(q) \cdot \xi\xi = \langle \mathcal{J}(q) \cdot \xi, \xi \rangle$$

where $\mathcal{J}(q)$ is the Jacobi operator on $T_q \mathcal{P}_{\mu,\nu} M^d$. Let $\{\Psi_k\}_k$ be a complete set of orthonormal eigenvectors of $\mathcal{J}(q)$:

(I.6)   $$\mathcal{J}(q) \cdot \Psi_k = \alpha_k \Psi_k , \, k \in \{0, 1, \cdots\} .$$

Let $\{u^k\}$ be the coordinates of $\xi$ in the $\{\Psi_k\}$ basis

(I.7)   $$\xi^\alpha(t) = \sum_{k=0}^{\infty} u^k \Psi_k^\alpha(t) ,$$

(I.8)   $$u^k = \int_{t_a}^{t_b} dt(\xi(t) | \Psi_k(t)) .$$

The $\{\Psi_k\}$ basis diagonalizes the hessian

(I.9)
$$S''(q) \cdot \xi\xi = \sum_{k=0}^{\infty} \alpha_k (u^k)^2 \, ;$$

Therefore the space of the $\{u^k\}$ is the space $l^2$ (hessian) of points $u$ such that $\sum \alpha_k (u^k)^2$ is finite.

2. *A basis for $\mathcal{U}^{2d}$: a complete set of linearly independent Jacobi fields.* Let $h$ be a Jacobi field, i.e. let $h$ satisfy the Jacobi equation

(I.10)
$$\mathcal{J}(q) \cdot h = 0$$

The Jacobi fields can be obtained without solving this equation: the critical point $q$ is a function of $t$ and of $2d$ constants of integration. The $2d$ derivatives of $q$ with respect to the constants of integration are Jacobi fields.[4]

If one (or several) eigenvalues of the Jacobi operator vanishes – assume only one, say $\alpha_0 = 0$, for simplicity – then $\Psi_0$ is a Jacobi field with vanishing boundary conditions

(I.11)
$$\mathcal{J}(q) \cdot \Psi_0 = 0 \, , \Psi_0 \in T_q \mathcal{P}_{\mu,\nu} M^d \, .$$

The $2d$ Jacobi fields are not linearly independent, $S''(q) \cdot \xi\xi$ is degenerate.

The finite dimensional case can be used as an example for classifying the various types of degeneracy of $S''(q) \cdot \xi\xi$.

(I.12)
$$\text{Let } S : \mathbf{R}^2 \to \mathbf{R}. \quad x = (x^1, x^2)$$

*Example 1:* $S(x) = x^1 + (x^2)^2$; relative critical points.

The first derivatives, $\frac{\partial S}{\partial x^1} = 1$ and $\frac{\partial S}{\partial x^2} = 2x^2$, do not define a critical point $S'(x_0) = 0$; however on any subspace $x^1 = $ constant, $x^2 = 0$ is a critical point. We say that $x^2 = 0$ is a relative critical point (relative to the subspace $x^1 = $ constant).

(I.13)
$$S''(x) = \begin{pmatrix} 0 & 0 \\ 0 & 2 \end{pmatrix}$$

$S''(x)$ is degenerate.

*Example 2:* $S(x) = (x^2)^2 + (x^1)^3$; double roots of $S'(x_0) = 0$

$$S'(x_0) = 0 \Rightarrow (x_0^1)^2 = 0, x_0^2 = 0$$

(I.14)
$$S''(x) = \begin{pmatrix} 6x^1 & 0 \\ 0 & 2 \end{pmatrix}, S''(x_0) = \begin{pmatrix} 0 & 0 \\ 0 & 2 \end{pmatrix}$$

$S''(x)$ is not degenerate, but $S''(x_0)$ is degenerate at the critical point. In this case we can say the critical point $x_0$ is degenerate.

The generalization to the infinite dimensional case is straightforward provided the action functional is stated in the variables which diagonalize the hessian (here the $u$-variables). In its diagonal form the hessian consists of a finite number of zeroes (say $\ell$) and a non-degenerate quadratic form (here $\sum_{k=\ell}^{\infty} \alpha_k (u^k)^2$) of codimension $\ell$.

If the critical points are relative (Example 1) $\mathcal{U}_{\mu,\nu}$ is a space of dimension $\ell > 0$. If $q \in \mathcal{U}_{\mu,\nu}$ is a multiple root (Example 2) the classical flow has an envelope. In both cases the strict $WKB$ approximation breaks down, but the semiclassical approximation with contributions from the first variation (Example 1) and/or from the third (possibly higher) variation (Example 2) gives finite, meaningful results.

We do not treat here the cases in which $\mathcal{U}_{\mu,\nu}$ consists of isolated points[3], nor the case in which $\mathcal{U}_{\mu,\nu}$ is empty. If $\mathcal{U}_{\mu,\nu}$ consists of isolated points $\{q_i\}_i$ (e.g. the anharmonic oscillator provided $(\mu, \nu)$ does not define a pair of conjugate points along $q_i$), the $WKB$ is a finite sum of finite contributions, each for a different $q_i$.

The case $\mathcal{U}_{\mu,\nu} = \emptyset$ is interesting but does not belong to the study of caustics. A typical example is the knife edge problem solved by L.S. Schulman [[5] and references therein].

# II   The intersection $\mathcal{U}_{\mu,\nu}$ is of dimension $\ell > 0$; conservation laws

It can happen, possibly after a change of variable in the space of paths $\mathcal{P}_{\mu,\nu} M^d$, that the Euler Lagrange equations which determine the critical points $q \in \mathcal{U}_{\mu,\nu}$ split into two sets

(II.1)          $S'_a = g_a$      for $a \in \{o, \ldots, \ell - 1\}$   a constant

(II.2)          $S'_A(q) = 0$     for $A \in \{\ell, \ldots, d\}$

with $S'_\alpha = \delta S(x)/\delta x^\alpha(t)$.

The $\ell$ equations (II.1) are constraints; the $d - \ell$ equations (II.2) determine $d - \ell$ coordinates $q^A$ of $q$. The space of critical points $\mathcal{U}_{\mu,\nu}$ is of dimension $\ell$.

The change of variable $\xi \leftrightarrow \{u^k\}$ given by (I.7) and (I.8) is not affected by the fact that $\ell$ eigenvectors $\Psi_k$ have zero eigenvalues. For the construction of eigenvectors with zero eigenvalues we refer to [1]. The dominating terms of the semiclassical expansion of the action functional in terms of the $\{u^k\}$ variables read, if we assume $\ell = 1$.

(II.3)                   $S(x) \simeq S(q) + c_0 u^0 + \frac{1}{2} \sum_{k=1}^{\infty} \alpha_k (u^k)^2$

with

(II.4)
$$c_0 = \int_T \mathrm{d}t \, \frac{\delta S}{\delta q^\alpha(t)} \psi_0^\alpha(t)$$

The probability amplitude for a transition from a state $(\mu, t_a)$ to a state $(\nu, t_b)$ of a system governed by the action functional $S$ can be written symbolically [$\vartheta$]

$$K(\nu, t_b; \mu, t_a) = \int_X \mathcal{D}x\chi \exp\left(\frac{2\pi i}{h} S(x)\right) \phi(x(t_a))$$

**X** is the space of pointed paths satisfying the conditions $(\nu)$ at $t_b$; we could write $\mathbf{X} = \mathcal{P}_\nu M^d$; the choice of initial wave function $\phi(x(t_a))$ is dictated by the conditions $(\mu)$ which define the state $(\mu, t_a)$.

The change of variables $x \mapsto \xi \to \{u^k\}$ are linear, and it is a straightforward matter to reexpress the functional integral over **X** as a functional integral of the space $\ell^2$ (hessian) of the variable $\{u^k\}$. See, for instance [1]. In the case $\ell = 1$, the novelty will be the integral over $u^0$, which does not appear in the quadratic form $\sum_{k=1}^{\infty} \alpha_k(u^k)^2$. It contributes a $\delta$-function to the propagator

$$\delta\left(\frac{1}{h} \int_T \mathrm{d}t \, \frac{\delta S}{\delta q^\alpha(t)} \psi_0^\alpha(t)\right)$$

which says that the propagator vanishes, unless the conservation law

$$\frac{1}{h} \int_T \mathrm{d}t \, \frac{\delta S}{\delta q^\alpha(t)} \psi_0^{\alpha(t)} = 0$$

is satisfied.

Explicit expressions for a variety of cases can be found in [6]. In conclusion, if the action is invariant under automorphisms of $\mathcal{U}_{\mu,\nu}$, then the dominating terms of the semi-classical expansion of $S(x)$ around $S(q)$ imply conservation laws. Conservation laws appear in the classical limit of quantum physics. It is not an anomaly for a quantum system to have less symmetry than its classical limit.

# III   The intersection $\mathcal{U}_{\mu,\nu}$ is a multiple root of $S'(q) \cdot \xi = 0$

In this situation the classical flows are caustic forming. Four examples are treated in references [10] and [7]. Two of them enter into well-known problems.

i) The soap bubble problem [11]. The "paths" are the curves defining (by rotation around an axis) the surface of a soap bubble held by two rings. The "classical flow" is a family of catenaries with one fixed point. The caustic is the envelope of the catenaries.

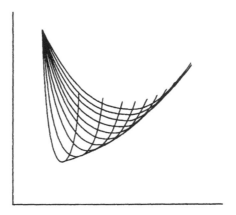

Figure 2: For a point in the "dark" side of the caustic there is no classical path; for a point on the "bright" side there are two classical paths which coalesce into a single one as the intersection of the two paths approaches the caustic. Note that the paths do not arrive at an intersection at the same time, the paths do not intersect in a space time diagram.

ii) The scattering of particles by a repulsive Coulomb potential. The flow is a family of Coulomb paths with fixed initial momentum. Its envelope is a parabola.

The two other examples are not readily identified as caustic problems because the flows do not have an envelope in the physical space. The vanishing boundary conditions of the Jacobi field at the caustic is the vanishing of its first derivative. In phase space the projection of the flow on the momentum space has an envelope.

iii) Rainbow scattering from a point source.

iv) Rainbow scattering from a source at infinity.

The relevant features can be analyzed on a specific example, for instance, the scattering of particles by a repulsive Coulomb potential. For other examples see [12].

Let $q$ and $q^\Delta$ be two solutions of the same Euler-Lagrange equation with slightly different boundary conditions at $t_b$, i.e. $q \in T_q \mathcal{P}_{\mu,\nu} M$ and $q^\Delta \in T_{q^\Delta} \mathcal{P}_{\mu^\Delta,\nu} M$.

$$p(t_a) = p_a \qquad q(t_b) = b$$
$$p^\Delta(t_a) = p_a \qquad q^\Delta(t_b) = b^\Delta .$$

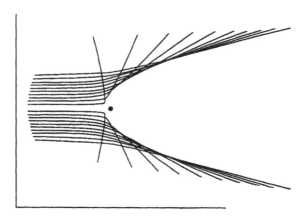

Figure 3: A flow on configuration space of charged particles in a repulsive Coulomb potential.

Assume $p_a$ and $b$ to be conjugate along $q$ with multiplicity 1; i.e. the Jacobi fields $h$ along $q$ such that

$$\dot{h}(t_a) = 0 , \quad h(t_b) = 0$$

form a one-dimensional space. Assume $(p_a, b^\Delta)$ not conjugate along $q^\Delta$.

We shall compute the probability amplitude $K(b^\Delta, t_b; p_a, t_a)$ when $b^\Delta$ is close to the caustic on the "bright" side or on the "dark" side. We shall *not* compute $K$ by expanding $S$ around $q^\Delta$ for the following reasons:

- If $b^\Delta$ is on the dark side, $q^\Delta$ does not exist.

- If $b^\Delta$ is on the bright side, one could consider $K$ to be the limit of the sum of two contributions corresponding to the two paths $q$ and $q^\Delta$ intersecting at $b^\Delta$

$$K(b, t_b; p_a, t_a) = \lim_{\Delta = 0} K_q(b^\Delta, t_b + \Delta t; p_a, t_a) + K_{q^\Delta}(b^\Delta, t_b; p_a, t_a)$$

but, at $b^\Delta$, $q$ has touched the caustic and "picked up" an additional phase equal to $-\pi/2$; both limits are infinite and their sum is not defined.

We compute $K(b^\Delta, t_b; p_a, t_a)$ by expanding $S$ around $q$, using (I.4) – and possibly higher derivatives if the third variation is singular. The calculation requires some care [see reference [7] for details] because $q(t_b) \neq b^\Delta$; in other words $q$ is not a critical point of the action restricted to the space of paths such that $x(t_b) = b^\Delta$. We approach the intersection in a direction other than a tangent to $\mathcal{U}$.

As before we make the change (I.8) of variable $\xi \mapsto u$ which diagonalizes $S''(q) \cdot \xi\xi$. Again we decompose the domain of integration in the $u$-variable

$$\ell^2(\text{hessian}) = \mathbf{X}^1 \times \mathbf{X}^\infty .$$

Again the second variation restricted to $\mathbf{X}^\infty$ is non singular, and calculating the integral over $\mathbf{X}^\infty$ proceeds as usual for the strict WKB approximation. The integral over $\mathbf{X}^1$ is

(III.1)   $I(\nu, c) = \displaystyle\int_{\mathbf{R}} du^0 \, \exp\left(i\left(cu^0 - \frac{\nu}{3}(u^0)^3\right)\right) = \nu^{-1/3} \, \text{Ai}(\nu^{-1/3} c)$

where

(III.2)   $\nu = \dfrac{\pi}{h} \displaystyle\int_{\mathbf{T}} dr \int_{\mathbf{T}} ds \int_{\mathbf{T}} dt \, \frac{\delta^3 S}{\delta q^\alpha(r) \, \delta q^\beta(s) \, \delta q^\gamma(t)} \, \psi_0^\alpha(r) \, \psi_0^\beta(s) \, \psi_0^\gamma(t)$

(III.3)                     $c = -\dfrac{2\pi}{h} \displaystyle\int_{\mathbf{T}} dt \, \frac{\delta S}{\delta q(t)} \cdot \psi_0(t) \, (b^\Delta - b)$

Ai is the Airy function. The leading contribution of the Airy function when $h$ tends to zero can be computed by the stationary phase method. At $v^2 = \nu^{-1/3} c$

(III.4)      $\text{Ai}(\nu^{-1/3} c) \simeq \begin{cases} 2\sqrt{\pi}\, v^{-1/4} \cos\left(\frac{2}{3} v^2 - \frac{\pi}{4}\right) & \text{for } v > 0 \\[2mm] \sqrt{\pi}\, (-v)^{-1/4} \exp\left(-\frac{2}{3} v^3\right) & \text{for } v < 0 \end{cases}$

$v$ is the critical point of the phase in the integrand of the Airy function; it is of order $h^{-1/3}$. For $v > 0$, $b^\Delta$ is in the illuminated region and the probability amplitude oscillates rapidly as $h$ tends to zero. For $v < 0$, $b^\Delta$ is in the shadow region and the probability amplitude decays exponentially.

The probability amplitude $K(b^\Delta, t_b; a, t_a)$ does not blow up when $b^\Delta$ tends to $b$. Quantum mechanics softens up the caustics.

*Remark.* The normalization and the argument of the Airy function can be expressed solely in terms of the Jacobi fields.

*Remark.*      Other cases, such as position-to-momentum, position-to-position, momentum- to-momentum, angular momentum transitions have been treated explicitly in references [7] and [12].

# IV   An example: glory scattering

Backward scattering of light, very close to the direction of the incoming rays has a long and interesting history (see for instance [13] and references therein). It creates a bright halo around one's shadow, and is usually called glory scattering. Early derivations of glory scattering were cumbersome, and used several approximations. It has been computed from first principles by functional integration using only the expansion in powers of the square root of Planck's constant [10], [7].

The classical cross-section for the scattering of a beam of particles in a solid angle $d\Omega = 2\pi \sin\theta \, d\theta$ by an axisymmetric potential is

(IV.1) $$d\sigma_{cl}(\Omega) = 2\pi \, B(\theta) \, dB(\theta)$$

where the deflection function $\Theta(B)$ giving the scattering angle $\theta$ as a function of the impact parameter $B$ is assumed to have a unique inverse $B(\Theta)$. We can write

(IV.2) $$d\sigma_{cl}(\Omega) = B(\Theta) \frac{dB(\Theta)}{d\Theta} \Big|_{\Theta=\theta} \frac{d\Omega}{\sin\theta}$$

abbreviated henceforth

(IV.3) $$d\sigma_{cl}(\Omega) = B(\theta) \frac{dB(\theta)}{d\theta} \frac{d\Omega}{\sin\theta} .$$

It can happen that for a certain value of $B$, say $B_g$ ($g$ for glory), the deflection function vanishes,

(IV.4) $$\theta = \Theta(B_g) \text{ is 0 or } \pi ,$$

implying $\sin\theta = 0$, and making (IV.3) useless.

The classical glory scattering cross-section is infinite because glory scattering is a caustic problem on two accounts.

  i) There is a conservation law: the final momentum $p_b = -p_a$ the initial momentum.

 ii) Near glory, particles with impact parameter $B_g + \delta B$ and $-B_g + \delta B$ exit with approximately the same angles, namely $\pi+$ terms of order $(\delta B)^3$.

The glory cross-section can be computed [10], [7] using the methods presented in sections II and III. The result is

(IV.5) $$d\sigma(\Omega) = 4\pi^2 \, h^{-1} \, |p_a| \, B^2(\theta) \frac{dB(\theta)}{d\theta} \, J_0(2\pi \, h^{-1} \, |p_a| \, B(\theta) \sin\theta)^2 \, d\Omega$$

where $J_0$ is the Bessel function of order 0.

A similar calculation [14], [7], [13] gives the WKB cross-section for polarized glories of massless waves in curved spacetimes

(IV.6) $$d\sigma(\Omega) = 4\pi^2 \, \lambda^{-1} \, B_g^2 \frac{dB}{d\theta} \, J_{2s}(2\pi \, \lambda^{-1} \, B_g \sin\theta)^2 \, d\Omega$$

$s = 0$ for scalar waves; at glory $J_0(0)^2 \neq 0$

$s = 1$ for electromagnetic waves; at glory $J_2(0)^2 = 0$

$s = 2$ for gravitational waves; at glory $J_4(0)^2 = 0$.

$\lambda$ is the wave length of the incoming wave.

Equation (IV.6) matches perfectly with the numerical calculations [13] of R. Matzner based on the partial wave decomposition method.

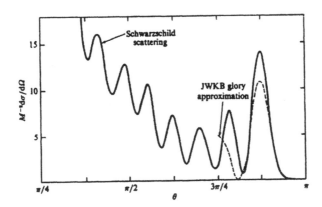

$$\frac{d\sigma}{d\Omega} = 2\pi\omega B_g^2 \left|\frac{dB}{d\theta}\right|_{\theta=\pi} J_{2s}(\omega B_g \sin\theta)^2$$
$$B_g = B(\pi) \qquad \text{glory impact parameter}$$
$$\omega = 2\pi\lambda^{-1}; \ s = 2 \text{ for gravitational wave}$$
analytic cross section: dashed line

numerical cross section: solid line

# References

[1] *Functional Integration; Basics and Applications.* Eds. C. DeWitt-Morette, P. Cartier, and A. Folacci; a NATO-ASI Series B (Physics), Vol. 361 (Plenum Press, New York, 1997).

[2] Functions which are square integrable as well as their first weak derivatives.

[3] The expansion of the action functional for computing the WKB approximation in functional integral was introduced in Cécile Morette, "On the definition and approximation of Ffeynman's path integral," *Physical Review* **81**, 848=852 (1951).

[4] The properties of Jacobi fields in configuration space and phase space, and the corresponding WKB approximations are scattered in several papers of Cécile DeWitt-Morette, John La Chapelle, Maurice Mizrahi, Bruce Nelson, Benny Sheeks, Alice Young, and Tian-Rong Zhang. For a summary of results obtained prior to 1984, see references [6], [7]. For more recent results, see reference [8].

[5] C. DeWitt-Morette, G. Low, L.S. Schulman, A.Y. Shiekh: "Wedges I," *Foundations of Physics* **16**, 311-349 (1986).

[6] Cécile DeWitt-Morette and Tian-Rong Zhang: "Path integrals and conservation laws," *Phys. Rev. D* **28**, 2503-2516 (1983).

[7] Cécile DeWitt-Morette: "Feynman path integrals, from the prodistribution definition to the calculation of glory scattering," *Acta Physica Austriaca Suppl.* **XXVI**, 101-170 (1984).

[8] John La Chapelle: "Functional Integration on Symplectic Manifolds," (Ph.D. Dissertation, The University of Texas at Austin, May 1995).

[9] Pierre Cartier and Cécile DeWitt-Morette: "A new perspective on functional integration," *J. Math. Phys.* **36**, 2237-2312 (1995). For strict WKB approximations, see the calculation leading to (III.55).

[10] Cécile DeWitt-Morette and Bruce L. Nelson: "Glories – and other degenerate points of the action," *Phys. Rev. D* **29**, 1663-1668 (1984).

[11] Cécile DeWitt-Morette: "Catastrophes in Lagrangian systems," and C. DeWitt-Morette and P. Tshumi: "Catastrophes in Lagrangian systems. An example," in *Long Time Prediction in Dynamics* Eds. V. Szebehely and B.D. Tapley (D. Reidel Pub. Co. 1976) pp. 57-69. Also Y. Choquet-Bruhat and C. DeWitt-Morette *Analysis, Manifolds and Physics* Vol. I pp. 105-109, and pp. 277-281 (North Holland 1982-1996).

[12] Cécile DeWitt-Morette, Bruce Nelson, and Tian-Rong Zhang: "Caustic problems in quantum mechanics with applications to scattering theory," *Phys. Rev. D* **28**, 2526-2546 (1983).

[13] R.A. Matzner, C. DeWitt-Morette, B. Nelson, and T.-R. Zhang: "Glory scattering by black holes," *Phys. Rev. D* **31**, 1869-1878 (1985).

[14] Tian-Rong Zhang and Cécile DeWitt-Morette: "WKB cross-section for polarized glories of massless waves in curved space-times," *Phys. Rev. Letters* **52**, 2313-2316 (1984).

# NORM ESTIMATE FOR KAC'S TRANSFER OPERATOR WITH APPLICATIONS TO THE LIE–TROTTER PRODUCT FORMULA

TAKASHI ICHINOSE*

Department of Mathematics, Kanazawa University
Kanazawa, 920–1192, Japan

*Dedicated to Professor Hiroshi Ezawa on the occasion
of his sixtyfifth birthday*

**Abstract.** A norm estimate of the difference between Kac's transfer operator and the Schrödinger semigroup with a power of small $t > 0$ greater than 1 is proved and it is applied to see the splitting of the first two eigenvalues of Kac's transfer operator as well as to establish the Lie–Trotter product formula for Schrödinger operators in *operator norm*.

## 1. INTRODUCTION

Kac's transfer operator we refer to in this note is an operator of the kind $K(t) = e^{-tV/2}e^{t\Delta/2}e^{-tV/2}$ with integral kernel

$$K(x,y;t) = e^{-tV(x)/2}\frac{\exp\left(-\frac{(x-y)^2}{2t}\right)}{(\sqrt{2\pi t})^d}e^{-tV(y)/2}, \qquad (1.1)$$

---

*) Talk at the *2nd Jagna International Workshop 'Mathematical Methods of Quantum Physics'*, January 4–8, 1998.

where $\Delta$ is the Laplacian and $V = V(x)$ a potential bounded below, in $d$–dimensional space $\mathbf{R}^d$.

Motivated by the result of B. Helffer[5,4,3] on a norm estimate for Kac's transfer operator $K(t)$, which we will call the Kac operator in the sequel, compared with the Schrödinger semigroup $e^{-t(-\frac{1}{2}\Delta+V)}$, we have extensively studied the problem in Ichinose–Takanobu[9,10], Doumeki–Ichinose–Tamura[2] and Ichinose–Tamura[13]. The aim of this note is to make a survey about these results together with applications, in particular, to the Lie–Trotter product formula.

The operator $K(t)$ can be regarded as a transfer matrix/operator for some lattice models in statistical mechanics which in his lecture[15] and also with C. J. Thompson[16,19], M. Kac studied to discuss a mathematical mechanism for a phase transition. To sketch how it comes out, consider a lattice model with exponential interaction, the simplest being a one-dimensional chain of $N$ spins $\sigma_i = \pm 1$ with interaction energy

$$E = -\tfrac{1}{2}J \sum_{1 \leq i < j \leq N} e^{-\frac{1}{2}|i-j|}\sigma_i\sigma_j, \tag{1.2}$$

where the interaction is assumed ferromagnetic, i.e. $J > 0$ and $t^{-1}$ is interpreted as the effective number of spins interacting with a given spin. The partition function $Q_N$ is given by

$$Q_N = e^{-\nu N t/4} \sum_{\{\sigma\}} \exp\big(\tfrac{\nu t}{4}\Sigma_{i=1}^N \Sigma_{j=1}^N e^{-\frac{1}{2}|i-j|}\sigma_i\sigma_j\big), \tag{1.3}$$

where $\nu = J/kT$ with $k$ Boltzmann's constant and $T$ the absolute temperature and the sum $\sum_{\{\sigma\}}$ is taken over all the spin configurations on the sites $i$ ($i = 1, 2, \cdots, N$) of the lattice. The transfer operator $\widetilde{K}(t)$ of this model turns out to be an integral operator with integral kernel

$$\widetilde{K}(x, y; t) = e^{t/4}e^{-tV(x)/2}\frac{\exp\big(-\frac{(x-y)^2}{4\sinh(t/2)}\big)}{\sqrt{4\pi\sinh(t/2)}}e^{-tV(y)/2}, \tag{1.4a}$$

where

$$tV(x) = \tfrac{1}{2}(\tanh\tfrac{t}{4})x^2 - \log\cosh(\sqrt{\tfrac{\nu t}{2}}x), \tag{1.4b}$$

so that for small $t > 0$

$$V(x) \sim \tfrac{1}{4}(\tfrac{1}{2} - \nu)x^2 + \tfrac{\nu^2}{48}tx^4 + \cdots.$$

This operator $\tilde{K}(t)$ plays a crucial role, because the partition function $Q_N$ is rewritten as

$$Q_N = e^{-\nu Nt/4} 2^N \iint \frac{1}{\sqrt{2\pi}} e^{-x^2/2} \tilde{K}^{(N-1)}(x,y;t) \frac{1}{\sqrt{2\pi}} e^{-y^2/2} dx\,dy, \quad (1.5)$$

where $\tilde{K}^{(N-1)}(x,y;t)$ is the integral kernel of the $(N-1)$ iterates $\tilde{K}(t)^{N-1}$ of $\tilde{K}(t)$. Moreover, for instance, the pair correlation $\rho(n)$ is given in terms of the eigenvalues $\tilde{\mu}_j(t)$ $(\tilde{\mu}_1(t) > \tilde{\mu}_2(t) \geq \cdots)$ of $\tilde{K}(t)$ and the corresponding eigenfunctions $\varphi_j(x)$ by

$$
\begin{aligned}
\rho(n) &= \lim_{N\to\infty} \langle \sigma_i \sigma_{i+n} \rangle_N \\
&\equiv \lim_{N\to\infty} Q_N^{-1} \sum_{\{\sigma\}} \sigma_i \sigma_{i+n} \exp\left(\tfrac{\nu t}{4} \Sigma_{i=1}^N \Sigma_{j=1}^N e^{-\frac{1}{2}|i-j|}\sigma_i\sigma_j\right) \\
&= \sum_{j=2}^{\infty} \left(\frac{\tilde{\mu}_j(t)}{\tilde{\mu}_1(t)}\right)^n \left(\int_{-\infty}^{\infty} \varphi_j(x)\varphi_1(x) \tanh\left(\sqrt{\tfrac{\nu t}{2}}x\right) dx\right)^2.
\end{aligned}
\quad (1.6)
$$

Therefore to know $\lim_{n\to\infty} \rho(n)$ is positive or zero, it is important to see whether the first eigenvalue $\tilde{\mu}_1(t)$ of $\tilde{K}(t)$ is asymptotically degenerate or not, or equivalently to estimate the quotient $\tilde{\mu}_2(t)/\tilde{\mu}_1(t)$ of the first two eigenvalues of $\tilde{K}(t)$. It is in fact based on this relation between asymptotic eigenvalue degeneracy and long-range order existence that M. Kac and C. J. Thompson[16,19] discussed a mathematical mechanism for a phase transition.

In this stage notice that the integral kernel (1.1) of our Kac operator $K(t)$ with $d = 1$ is nearly equal to the integral kernel (1.4a) of this $\tilde{K}(t)$ for sufficiently small $t > 0$, so that as $t \downarrow 0$, both the operators $K(t)$ and $\tilde{K}(t)$ will have analogous asymptotic behavior. In fact, in order to estimate the quotient $\tilde{\mu}_2(t)/\tilde{\mu}_1(t)$ of the first two eigenvalues of $\tilde{K}(t)$, Kac himself took this approximation for granted and analyzed the eigenvalues $\lambda$ of the Schrödinger equation $(-\frac{1}{2}\Delta + V(x))\varphi(x) = \lambda\varphi(x)$ associated with our Kac operator $K(t)$.

In this note we shall consider both the nonrelativistic and relativistic Schrödinger operators

$$H = H_0 + V \equiv -\tfrac{1}{2}\Delta + V(x), \quad (1.7a)$$

$$H^r = H_0^r + V \equiv \sqrt{-\Delta + 1} - 1 + V(x), \quad (1.7b)$$

with mass 1, and the associated the Kac operators

$$K(t) = e^{-tV/2} e^{-tH_0} e^{-tV/2}, \quad (1.8a)$$

$$K^r(t) = e^{-tV/2} e^{-tH_0^r} e^{-tV/2}, \quad (1.8b)$$

where $V(x)$ is a real-valued continuous function in $\mathbf{R}^d$ bounded below.

It is seen through the Feynman–Kac formula that both $e^{-tH}$ and $e^{-tH^r}, t \geq 0$, define strongly continuous semigroups not only on $L^2(\mathbf{R}^d)$ but also on all $L^p(\mathbf{R}^d)$, $1 \leq p < \infty$, and on the Banach space $C_\infty(\mathbf{R}^d)$ of the continuous functions vanishing at infinity.

In Section 2 we shall describe results on estimates in $L^p$ operator norm for the difference between the Kac operator and the Schrödinger semigroup by a power $O(t^{1+a})$ of small $t > 0$ with $a \geq 0$:

$$\|K(t) - e^{-tH}\|_p, \quad \|K^r(t) - e^{-tH^r}\|_p = O(t^{1+a}), \qquad (1.9)$$

where $\|\cdot\|_p$ stands for the $L^p$ operator norm on $L^p(\mathbf{R}^d)$ for $1 \leq p < \infty$ and on $C_\infty(\mathbf{R}^d)$ for $p = \infty$. Section 3 gives an idea of the proof of (1.9).

Finally in this section, we mention two applications, when $a > 0$. This operator norm estimate, as one application, may be used with Min-max principle to get an estimate of the quotient of the first two eigenvalues $\mu_1(t)$ and $\mu_2(t)$ of the Kac operator compared with the first two eigenvalues $\lambda_1$ and $\lambda_2$ of the Schrödinger operator such as

$$\frac{\mu_2(t)}{\mu_1(t)} = 1 - (\lambda_2 - \lambda_1)t + O(t^{1+a}).$$

As a second application, it can give the Lie–Trotter product formula in $L^p$ operator norm.

In his papers[6,7] B. Helffer has elaborated some of the ideas of M. Kac and also dealt with the Kac operator of the kinds $e^{-V/2}e^{-h^2H_0}e^{-V/2}$ and $e^{-V/2h}e^{-hH_0}e^{-V/2h}$ with semiclassical parameter $h > 0$, by using semiclassical analysis.

## 2. NORM ESTIMATES

First we describe the nonrelativistic result on $K(t)$ in (1.8a). B. Helffer[5,4,3] (also see his treatise Ref. 6) is the first who proved in $L^2$ operator norm

$$\|K(t) - e^{-tH}\|_2 = O(t^2), \qquad (2.1)$$

when $V(x)$ is a $C^\infty$-function bounded below by a constant $b$ and satisfying $|\partial^\alpha V(x)| \leq C_\alpha (1 + x^2)^{(2-|\alpha|)+/2}$ for every multi-index $\alpha$ with constant $C_\alpha$. To do so he used pseudo-differential operator calculus. Dia–Schatzman[1] gave another operator-theoretical proof.

Then in Ref. 9 we have extended this result to the case of more general potentials $V(x)$ by a probabilistic method with the Feynman–Kac formula and even in $L^p$ operator norm as in the following theorem. An alternative proof of this result is given with an operator-theoretical method in Ref. 2.

**THEOREM 2.1.** *(Nonrelativistic case) Let $0 < \delta \leq 1$ and $m$ a nonnegative integer such that $m\delta \leq 1$. Suppose that $V(x)$ is a $C^m$-function in $\mathbf{R}^d$ bounded below by a constant $b$ satisfying that*

$$|\partial^\alpha V(x)| \leq C(V(x) - b + 1)^{1-|\alpha|\delta}, \quad 0 \leq |\alpha| \leq m, \qquad (2.2a)$$

*and further that $\partial^\alpha V(x), |\alpha| = m$, are Hölder-continuous:*

$$|\partial^\alpha V(x) - \partial^\alpha V(y)| \leq C|x - y|^\kappa, \quad x, y \in \mathbf{R}^d, \qquad (2.2b)$$

*with constants $C > 0$ and $0 \leq \kappa \leq 1$ (By $\kappa = 0$ we understand $\partial^\alpha V(x), |\alpha| = m$, bounded). Then it holds that, as $t \downarrow 0$,*

$$\|K(t) - e^{-tH}\|_p = \begin{cases} O(t^{1+\kappa/2}), & m = 0, \\ O(t^{1+2\delta \wedge \frac{1+\kappa}{2}}), & m = 1, \\ O(t^{1+2\delta}), & m \geq 2. \end{cases} \qquad (2.3)$$

Here note that condition (2.2b) with $\kappa = 1$ is equivalent to that $\partial^\alpha V(x)$, $|\alpha| = m + 1$, are essentially bounded.

From Theorem 2.1 we can prove the following the Lie–Trotter product formula in $L^p$ operator norm. We refer to Ref. 18 for more general potentials.

**THEOREM 2.2.** *(Nonrelativistic case) For the same function $V(x)$ as in Theorem 2.1, it holds uniformly on each finite $t$-interval in $[0, \infty)$, as $n \to \infty$, that*

$$\|(e^{-tV/2n}e^{-tH_0/n}e^{-tV/2n})^n - e^{-tH}\|_p, \quad \|(e^{-tV/n}e^{-tH_0/n})^n - e^{-tH}\|_p$$

$$= \begin{cases} O(n^{-\kappa/2}), & m = 0, 0 < \kappa \leq 1, \\ O(n^{-2\delta \wedge \frac{1+\kappa}{2}}), & m = 1, 0 \leq \kappa \leq 1, \\ O(n^{-2\delta}), & m \geq 2. \end{cases}$$

$$(2.4)$$

*Proof.* We may assume that $V(x) \geq 0$, so that both $K(t)$ and $e^{-tH}$ are contractions.

The symmetric product case, i.e. for $K(\frac{t}{n})^n$, is an immediate consequence of Theorem 2.1 (2.3). In fact, once we have an estimate like (1.9), we have by telescoping

$$\|K(\tfrac{t}{n})^n - e^{-tH}\|_p = \|\sum_{j=1}^{n} K(\tfrac{t}{n})^{n-j}(K(\tfrac{t}{n}) - e^{-tH/n})e^{-(j-1)tH/n}\|_p$$

$$\leq \sum_{j=1}^{n} \|K(\tfrac{t}{n}) - e^{-tH/n}\|_p \leq nO(\tfrac{t}{n})^{1+a}) = n^{-a}O(t^{1+a}).$$

In the non-symmetric product case, we have

$$(e^{-tV/n}e^{-tH_0/n})^n - e^{-tH} = \big((e^{-tV/n}e^{-tH_0/n})^n - K(\tfrac{t}{n})^n\big) + \big(K(\tfrac{t}{n})^n - e^{-tH}\big).$$

The second term is nothing but the symmetric product case. The first term is rewritten as

$$(e^{-tV/n}e^{-tH_0/n})^n - K(\tfrac{t}{n})^n$$
$$= [(e^{-tV/2n}, K(\tfrac{t}{n})^{n-1}]e^{-tV/2n}e^{-tH_0/n} + K(\tfrac{t}{n})^{n-1}[e^{-tV/2n}, e^{-tH_0/n}],$$

where $[A, B] = AB - BA$ stands for the commutator for operators $A$ and $B$. So estimating these two commutators, we have the desired estimate.

*Examples.* The following functions satisfy condition (2.2ab) for $V(x)$:

(i) $|x|^2$ (harmonic oscillator potential) with $(\delta, m, \kappa) = (\tfrac{1}{2}, 1, 1)$ or $(\tfrac{1}{2}, 2, 0)$.

(ii) $|x|^4 - |x|^2$ (double well potential) with $(\delta, m, \kappa) = (\tfrac{1}{4}, 3, 1)$ or $(\tfrac{1}{4}, 4, 0)$.

(iii) $|x|^{1998}$ with $(\delta, m, \kappa) = (\tfrac{1}{1998}, 1997, 1)$ or $(\tfrac{1}{1998}, 1998, 0)$.

(iv) More generally, $|x|^\rho - c|x|^\sigma$, where $c \geq 0$ and $0 \leq \sigma < \rho$, with $(\delta, m, \kappa) = (1, 0, \rho)$ for $0 < \rho \leq 1$ and $(\delta, m, \kappa) = (1/\rho, [\rho], \rho - [\rho])$ for $\rho > 1$, $[\rho]$ being the maximal integer that is not greater than $\rho$.

However, for instance, $\exp(|x|^2 + 1)^a$, $a > 0$, and $\exp|x|^2$ do not satisfy the condition.

*Remark 1.* Helffer's result[5] (2.1) is included in Theorem 2.1 with $p = 2$ and $(\delta, m, \kappa) = (\tfrac{1}{2}, 1, 1)$ or $(\delta, m, \kappa) = (\tfrac{1}{2}, 2, 0)$. In fact, as his condition on $V(x)$ implies that for every multi-index $\alpha$

$$|\partial^\alpha V(x)| \leq C(V(x) - b + 1)^{(1-|\alpha|/2)_+}. \tag{2.5}$$

*Remark 2.* Theorems 2.1 and 2.2 are valid with the operator $H_0$ replaced by the magnetic Schrödinger operator $H_0(A) = \tfrac{1}{2}(-i\partial - A(x))^2$ with vector potential $A(x)$ including the case of constant magnetic fields (See Ref. 9, 2).

*Remark 3.* As concerns the Lie–Trotter product formula in operator norm, Rogava[17] proved for nonnegative selfadjoint operators $A$ and $B$ in a Hilbert space that, if the domain $D[A]$ of $A$ is a subset of the domain $D[B]$ of $B$ and $C = A + B$ is selfadjoint on $D[C] = D[A]$, then, as $n \to \infty$,

$$\|(e^{-tA/2n}e^{-tB/n}e^{-tA/2n})^n - e^{-tC}\| = \|(e^{-tB/n}e^{-tA/n})^n - e^{-tC}\|$$
$$= O(n^{-1/2}\ln n).$$

In this case, $B$ is $A$–bounded. Notice that in our Theorems 2.1 and 2.2, neither $V$ is $H_0$–bounded nor $H_0$ is $V$–bounded.

We refer to Ref. 12, 14 for some results complementary to Rogava's with an extension to the time-dependent operators and to Ref. 13 for the Lie–Trotter product formula in trace norm, both of which were proved by operator-theoretic methods.

Next we come to the relativistic result on $K^r(t)$ in (1.8b). We have obtained in Ref. 10 the following result. We refer to Ref. 11 for more general operators $H_0^r$ associated with the Lévy process and for slightly more general potentials $V(x)$.

**THEOREM 2.3.** *(Relativistic case) Let $V(x)$ be the same function as in Theorem 2.1. Then it holds that, as $t \downarrow 0$,*

$$\|K^r(t) - e^{-tH^r}\|_p = \begin{cases} O(t^{1+\kappa}), & m = 0, 0 \le \kappa < 1, \\ O(t^2|\ln t|), & (m,\kappa) = (0,1), \\ O(t(t^{2\delta} \vee t|\ln t|)), & (m,\kappa) = (1,0), \\ O(t^{1+2\delta\wedge 1}), & m = 1, 0 < \kappa \le 1, \\ O(t^{1+2\delta}), & m \ge 2. \end{cases} \qquad (2.6)$$

With Theorem 2.3 we can prove, similary to Theorem 2.2, the Lie-Trotter product formula in $L^p$ operator norm.

**THEOREM 2.4.** *(Relativistic case) For the same function $V(x)$ as in Theorem 2.1, it holds uniformly on each finite $t$–interval in $[0, \infty)$, as $n \to \infty$, that*

$$\|(e^{-tV/2n}e^{-tH_0^r/n}e^{-tV/2n})^n - e^{-tH^r}\|_p, \quad \|(e^{-tV/n}e^{-tH_0^r/n})^n - e^{-tH^r}\|_p$$

$$= \begin{cases} O(n^{-\kappa}), & m = 0, 0 < \kappa < 1, \\ O(n^{-1}\ln n), & (m,\kappa) = (0,1), \\ O(n^{-2\delta}) \vee O(n^{-1}\ln n), & (m,\kappa) = (1,0), \\ O(n^{-2\delta\wedge 1}), & m = 1, 0 < \kappa \le 1, \\ O(n^{-2\delta}), & m \ge 2. \end{cases}$$
$$(2.7)$$

Part of these results has been briefly announced in Ref. 8.

## 3. IDEA OF PROOF OF THEOREM 2.1

The proof will be done in two ways; a) one[9,10] is probabilistic in $L^p$ with the Feynman–Kac formula and b) the other[2] $L^2$ operator-theoretical. We only sketch it for Theorem 2.1.

a) By the Feynman–Kac formula we have for $f \in C_0^\infty(\mathbf{R}^d)$, $\|f\|_p = 1$

$$([K(t) - e^{-tH}]f)(x)$$

$$= \mathbf{E}_x\left[\left(e^{-\frac{t}{2}(V(x)+V(X(t)))} - e^{-\int_0^t V(X(s))ds}\right)f(X(t))\right]$$

$$= \int f(y)p(t, x - y)\mathbf{E}_x[v(t, x, y)|X(t) = y]dy$$

with $v(t, x, y) = e^{-\frac{t}{2}(V(x)+V(X(t)))} - e^{-\int_0^t V(X(s))ds}$, where $p(t, x - y)$ is the integral kernel of $e^{-tH_0}$ and $\mathbf{E}_x$ is the expectation with respect to the Wiener measure on the space of the continuous paths $X : [0, \infty) \to \mathbf{R}^d$ starting at $X(0) = x$. Then we have by Taylor's theorem

$$v(t, x, y)$$

$$= -w(t, x, y)e^{-\frac{t}{2}(V(x)+V(y))} - \sum_{j=2}^m \frac{1}{j!}w(t, x, y)^j e^{-\frac{t}{2}(V(x)+V(y))}$$

$$- \frac{1}{m!}w(t, x, y)^{m+1}\int_0^1 d\theta(1 - \theta)^m e^{-(1-\theta)\frac{t}{2}(V(x)+V(y))-\theta\int_0^t V(X(s))ds}$$

$$\equiv \sum_{i=1}^3 v_i(t, x, y),$$

with

$$w(t, x, y) = \frac{t}{2}\Big(V(x) + V(y)\Big) - \int_0^t V(X(s))ds.$$

The main task is to estimate the conditional expectations of the $v_i$

$$\mathbf{E}_x[v_i(t, x, y)|\ X(t) = y], \ i = 1, 2, 3,$$

by a sum of powers of $|x-y|$ and $t$.

b) We can see $K(t)$ satisfies

$$K'(t) = -HK(t) + R(t), \ R(t) = R_1(t) + R_2(t),$$

where

$$R_1(t) = [H_0, e^{-tV/2}]e^{-tH_0}e^{-tV/2}, \quad R_2(t) = e^{-tV/2}[V/2, e^{-tH_0}]e^{-tV/2}.$$

We have

$$K(t) - e^{-tH} = \int_0^t e^{-(t-s)H} R(s) ds.$$

Then the key is to prove an estimate of the kind:

$$\|R(s)\| = O(s^a), \quad 0 < s < t << 1, \quad a > 0.$$

Hence, on integrating it, we can get

$$\|K(t) - e^{-tH}\|_2 = O(t^{1+a}).$$

## ACKNOWLEDGEMENT

The author should like to thank the organizers for their kind invitation to this workshop in Jagna and in particular, Professors V. and Ch. Bernido for their warm hospitality.

## REFERENCES

1. Boun O. Dia and Michelle Schatzman, An estimate on the Kac transfer operator, *J. Functional Analysis*, **145**, 108–135 (1997).

2. Atsushi Doumeki, Takashi Ichinose and Hideo Tamura, Error bound on exponential product formulas for Schrödinger operators, to appear in *J. Math. Soc. Japan*, **50**, 1998.

3. Bernard Helffer, Spectral properties of the Kac operator in large dimension, *Centre de Recherches Math., CRM Proceedings and Lecture Notes*, **8**, 179–211 (1995).

4. Bernard Helffer, Correlation decay and gap of the transfer operator, *Algebra i Analiz (St. Petersburg Math. J.)*, **8**, 192–210 (1996).

5. Bernard Helffer, Around the transfer operator and the Trotter–Kato formula, *Operator Theory: Advances and Appl.*, **78**, 161–174 (1995).

6. Bernard Helffer, Recent results and open problems on Schrödinger operators, Laplace integrals, and transfer operators in large dimension, *Schrödinger Operators, Markov Semigroups, Wavelet Analysis, Operator Algebras, Math. topics*, **11**, edited by M. Demuth, E. Schrohe, B.-W. Schulze and J. Sjöstrand, pp. 11–162, Berlin: Akademie Verlag 1996.

7. Bernard Helffer, Semi-classical analysis for the transfer operator: Formal WKB constructions in large dimension, *Commun. Math. Phys.*, **187**, 81–113 (1997).

8. Takashi Ichinose, Norm convergence of the Trotter product formula for Schrödinger operators via the Feynman–Kac formula, *Proc. of Dubna Joint Meeting of International Seminar "Path Integrals: Theory & Applications" and 5th Internatioal Conference "Path Integrals from meV to MeV"*, edited by V. S. Yarunin and M. K. Smondyrev, pp. 341–346, Dubna 1996.

9. Takashi Ichinose and Satoshi Takanobu, Estimate of the difference between the Kac operator and the Schrödinger semigroup, *Commun. Math. Phys.*, **186**, 167–197 (1997).

10. Takashi Ichinose and Satoshi Takanobu, The norm estimate of the difference between the Kac operator and the Schrödinger semigroup: A unified approach to the nonrelativistic and relativistic cases, to appear in *Nagoya Math. J.*, **149**, 1998.

11. Takashi Ichinose and Satoshi Takanobu, The norm estimate of the difference between the Kac operator and the Schrödinger semigroup II: Supplement to the relativistic case, *Preprint* 1997.

12. Takashi Ichinose and Hideo Tamura, Error estimates in operator norm for Trotter–Kato product formula, *Integr. Equat. Oper. Theory*, **27**, 195–207 (1997).

13. Takashi Ichinose and Hideo Tamura, Error bound in trace norm for Trotter–Kato product formula of Gibbs semigroups, to appear in *Asymptotic Analysis*.

14. Takashi Ichinose and Hideo Tamura, Error estimates in operator norm of exponential product formulas for propagators of parabolic evolution equations, to appear in *Osaka J. Math.*

15. Mark Kac, Mathematical mechanism of phase transitions, 1966 *Brandeis Lecture*, pp. 242–305, New York: Gordon and Breach 1968.

16. Mark Kac and Colin J. Thompson, On the mathematical mechanism of phase transition, *Proc. N. A. S.*, **55**, 676–683 (1966); Erratum, *ibid.* **56**, 1625 (1966).

17. Dzh. L. Rogava, Error bounds for Trotter–type formulas for self-adjoint operators, *Functional Analysis and Its Applications*, **27**, 217–219 (1993).

18. Satoshi Takanobu, On the error estimate of the integral kernel for the Trotter product formula for Schrödinger operators, to appear in *Ann. Probab.*

19. Colin J. Thompson and Mark Kac, Phase transition and eigenvalue degeneracy of a one dimensional anharmonic oscillator, *Studies in Appl. Math.*, **48**, 257–264 (1969).

# NONSTANDARD ANALYTICAL CONSTRUCTION OF A PATH SPACE MEASURE FOR THE 4-D DIRAC EQUATION

TORU NAKAMURA

Department of Mathematics, Sundai Preparatory School,
Kanda-Surugadai, Chiyoda-ku, Tokyo 101-8313, Japan

KEIJI WATANABE

Department of Physics, Meisei University,
Hino, Tokyo 191-8506, Japan

HIROSHI EZAWA

Department of Physics, Gakushuin University,
Mejiro, Toshima-ku, Tokyo 171-8588, Japan

Abstract  Nonstandard analysis is used to construct a measure over paths in momentum space for the path integral solution to the 4-D Dirac equation. In this approach, it is sufficient simply to assign a 4 × 4 matrix to each path in a discretized momentum space over discretized time. The standard part of our measure is the same as the measure given by B.Gaveau[1].

Key Words   nonstandard analysis, Dirac equation, path integral, path-space measure.

## 1. INTRODUCTION

The Green function for an initial data problem in quantum mechanics can be obtained as a repeated integral of a product of Green functions for time slices $\Delta t$. The limit $\Delta t \to 0$ of the repeated integral is often called the path integral, but actually it is not an integral over the particle paths. In order to have a genuine integral, we need a measure over the infinite dimensional space of paths. However, it was proved

155

by R.H.Cameron[2] in 1960 that such a measure does not exist for the case of the Schrödinger equation.

To our surprise, existence of a measure was established by T.Ichinose[3] in 1982 for the case of the Dirac equation in 1+1-dimensions. Then in 1985, Ph.Blanchard et al.[4] constructed the measure. In 1991-1997, one of the present authors (T.N.)[5,6] pointed out that the nonstandard analysis provides another construction more concrete and visible; he studied also the nonstandard measure for the Schrödinger case as the limit of the light velocity $\rightarrow \infty$ of the measure for the Dirac case.

Ichinose's discovery could not be extended to the 3+1-dimensions, for which Zastawniak[7] proved in 1989 that the path space measure does not exist, though preceding him by five years, B.Gaveau[1] recognized that a measure could be found for paths if one turned to the momentum space. The present work establishes that the measure is obtained in a transparent way if one uses the Loeb theory[8] of the nonstandard analysis.

In Sec.2, we shall review the construction by T.N.[5,6] of a path space measure for a Dirac particle moving in the 1+1-dimensional space-time with a given vector potential. The paths are treated as those of a massless Dirac particle in a product space of the coordinate- and the helicity-spaces which are perturbed by the mass and the potential. Then, the path integral is obtained as the standard part of a nonstandard analytic sum over sample paths of a Poisson process in which the helicity is flipped by the mass.

This picture is found helpful in Sec.4 where a free Dirac particle in 3+1-dimensions is treated as a particular case of the path integral generally formulated in Sec.3. In this main part of the present paper, a path space measure will be constructed in a transparent way for the paths in a product space of the momentum- and the helicity-spaces, in which the particle is subjected to a compound Poisson processes with imaginary parameters as given by the potential and the mass. The standard part of the measure thus obtained can be identified with the Dirac case of the measures given by Gaveau[1].

## 2. PATH SPACE MEASURE FOR THE DIRAC EQUATION IN 1+1-DIMENSIONS

Let us first survey the results for the 1+1-dimensional Dirac equation. The results in this section were obtained by the first author partly in 1991[5] and finally in 1997[6].

We begin with a heuristic argument for orientation, and then give a rigorous formulation after equation (7) below in terms of nonstandard analysis.

The Green function for initial value problem can be expressed as a repeated integral:

$$
\begin{aligned}
G(t,x\,;0,y) &= \int \cdots \int \langle x|U(\Delta t)|x_{n-1}\rangle \langle x_{n-1}|U(\Delta t)|x_{n-2}\rangle \cdots \langle x_1|U(\Delta t)|y\rangle \prod_{j=1}^{n-1} dx_j \\
&= \int \cdots \int \langle x|U(\Delta t)|p_{n-1}\rangle \langle p_{n-1}|x_{n-1}\rangle \langle x_{n-1}|U(\Delta t)|p_{n-2}\rangle \cdots
\end{aligned}
$$

$$\cdots \langle x_1 | U(\Delta t) | p_0 \rangle \langle p_0 | y \rangle \prod_{j=1}^{n-1} dx_j \prod_{j=0}^{n-1} dp_j, \qquad (1)$$

where $t/n = \Delta t$. When the time interval $\Delta t$ is short, $\langle x_{k+1} | U(\Delta t) | p_k \rangle$ in (1) can be approximated by

$$\exp\left[-\frac{i\Delta t}{\hbar} mc^2 \beta\right] \exp\left[-\frac{i\Delta t}{\hbar} cp_k \alpha\right] \exp\left[\frac{ie\Delta t}{\hbar} A(k\Delta t, x_k)\right] \frac{1}{\sqrt{2\pi\hbar}} \exp\left[-\frac{i}{\hbar} p_k x_{k+1}\right]$$

where $\alpha$ and $\beta$ are Dirac matrices, $c$ is the light velocity and

$$A(t, x_k) = -A_0(t, x_k) + A_1(t, x_k)\alpha. \qquad (2)$$

Making use of the projections $P_{\pm 1}$ to the eigenspaces of the eigenvalues $\pm 1$ of $\alpha$, we can carry out the $p$-integrations, to obtain

$$G(t, x; 0, y) \simeq \int \cdots \int \prod_{k=0}^{n-1}{}' \Bigg\{ \Big( \delta(x_{k+1} - x_k - c\Delta t)P_1 + \delta(x_{k+1} - x_k + c\Delta t)P_{-1} \Big)$$

$$\times \exp\left[\frac{ie\Delta t}{\hbar} A(k\Delta t, x_k)\right] \left(1 - \frac{i\Delta t}{\hbar} mc^2 \beta\right) \Bigg\} \prod_{k=1}^{n-1} dx_k \qquad (3)$$

with $x_0 = y$ and $x_n = x$, where $\prod'$ denotes the time-ordered product. In order to expand the product $\prod_{k=0}^{n-1}{}' (\cdots\cdots)$, we introduce the functions $\lambda$ and $\omega$ defined on the set $\{0, \cdots, n-1\}$ with values in $\{-1, 1\}$, the former standing for the choice

$$M_{\lambda(k)} = \begin{cases} -imc^2 \beta\Delta t/\hbar & \text{if } \lambda(k) = -1 \\ 1 & \text{if } \lambda(k) = 1, \end{cases}$$

and the latter for

$$Q_{\omega(k)} = \begin{cases} \delta(x_{k+1} - x_k + c\Delta t)P_{-1} & \text{if } \omega(k) = -1 \\ \delta(x_{k+1} - x_k - c\Delta t)P_1 & \text{if } \omega(k) = 1. \end{cases}$$

The Green function is expressed as a double sum over $\lambda$ and $\omega$.

$$G(t, x; 0, y) \simeq \int \cdots \int \sum_{\lambda, \omega} \left(\prod_{k=0}^{n-1}{}' M_{\lambda(k)} Q_{\omega(k)} \exp\left[\frac{ie\Delta t}{\hbar} A(k\Delta t, x_k)\right]\right) \prod_{k=1}^{n-1} dx_k. \qquad (4)$$

Because of the property of the projections

$$P_1 P_{-1} = P_{-1} P_1 = P_1 \beta P_1 = P_{-1} \beta P_{-1} = 0,$$

all the products of $M$'s and $Q$'s vanish except for the $\lambda$ uniquely determined for each $\omega$. For example, let $n = 4$ and consider the case where terms in (4) are such that

$$\omega(0) = 1, \quad \omega(1) = -1, \quad \omega(2) = -1, \quad \omega(3) = 1,$$

then the product is

$$M_{\lambda(3)}P_1 M_{\lambda(2)}P_{-1}M_{\lambda(1)}P_{-1}M_{\lambda(0)}P_1,$$

where we have omitted the $\delta$-functions and $A_k$ for simplicity. Then, the function $\lambda$ has to be such that

$$\lambda(0) = -1, \quad \lambda(1) = 1, \quad \lambda(2) = -1$$

in order for the product to be non-vanishing. We are still left with two alternatives $\lambda(3) = \pm 1$ over which we have to sum, but we can take only one $\lambda(3) = 1$ because $M_\lambda$ for $\lambda(3) = -1$ carries the factor $\Delta t$.

Thus, in effect, we can write $M_\omega$ for $M_\lambda$ such that

$$M_{\omega(k)} = \begin{cases} -imc^2\beta\Delta t/\hbar & \text{if } \omega(k) \neq \omega(k+1) \\ 1 & \text{if } \omega(k) = \omega(k+1). \end{cases} \tag{5}$$

Then, the double sum in (4) reduces to a single sum.

$$G(t,x;0,y) \simeq \int \cdots \int \sum_\omega \left( \prod_{k=0}^{n-1}{}' M_{\omega(k)} Q_{\omega(k)} \exp\left[ \frac{ie\Delta t}{\hbar} A(k\Delta t, x_k) \right] \right) \prod_{k=1}^{n-1} dx_k. \tag{6}$$

We can carry out the $x_k$-integrations in (6) with the help of the $\delta$-functions in $Q_{\omega(k)}$'s, to obtain

$$G(t,x;0,y) \simeq \sum_{X_\omega} \left( \prod_{k=0}^{n-1}{}' M_{\omega(k)} P_{\omega(k)} \exp\left[ \frac{ie\Delta t}{\hbar} A(k\Delta t, X_\omega(k)) \right] \right) \tag{7}$$

where $X_\omega(k) = y + \sum_{l=0}^{k-1} c\Delta t\, \omega(l)$. Notice that the sum over $\omega$ in (6) has been replaced in (7) by the sum over zig-zag paths $X_\omega$ such that $X_\omega(n) = x$, each path carrying the term $\prod_{k=0}^{n-1}{}' M_{\omega(k)} P_{\omega(k)} \exp\left[ \frac{ie\Delta t}{\hbar} A(k\Delta t, X_\omega(k)) \right]$.

The heuristic argument has thus led to the form of the Green function (7) to which we shall now give a rigorous meaning by the nonstandard analysis. For this purpose, we discretize the time and the space with infinitesimal spacing $\varepsilon$ and $c\varepsilon$, respectively where $\varepsilon$ is an infinitesimal number in nonstandard analysis. Denote $k\varepsilon \in \varepsilon^*\mathbb{N}$ and $lc\varepsilon \in c\varepsilon^*\mathbb{Z}$ closest to $t$ and $x$ by $\underline{t}$ and $\underline{x}$, respectively. For each $\omega$, let us define a path $X_\omega$ in the discretized space-time by

$$X_\omega(k) = X_\omega(0) + \sum_{l=0}^{k-1} c\varepsilon\, \omega(l) \quad (k = 1, \cdots, n_t).$$

Here, $\omega$ is to be regarded as an internal function in the sense of nonstandard analysis, defined on the set $\{0, \cdots, n_t - 1\}$ with values in $\{-1, 1\}$, where $n_t\varepsilon = \underline{t}$.

Let us also define a measure $\mu(X_\omega)$ for each path $X_\omega$ by

$$\mu(X_\omega) = \prod_{k=0}^{n_t-1} {}' M_{\omega(k)} P_{\omega(k)} \exp\left[\frac{i e \varepsilon}{\hbar} A(k\varepsilon, X_\omega(k))\right]. \tag{8}$$

In the free case where $A_0 = A_1 = 0$, $\mu(X_\omega)$ reduces to

$$\mu_0(X_\omega) = \prod_{k=0}^{n_t-1} {}' M_{\omega(k)} P_{\omega(k)},$$

which, despite of its appearance as a chronological product of matrices, is in fact determined simply by $\omega(0)$ and $\omega(n_t - 1)$, or in other words by $X_\omega(1) - X_\omega(0)$ and $X_\omega(n_t) - X_\omega(n_t - 1)$, the initial and final steps being forward or backward. In $\mu(X_\omega)$, on the other hand, $A(k\varepsilon, X_\omega(k))$'s can be factored out to be multiplied by $\mu_0(X_\omega)$. Namely:

**Proposition 1** *Let $r$ denote the number of $k$'s for which $\omega(k) \neq \omega(k-1)$.*

(1) *If $\omega(0) = \omega(n_t - 1)$, then $\mu_0(X_\omega) = \left(-\dfrac{i\varepsilon}{\hbar} mc^2\right)^r P_{\omega(n_t-1)}$.*

(2) *If $\omega(0) = -\omega(n_t - 1)$, then $\mu_0(X_\omega) = \left(-\dfrac{i\varepsilon}{\hbar} mc^2\right)^r P_{\omega(n_t-1)}\beta$.*

(3) *The effects of the potentials are factorized if $O(\varepsilon^{n_t+1})$ is disregarded:*

$$\begin{aligned}
\mu(X_\omega) &= \exp\left[\int_0^t \frac{ie}{c\hbar} A_1(s, X_\omega(s)) dX_\omega(s) - \int_0^t \frac{ie}{c\hbar} A_0(s, X_\omega(s)) ds\right] \mu_0(X_\omega) \\
&\quad + O(\varepsilon^{n_t+1}).
\end{aligned}$$

We have the following theorems.

**Theorem 1** *Assume $A_0(s,y)$, $A_1(s,y)$ to be $C^2(\mathbb{R}^2)$-functions and $\psi(0,y)$ to be a $C^2(\mathbb{R})$-function. Then, the nonstandard path integral*

$$\sum_{X_\omega \in \mathcal{P}_{\underline{t},\underline{x}}} \mu(X_\omega) \, {}^*\psi(0, X_\omega(0)) \quad \text{with} \quad \mathcal{P}_{\underline{t},\underline{x}} = \{\, X_\omega : X_\omega(\underline{t}) = \underline{x} \,\} \tag{9}$$

*with the measure $\mu$ over the path space $\mathcal{P}_{\underline{t},\underline{x}}$ has the standard part that satisfies the $1 + 1$-dimensional Dirac equation with potentials $A_0$ and $A_1$ for given initial data $\psi(0,y)$.*

The finitely additive nonstandard measure space $\{\, \mathcal{P}_{\underline{t},\underline{x}}, \, \mu_0 \,\}$ has a property of the Poisson process[6] as represented by the factor $(-imc^2\varepsilon/\hbar)^r$. Consequently it is of bounded variation permitting us to take its standard part by the Loeb measure theory [8].

**Theorem 2** *A standard path space $\mathsf{P}_{t,x}$ and a completely additive standard measure $m_L$ over $\mathsf{P}_{t,x}$ are defined as the standard parts of the nonstandard $\mathcal{P}_{\underline{t},\underline{x}}$ and $\mu_0$, respectively. The standard path integral with respect to the former is the same as the standard part of the path integral with respect to the latter.*

## 3. PATH SPACE MEASURE FOR THE DIRAC EQUATION IN 3+1-DIMENSIONS

In the 3+1-dimensional case, (1) turns out heuristically to be

$$
G(t,\vec{x};0,\vec{y}) \simeq \int \cdots \int \prod_{k=0}^{n-1}{}' \left\{ \exp\left[ -\frac{ie\Delta t}{\hbar}\Big(A_0(k\Delta t,\vec{x}_k) - \vec{A}(k\Delta t,\vec{x}_k)\cdot\vec{\alpha}\Big) \right] \right.
$$

$$
\left. \times \frac{1}{(2\pi\hbar)^3}\exp\left[ -\frac{i\Delta t}{\hbar}mc^2\beta \right]\exp\left[ \frac{i}{\hbar}\vec{p}_k\cdot(\vec{x}_{k+1}-\vec{x}_k - c\Delta t\vec{\alpha}) \right] \right\} \prod_{k=1}^{n-1} d^3x_k \prod_{k=0}^{n-1} d^3p_k . \quad (10)
$$

If $\vec{p}_k$-integrations are carried out, then the result contains not only the $\delta$-functions as in (3) for the 1+1-dimensional case but also their derivatives, as illustrated by a simple example where $m = A_0 = \vec{A} = 0$,

$$
G(t,\vec{x};0,\vec{y}) \simeq \int\cdots\int \prod_{k=1}^{n-1} d^3x_k \prod_{k=0}^{n-1}{}' \left\{ \frac{-\vec{\alpha}\cdot(\vec{x}_k - \vec{x}_{k+1})}{4\pi|\vec{x}_k - \vec{x}_{k+1}|^3}\delta(|\vec{x}_{k+1}-\vec{x}_k| - c\Delta t) \right.
$$

$$
\left. +\frac{1}{2\pi|\vec{x}_k - \vec{x}_{k+1}|}\Big(-\frac{1}{2} + \frac{\vec{\alpha}\cdot(\vec{x}_k - \vec{x}_{k+1})}{2|\vec{x}_k - \vec{x}_{k+1}|}\Big)\delta'(|\vec{x}_{k+1}-\vec{x}_k| - c\Delta t) \right\}.
$$

The derivatives of the delta-functions have to be interpreted somehow by a difference of delta-functions on the discretized space-time, and therefore keep us from assigning a measure to each path separately. Thus, we choose to carry out the $\vec{x}_k$-integrations in (10) rather than the $\vec{p}_k$-integration we did in the 1+1-dimensional case.

$$
G(t,\vec{p};0,\vec{q}) = \frac{1}{(2\pi\hbar)^3}\int \exp\left[ -\frac{i}{\hbar}\vec{p}\cdot\vec{x} \right]\exp\left[ \frac{i}{\hbar}\vec{q}\cdot\vec{y} \right]G(t,\vec{x};0,\vec{y})d^3x\, d^3y
$$

$$
\simeq \int\cdots\int \langle\vec{p}|U_I(\Delta t)|\vec{k}_{n-1}\rangle\langle\vec{k}_{n-1}|U_0(\Delta t)|\vec{p}_{n-1}\rangle\langle\vec{p}_{n-1}|U_I(\Delta t)|\vec{k}_{n-2}\rangle
$$

$$
\times \cdots \times \langle\vec{p}_1|U_I(\Delta t)|\vec{k}_0\rangle\langle\vec{k}_0|U_0(\Delta t)|\vec{q}\rangle \prod_{j=1}^{n-1} d^3p_j \prod_{j=0}^{n-1} d^3k_j, \quad (11)
$$

where

$$
\langle\vec{k}_j|U_0(\Delta t)|\vec{p}_j\rangle \simeq \Big(1 - \frac{i\Delta t}{\hbar}mc^2\beta\Big)\exp\left[ -\frac{ic\Delta t}{\hbar}\vec{k}_j\cdot\vec{\alpha} \right]\delta(\vec{k}_j - \vec{p}_j), \quad (12)
$$

$$
\langle\vec{p}_j|U_I(\Delta t)|\vec{k}_{j-1}\rangle \simeq \frac{1}{(2\pi\hbar)^3}\int \exp\left[ -\frac{i}{\hbar}(\vec{p}_j - \vec{k}_{j-1})\cdot\vec{x} \right]
$$

$$
\times \left\{ 1 - \frac{ie\Delta t}{\hbar}\Big(A_0(j\Delta t,\vec{x}) - \vec{A}(j\Delta t,\vec{x})\cdot\vec{\alpha}\Big) \right\} d^3x
$$

$$
= \delta(\vec{p}_j - \vec{k}_{j-1}) + \frac{ie\Delta t}{\hbar}\widehat{\vec{A}}(j\Delta t,\vec{p}_j - \vec{k}_{j-1}), \quad (13)
$$

and

$$
\widehat{\vec{A}}(t,\vec{k}) = \frac{1}{(2\pi\hbar)^3}\int \exp\left[ -\frac{i}{\hbar}\vec{k}\cdot\vec{x} \right]\Big(-A_0(t,\vec{x}) + \vec{A}(t,\vec{x})\cdot\vec{\alpha}\Big)d^3x. \quad (14)
$$

Putting (12) and (13) into (11), we obtain

$$G(t, \vec{p}; 0, \vec{q}) \simeq \int \cdots \int \prod_{k=1}^{n}{}' \left\{ \left( \delta(\vec{p}_k - \vec{p}_{k-1}) + \frac{ie\Delta t}{\hbar} \widehat{A}(k\Delta t, \vec{p}_k - \vec{p}_{k-1}) \right) \right.$$

$$\left. \times \left( 1 - \frac{i\Delta t}{\hbar} mc^2 \beta \right) \exp\left[ -\frac{ic\Delta t}{\hbar} \vec{p}_{k-1} \cdot \vec{\alpha} \right] \right\} \prod_{k=1}^{n-1} d^3 p_k \quad (15)$$

where $\vec{p}_0 = \vec{q}$ and $\vec{p}_n = \vec{p}$.

Let us introduce the functions $\omega$ and $\lambda$ defined on the set $\{1, \cdots, n\}$ with values in $\{-1, 1\}$ as before, the former standing for

$$M_{\omega(k)} = \begin{cases} -imc^2\beta\Delta t/\hbar & \text{(helicity flip)} & \text{if } \omega(k) = -1 \\ 1 & \text{(no flip)} & \text{if } \omega(k) = 1, \end{cases}$$

and the latter for

$$Q_{\lambda(k)} = \begin{cases} ie\Delta t \, \widehat{A}(k\Delta t, \vec{p}_k - \vec{p}_{k-1})/\hbar & \text{if } \lambda(k) = -1 \\ \delta(\vec{p}_k - \vec{p}_{k-1}) & \text{if } \lambda(k) = 1. \end{cases}$$

Then, the Green function is expressed as a sum over $\omega$ and $\lambda$.

$$G(t, \vec{p}; 0, \vec{q}) \simeq \sum_{\omega, \lambda} \int \cdots \int \prod_{k=1}^{n}{}' \left( Q_{\lambda(k)} M_{\omega(k)} \exp\left[ -\frac{ic\Delta t}{\hbar} \vec{p}_{k-1} \cdot \vec{\alpha} \right] \right) \prod_{k=1}^{n-1} d^3 p_k. \quad (16)$$

For a given $\lambda$, we can carry out the $\vec{p}_k$-integrations if $\lambda(k) = 1$ with the help of the $\delta$-functions in $Q_{\lambda(k)}$, leaving $\vec{p}_k$-integrals untouched if $\lambda(k) = -1$. Let $\lambda(n_j) = -1$ $(j = 1, \cdots, l)$ and $\vec{p}_j$ be the untouched momentum for $k = n_j$. Then, the sum over $\lambda$ reduces to the sum over $\rho = \langle l; n_1, \cdots, n_l; \vec{p}_1, \cdots, \vec{p}_l \rangle$ $(\vec{p}_l = \vec{p})$ representing the path

$$\vec{p}_\rho(s) = \vec{p}_j \quad (n_j \Delta t < s < n_{j+1} \Delta t) \quad \text{(see Fig.1)}.$$

If $l = 0$, however, $\rho = \langle 0 \rangle$ and $\vec{p}_\rho(s) = \vec{p} \;(0 < s < t)$.

Then,

$$G(t, \vec{p}; 0, \vec{q})$$

$$\simeq \sum_{\omega, \rho} \prod_{j=1}^{l}{}' \left\{ \frac{ie\Delta t}{\hbar} \widehat{A}(n_j \Delta t, \vec{p}_j - \vec{p}_{j-1}) \prod_{k=n_j}^{n_{j+1}-1}{}' \left( M_{\omega(k)} \exp\left[ -\frac{ic\Delta t}{\hbar} \vec{p}_{j-1} \cdot \vec{\alpha} \right] \right) \right\} (\Delta v)^l (17)$$

where $\vec{p}_0 = \vec{q}$ and $\Delta v$ is the volume element of the momentum space. The summand for $\rho$ with $l = 0$ should read

$$\prod_{k=1}^{n}{}' \left( M_{\omega(k)} \exp\left[ -\frac{ic\Delta t}{\hbar} \vec{p} \cdot \vec{\alpha} \right] \right).$$

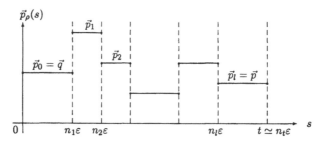

Figure 1 The path represented by $\rho$

We have been using heuristic arguments so far. We shall make it rigorous by nonstandard analysis. We shall discretize the time with spacing $\varepsilon$ as before and the momentum space into cubes of infinitesimal volume $\tau$. Then, a pair $(\omega, \rho)$ represents a path in the product space of the helicity- and the discretized momentum-spaces, which we shall denote by $Y_{\omega,\rho}$. Let us define a measure $\mu(Y_{\omega,\rho})$ for each path $Y_{\omega,\rho}$ by

(a) for $\rho$ with $l \neq 0$

$$\mu(Y_{\omega,\rho}) = \tau^l \prod_{j=1}^{l}{}' \left\{ \frac{ie\varepsilon}{\hbar} \, {}^*\widehat{A}(n_j\varepsilon, \vec{p}_j - \vec{p}_{j-1}) \prod_{k=n_j}^{n_{j+1}-1}{}' \left( M_{\omega(k)} \exp\left[ -\frac{ic\varepsilon}{\hbar} \vec{p}_{j-1} \cdot \vec{\alpha} \right] \right) \right\} \quad (18)$$

(b) for $\rho$ with $l = 0$

$$\mu(Y_{\omega,\rho}) = \prod_{k=1}^{n_t}{}' \left( M_{\omega(k)} \exp\left[ -\frac{ic\varepsilon}{\hbar} \vec{p} \cdot \vec{\alpha} \right] \right) \quad (19)$$

where ${}^*\widehat{A}$ is the nonstandard extension of $\widehat{A}$.

**Theorem 3**    (1) *Assume that the matrix elements of $\widehat{A}(s, \vec{q})$ are in $L^1(\mathsf{R}^3) \cap C(\mathsf{R}^3)$ as a function of $\vec{q}$, satisfying*

$$\sup_{0 \leq s \leq t} \|\widehat{A}(s, \cdot)\|_1 < \infty,$$

*where $\| \quad \|_1$ denotes the maximum of the $L^1(\mathsf{R}^3)$-norms of matrix-elements multiplied by 4.*

*Assume in addition that the elements of $\widehat{\psi}(0, \vec{q})$ are in $L^\infty(\mathsf{R}^3) \cap C(\mathsf{R}^3)$.*

*Then, the nonstandard path integral*

$$\widehat{\Psi}(\underline{t}, \vec{p}) = \sum_{Y_{\omega,\rho} \in \mathcal{P}_{\underline{t},\vec{p}}} \mu(Y_{\omega,\rho}) \, {}^*\widehat{\psi}(0, Y_{\omega,\rho}(0)) \quad \text{with} \quad \mathcal{P}_{\underline{t},\vec{p}} = \{ Y_{\omega,\rho} : Y_{\omega,\rho}(\underline{t}) = \vec{p} \}$$

$$(20)$$

has the standard part $\widehat{\psi}(t, \vec{p})$ that satisfies the $3+1$-dimensional Dirac equation in the momentum space for given initial data $\widehat{\psi}(0, \vec{p})$:

$$i\hbar\frac{\partial}{\partial t}\widehat{\psi}(t, \vec{p}) = (c\vec{p}\cdot\vec{\alpha} + mc^2\beta)\widehat{\psi}(t, \vec{p}) - (e\widehat{A} * \widehat{\psi})(t, \vec{p}).$$

Here, $f * g$ is the convolution of $f$ and $g$.

(2) A standard measure is obtained from $\mu$ as its Loeb measure, and the standard path integral with respect to the former is the same as the standard part of the nonstandard path integral with respect to the latter.

It can be shown that the measure in Theorem 3 is the same as that given by B.Gaveau[1].

*Short sketch of the proof*

The absolute value of $\mu(Y_{\omega,\rho})\,{}^*\widehat{\psi}(0, Y_{\omega,\rho}(0))$ is bounded by an infinitesimal

$$\tau^l\left(\frac{mc^2\varepsilon}{\hbar}\right)^{r(\omega)}\prod_{j=1}^{l}\frac{\varepsilon e M}{\hbar}|4\,{}^*\widehat{A}(j\varepsilon, \vec{q}_j)|_{\infty} \tag{21}$$

where $\vec{q}_j = \vec{p}_j - \vec{p}_{j-1}$, $r(\omega)$ is the number of $k$'s satisfying $\omega(k) = -1$, and

$$|{}^*\widehat{A}(s, \vec{q})|_{\infty} = \max_{\alpha,\beta}|{}^*\widehat{A}_{\alpha,\beta}(s, \vec{q})|, \quad M = \sup_{\vec{q}}|\widehat{\psi}(0, \vec{q})|.$$

The factor $(mc^2\varepsilon/\hbar)^{r(\omega)}$ is appropriate to control the number of $\omega$'s in the following way.

$$\sum_{r=0}^{n_t}\left(\frac{mc^2\varepsilon}{\hbar}\right)^r\binom{n_t}{r} = \left(1 + \frac{mc^2\varepsilon}{\hbar}\right)^{n_t} \simeq e^{mc^2t/\hbar}. \tag{22}$$

Since $\rho$ represents both $(\vec{p}_0, \cdots, \vec{p}_{l-1})$ and $(l; n_1, \cdots, n_l)$, the sum over $\rho$ in (20) is actually a double sum. The first sum over $(\vec{p}_0, \cdots, \vec{p}_{l-1})$ is bounded by $M(\|\widehat{A}\|_1)^l$ where

$$\|\widehat{A}\|_1 = \sup_{0\leq s\leq t}\|\widehat{A}(s, \cdot)\|_1,$$

because the sum over $\vec{p}_j$ together with $\tau$ can be replaced by an integral over $\vec{p}_j$. The second sum over $(l; n_1, \cdots, n_l)$ is controlled by $\varepsilon^l$ in the same way as in (22). From these, we can see that $\widehat{\Psi}(\underline{t}, \vec{p})$ has the standard part.

The following is a direct result from the definition of $\widehat{\Psi}(\underline{t}, \vec{p})$.

$$i\hbar\frac{1}{\varepsilon}\left(\widehat{\Psi}(\underline{t} + \varepsilon, \vec{p}) - \widehat{\Psi}(\underline{t}, \vec{p})\right) \simeq (c\vec{p}\cdot\vec{\alpha} + mc^2\beta)\widehat{\Psi}(\underline{t}, \vec{p}) - \sum_{\vec{q}}\tau\,e^*\widehat{A}(\underline{t}, \vec{q} - \vec{p})\widehat{\Psi}(\underline{t}, \vec{q}).$$

By these facts, we can prove the theorem in a similar way to the case of $1+1$-dimensions[5], and estimations are simpler since the derivative with respect to $x$ is absent.

## 4. CALCULATIONS FOR THE FREE CASE

If the potentials are zero, then $Q_{-1} = 0$ so that the double sum over the helicity space and the momentum spaces reduces to the single sum over the helicity space. Therefore, we only have to consider the case of (19):

$$\mu_0(Y_\omega) = \prod_{k=1}^{n_t} {}' \left( M_{\omega(k)} \exp\left[ -\frac{ic\varepsilon}{\hbar} \vec{p} \cdot \vec{\alpha} \right] \right).$$

Let $P_a(\vec{p})$ $(a = 1, -1)$ be the projection to the eigenspace of the eigenvalue $a$ of the helicity $\frac{\vec{p} \cdot \vec{\alpha}}{|\vec{p}|}$, that is $P_a(\vec{p}) = \frac{1}{2} \left( 1 + a \frac{\vec{p} \cdot \vec{\alpha}}{|\vec{p}|} \right)$.

**Proposition 2** *When $\widehat{A}$ is zero, the path space measure is given by a formula similar to the one in the 1+1-dimensional case ( cf. Prop.1).*

(1) *For $\omega$ such that $r(\omega)$ is even, and $\omega(j) = -1$ for*

$$j = a_1, \ a_1 + b_1, \ a_1 + b_1 + a_2, \ \cdots, \ \sum_{j=1}^{k}(a_j + b_j),$$

one has

$$\mu_0(Y_\omega) = \left( -\frac{i\varepsilon}{\hbar} mc^2 \right)^{r(\omega)} \sum_{a=\pm 1} P_a(\vec{p}) \exp\left[ -\frac{ic\varepsilon}{\hbar} a p f(\omega) \right].$$

(2) *For $\omega$ such that $r(\omega)$ is odd, and $\omega(j) = -1$ for*

$$j = a_1, \ a_1 + b_1, \ a_1 + b_1 + a_2, \ \cdots, \ \sum_{j=1}^{k-1}(a_j + b_j) + a_k,$$

one has

$$\mu_0(Y_\omega) = \left( -\frac{i\varepsilon}{\hbar} mc^2 \right)^{r(\omega)} \sum_{a=\pm 1} P_a(\vec{p}) \exp\left[ -\frac{ic\varepsilon}{\hbar} a p f(\omega) \right] \beta.$$

*Here, $p = |\vec{p}|$, and $f(\omega)$ is $\sum_{j=1}^{k}(a_j - b_j) + a_{k+1}$ in (1) and $\sum_{j=1}^{k}(a_j - b_j)$ in (2).*

**Theorem 4** *The Green function for the initial value problem of the free Dirac equation is obtained as the standard part "st" of the nonstandard path integral over the helicity space:*

$$\frac{1}{(2\pi\hbar)^3} \int \exp\left[ -\frac{i}{\hbar} \vec{p} \cdot (\vec{y} - \vec{x}) \right] \text{st} \left( \sum_\omega \mu_0(Y_\omega) \right) d^3 p$$

$$= \left( \frac{1}{c} \frac{\partial}{\partial t} - \vec{\alpha} \cdot \vec{\nabla}_x - \frac{imc}{\hbar} \beta \right) \left( \frac{1}{2\pi} \delta(s) - \frac{mc}{4\pi\hbar} \frac{\theta(s)}{\sqrt{s}} J_1\left( \frac{mc}{\hbar} \sqrt{s} \right) \right) \qquad (23)$$

*where $s = c^2 t^2 - |\vec{y} - \vec{x}|^2$.*

The right-hand side agrees with the Green function one obtains from the S-function in quantum field theory[9].

*Short sketch of the proof*

Write the left-hand side of (23) as sum of the three parts, $\mathcal{K}_1$, $\mathcal{K}_2$ and $\mathcal{K}_3$ the contributions from paths with $r(Y_\omega) = 0$, odd and even, respectively where $r(\omega)$ is defined below (21).

To $\mathcal{K}_1$, only one $\omega$ contributes and

$$\mu_0(Y_\omega) = \exp\left[-\frac{ict}{\hbar}\vec{p}\cdot\vec{\alpha}\right] = \sum_{a=\pm 1}\exp\left[-\frac{ict}{\hbar}ap\right]P_a.$$

Hence

$$\mathcal{K}_1 \simeq \sum_{a=\pm 1}\frac{1}{(2\pi\hbar)^3}\int\exp\left[-\frac{i}{\hbar}\vec{p}\cdot(\vec{y}-\vec{x})\right]\exp\left[-\frac{ict}{\hbar}ap\right]P_a(\vec{p})d^3p$$

$$= \frac{1}{4\pi|\vec{y}-\vec{x}|}\delta'(ct-|\vec{y}-\vec{x}|) - \frac{\vec{\alpha}\cdot(\vec{y}-\vec{x})}{4\pi|\vec{y}-\vec{x}|^2}\delta(ct-|\vec{y}-\vec{x}|)$$

$$\qquad - \frac{\vec{\alpha}\cdot(\vec{y}-\vec{x})}{4\pi|\vec{y}-\vec{x}|}\delta'(ct-|\vec{y}-\vec{x}|)$$

$$= \left(\frac{1}{c}\frac{\partial}{\partial t} - \vec{\alpha}\cdot\vec{\nabla}_x\right)\frac{1}{2\pi}\delta(s). \tag{24}$$

For $\omega$'s contributing to $\mathcal{K}_2$, $\mu_0(Y_\omega)$ is given by (2) of Prop.2. Since the number of $\omega$ for which $f(\omega) = z$ and $r(\omega) = 2l - 1$ can be shown approximately to be

$$\frac{u^{l-1}v^{l-1}}{(l-1)!(l-1)!} \quad \text{with} \quad u = \frac{n_t+z}{2}, \; v = \frac{n_t-z}{2},$$

so that

$$\mathcal{K}_2 \simeq \frac{imc^2}{2\hbar(2\pi\hbar)^3}\int d^3p\exp\left[-\frac{i}{\hbar}\vec{p}\cdot(\vec{y}-\vec{x})\right]\sum_{|z|\leq n_t}\varepsilon\left\{\left(\exp\left[-\frac{ic\varepsilon}{\hbar}zp\right] + \exp\left[\frac{ic\varepsilon}{\hbar}zp\right]\right)\right.$$

$$\left. \times\sum_l\left(-\frac{imc^2\varepsilon}{\hbar}\right)^{2l-2}\frac{u^{l-1}v^{l-1}}{(l-1)!(l-1)!}\right\}\beta.$$

Apart from an infinitesimal in the nonstandard analysis, the sum over $l$ is the Bessel function and the sum over $z$ is nothing but the integral. Then,

$$\mathcal{K}_2 \simeq -\frac{imc^2\beta}{4\hbar}\frac{1}{2(2\pi\hbar)^3}\int d^3p\exp\left[-\frac{i}{\hbar}\vec{p}\cdot(\vec{y}-\vec{x})\right]$$

$$\times\int_{-t}^{t}\left(\exp\left[-\frac{ic}{\hbar}\xi p\right] + \exp\left[\frac{ic}{\hbar}\xi p\right]\right)J_0\left(\frac{mc}{\hbar}\sqrt{c^2t^2 - c^2\xi^2}\right)d\xi$$

$$= -\frac{imc\beta}{2\pi\hbar}\delta(s) + \frac{im^2c^2\beta}{4\pi\hbar^2}\frac{\theta(s)}{\sqrt{s}}J_1\left(\frac{mc}{\hbar}\sqrt{s}\right). \tag{25}$$

The calculation of $\mathcal{K}_3$ is similar to that of $\mathcal{K}_2$ and the result is

$$\mathcal{K}_3 \simeq \left(\frac{1}{c}\frac{\partial}{\partial t} - \vec{\alpha}\cdot\vec{\nabla}_x\right)\left(-\frac{mc}{4\pi\hbar}\frac{\theta(s)}{\sqrt{s}}J_1\left(\frac{mc}{\hbar}\sqrt{s}\right)\right). \tag{26}$$

From (24), (25) and (26), we get the conclusion of the theorem.

## ACKNOWLEDGEMENTS

We are grateful to Prof. M.Hirokawa of Tokyo Gakugei Univ., Prof. M.Yamaguchi and Prof. W.Matsumoto of Ryukoku Univ. for discussion and valuable comments.

## REFERENCES

1. Gaveau,B. (1984) Representation Formulas of the Cauchy Problem for Hyperbolic Systems Generalizing Dirac System. *Journal of Functional Analysis*, **58**, 310–319.

2. Cameron,R.H. (1960) A family of integrals serving to connect the Wiener and Feynman integrals. *Journal of Mathematics and Physics*, **39**, 126–140.

3. Ichinose,T. (1982) Path integral for the Dirac equation in two space-time dimensions. Proceedings of the Japan Academy, **58**, 290–293.

4. Blanchard.Ph., Combe,Ph., Sirugue,M. & Collin, S. (1985) Probabilistic solution of the Dirac equation, Path integral representation for the solution of the Dirac equation in presence of an electromagnetic field. *Bielefeld BiBos,* **44**.

5. Nakamura,T. (1991) A nonstandard representation of Feynman's path integrals. *Journal of Mathematical Physics*, **32**, 457–463.

6. Nakamura,T. (1997) Path space measures for Dirac and Schrödinger equations: Nonstandard analytical approach. *Journal of Mathematical Physics*, **38**(8), 4052–4072.

7. Zastawniak,T. (1989) The nonexistence of the path-space measure for the Dirac equation in four space-time dimensions. *Journal of Mathematical Physics*, **30**, 1354–1358.

8. Loeb,P.A. (1975) Conversion from nonstandard to standard measure spaces and applications in probability theory. *Transactions of the American Mathematical Society*, **211**, 113-122.

9. Bogoliubov,N.N. & Shirkov,D.V. (1980) *Introduction to the theory of quantum fields*. Wiley-Interscience, John Wiley & Sons.

# RELATIVISTIC QUANTUM MECHANICS IN SPACES WITH A RING-SHAPED TOPOLOGICAL DEFECT

CHRISTOPHER C. BERNIDO

National Institute of Physics, University of the Philippines
Diliman, Quezon City 1101, Philippines, and
Research Center for Theoretical Physics, CVI
Jagna, Bohol 6308, Philippines

**Abstract**    The path integral method is used to solve the Klein-Gordon and Dirac equations for a relativistic particle in a space with a ring-shaped topological defect.

## 1  Introduction

Quantum mechanics in spaces with a topological defect has various interesting applications in polymer science, condensed matter physics, elementary particles, and cosmology. Two examples of such a space which have received much attention are a punctured plane for two-dimensional systems, and a ring-shaped object, or a torus, in three dimensions. These topological defects effectively make the space of interest multiply-connected, and a convenient tool for investigating these types of problem is Feynman's sum-over-histories approach to quantum mechanics.

The Feynman path integral method has, in fact, been used to study spaces with topological defects for the nonrelativistic case[1-8]. For a ring in three dimensions, the solution yields in toroidal coordinates an explicit form for the propagator and applications to the toroidal Aharonov-Bohm set-up[8,9] and the entanglement problem involving a torus-shaped molecule[10] have been considered. In this paper, we now extend this investigation to a relativistic particle of mass $\mu$ moving in a space with a ring-shaped defect in the presence of an external electromagnetic field, $A^\mu = (\Phi, \vec{A})$. In particular, we seek a solution for the Dirac equation ($\hbar = c = 1$),

$$(\mu - \widehat{M})G(\vec{r}\,", \vec{r}\,') = \delta(\vec{r}\," - \vec{r}\,), \tag{1}$$

where,

$$\widehat{M} = -\beta\alpha \cdot (\vec{p} - e\,\vec{A}) - \beta V + \beta E, \tag{2}$$

and the Klein-Gordon equation,

$$(\Box^2 - \mu^2)G(x", x') = -\delta(x", x'). \tag{3}$$

167

Here, $\vec{\alpha}$ and $\beta$ are the Dirac matrices, $V = e\Phi$, $\Box^2 = (\partial_\mu - ieA_\mu)(\partial^\mu - ieA^\mu)$, $\delta(x", x') = \langle x" \mid x' \rangle$, and, $x = (t, \vec{r})$. In Eqs. (1) and (3), the Green functions are treated as matrix elements, i.e., $G(\vec{r}\,", \vec{r}') = \langle \vec{r}\," |G| \vec{r}' \rangle$, and, $G(x", x') = \langle x" |G| x' \rangle$. We present in sections 2 and 3 a method of expressing these Green functions as a sum-over-histories, and in section 4, we introduce a ring-shaped defect using toroidal coordinates. Section 5 discusses a path integral evaluation leading to an explicit form of the Green function for Eqs. (1) and (3).

## 2   Path Summation for the Dirac Particle

Let us first consider the Dirac equation by expressing the Green function as,

$$G(\vec{r}\,", \vec{r}') = (\mu + \widehat{M})g(\vec{r}\,", \vec{r}'), \tag{4}$$

where, $g(\vec{r}\,", \vec{r}') = \langle \vec{r}\," |g| \vec{r}' \rangle$, which allows us to obtain from Eq. (1) the iterated Dirac equation,

$$\left(\mu^2 - \widehat{M}^2\right) g(\vec{r}\,", \vec{r}') = \delta(\vec{r}\," - \vec{r}). \tag{5}$$

The operator $g$ in Eq. (5) can be written as,[11]

$$g = \left(\mu^2 - \widehat{M}^2\right)^{-1} = (i/2\mu) \int_0^\infty \exp[-iH\Lambda] \, d\Lambda, \tag{6}$$

where we defined,

$$H = (1/2\mu) \left(\mu^2 - \widehat{M}^2\right). \tag{7}$$

Note that if we take the matrix element of Eq. (6), such that,

$$\langle \vec{r}\," |g| \vec{r}' \rangle = (i/2\mu) \int_0^\infty \langle \vec{r}\," |\exp[-iH\Lambda]| \vec{r}' \rangle \, d\Lambda, \tag{8}$$

then the integrand, $\langle \vec{r}\," |\exp[-iH\Lambda]| \vec{r}' \rangle$, is similar in form to the propagator of a system evolving in a time-like parameter $\Lambda$ with an effective Hamiltonian $H$. The $H$, Eq. (7), obtained by squaring Eq. (2), has the form,

$$H = (1/2\mu)(\vec{p} - e\vec{A})^2 - (e/2\mu)(\vec{\sigma} \cdot \vec{B} - i\vec{\alpha} \cdot \vec{E}) - (1/2\mu)\left[(E - V)^2 - \mu^2\right]. \tag{9}$$

In the absence of the external electric and magnetic fields, $\vec{E}$ and $\vec{B}$, the effective Hamiltonian reduces to

$$H = (1/2\mu)\left(\vec{p}^{\,2} - k^2\right), \tag{10}$$

where, $k^2 = E^2 - \mu^2$.

A procedure for solving the Dirac equation may, therefore, start by calculating, $\left\langle \vec{r}\,'' \left| \exp[-iH\Lambda] \right| \vec{r}\,' \right\rangle$, from which we obtain the Green function, $g(\vec{r}\,'', \vec{r}\,')$, Eq.(8), that satisfies the iterated Dirac equation (5). Once we obtain, $g(\vec{r}\,'', \vec{r}\,')$, we can then use Eq. (4), to evaluate the Green function $G(\vec{r}\,'', \vec{r}\,')$, which is the solution of the Dirac equation (1). To solve the integrand in Eq. (8), we shall express it as a sum-over-histories of the particle, or the path integral,

$$\left\langle \vec{r}\,'' \left| \exp[-iH\Lambda] \right| \vec{r}\,' \right\rangle = \int \exp[iS] \, D[\vec{r}], \tag{11}$$

where $S$ is the corresponding action of the form,

$$S = \int \left\{ \frac{1}{2}\mu \, \dot{\vec{r}}^2 + e\,\vec{A} \cdot \dot{\vec{r}} + (e/2\mu)(\vec{\sigma} \cdot \vec{B} - i\,\vec{\alpha} \cdot \vec{E}) \right.$$
$$\left. + (1/2\mu) \left[ (E - V)^2 - \mu^2 \right] \right\} d\lambda. \tag{12}$$

Here, $\lambda$ is a time-like parameter, $0 \leq \lambda < \Lambda$, and, $\dot{\vec{r}} = d\,\vec{r}/d\lambda$. We shall apply this method of solving for the Dirac equation in section 5.

## 3   Path Summation for the Klein-Gordon Particle

The Green function, $G(x'', x') = \langle x'' |G| x' \rangle$, for the Klein-Gordon equation (3), can likewise be obtained via the path integral method. We can express the operator $G$ as,[11]

$$G = -(\Box^2 - \mu^2)^{-1} = (i/2) \int_0^\infty \exp[-i\tilde{H}\Lambda] \, d\Lambda, \tag{13}$$

where we defined $\tilde{H}$ as,

$$\tilde{H} = \tfrac{1}{2}(-\Box^2 + \mu^2). \tag{14}$$

The matrix element of Eq. (13),

$$G(x'', x') = (i/2) \int_0^\infty \left\langle x'' \left| \exp[-i\tilde{H}\Lambda] \right| x' \right\rangle \, d\Lambda, \tag{15}$$

leads to an integrand $\left\langle x'' \left| \exp[-i\tilde{H}\Lambda] \right| x' \right\rangle$ which is similar to the propagator of a quantum system evolving in $\Lambda$-time with an effective Hamiltonian $\tilde{H}$. The integrand can then be expressed as a path integral,

$$\left\langle x'' \left| \exp[-i\tilde{H}\Lambda] \right| x' \right\rangle = \int \exp[iS] \, D[x], \tag{16}$$

where $S$ is the corresponding action of the form,

$$S = \int \left[ \tfrac{1}{2}\left( -\dot{t}^2 + \dot{\vec{r}}^2 \right) + e\,\dot{x}^\mu A_\mu - (\mu^2/2) \right] d\lambda. \qquad (17)$$

Here, we have, $\dot{x}^\mu = dx^\mu/d\lambda$, $0 \leq \lambda < \Lambda$, and $A_\mu$ are external electromagnetic fields.

In the next section, we consider the presence of a ring-shaped defect using toroidal coordinates.

## 4    Path Summation with a Ring-Shaped Defect

A ring-shaped topological defect is best described using the toroidal coordinates $(\eta, \xi, \phi)$, where $(0 < \eta < +\infty)$, $(0 \leq \xi < 2\pi)$, and $(0 \leq \phi < 2\pi)$. When the variable $\eta$ is fixed, i.e., $\eta = \eta_0$, a two-dimensional surface shaped like a torus is generated. This toroidal object has an axial circle in the $x - y$ plane, and centered at the origin of radius $a \coth \eta_0$, with a circular cross-section of radius $a \operatorname{cosech} \eta_0$. In the limit $\eta \to \infty$, one obtains an infinitesimally thin ring of radius $a$, and the limit $\eta \to 0$ corresponds to the $z$-axis.

If we now consider a quantum particle in three-dimensional space which moves outside an impenetrable ring described by the toroidal surface, $\eta = \bar{\eta} = $ constant, a term $S^c$ can be added to the effective action of the path integral, such that,

$$\exp(iS^c) = \Theta(\bar{\eta} - \eta). \qquad (18)$$

Here, $\Theta(\bar{\eta} - \eta)$ is a step function with values of unity for $\eta < \bar{\eta}$ (outside the ring), and zero for $\eta > \bar{\eta}$ (inside the ring). Eq. (18), therefore, wipes out the particle propagator when the path of the particle hits the impenetrable ring barrier. The ring, in fact, makes the physical space multiply-connected such that the paths, or trajectories, of the particle that wind around the ring in going from $\vec{r}'$ to $\vec{r}''$ can no longer be deformed into others. If we consider, for instance, Eq. (11), the sum over all paths becomes a sum of partial propagators $K_n(\vec{r}'', \vec{r}'; \Lambda)$, each corresponding to homotopically inequivalent paths where $n$ is a winding number. Eq. (11) can, therefore, be written as,[1-3]

$$\left\langle \vec{r}'' \left| \exp[-iH\Lambda] \right| \vec{r}' \right\rangle = \int \exp[iS]\, D[\vec{r}] = \sum_{n=-\infty}^{+\infty} K_n(\vec{r}'', \vec{r}'; \Lambda). \qquad (19)$$

The winding number $n$ signifies, if positive, a particle's path that first goes inside the ring and then loops $n$ times around it; and if negative, a path that first passes outside the ring and then loops $|n| - 1$ times. The winding numbers $n = -1$ and $n = 0$, for example, correspond to paths passing outside and inside the ring, respectively, but with no winding.

In the following section, we evaluate the path integral for a relativistic particle in spaces with a ring-shaped defect by taking the Aharonov-Bohm set-up as an example.

## 5   The Toroidal Aharonov-Bohm Experiment

### 5.1   Spin $\frac{1}{2}$ Particle:

As an example of a Dirac particle in a space with a ring-shaped topological defect, we consider the toroidal Aharonov-Bohm experiment[9] where an impenetrable ring confines the magnetic field $\vec{B}$. Since the particle moves outside the ring where the magnetic field is zero, the effective action, Eq. (12), becomes (note, $\vec{E} = V = 0$),

$$S = \int \left\{ \frac{1}{2}\mu \, \dot{\vec{r}}^{\,2} + e\,\vec{A} \cdot \dot{\vec{r}} + (k^2/2\mu) \right\} d\lambda + S^c \tag{20}$$

where, $S^c$ is given by Eq. (18), and $k^2 = E^2 - \mu^2$. In toroidal coordinates, the vector potential $\vec{A}$ for a magnetic flux $\Phi_0$ enclosed by the toroid is given by, $\vec{A} = (\Phi_0/2\pi) \, \vec{\nabla} \, \xi$. To evaluate the propagator $K_n(\vec{r}\,",\vec{r}';\Lambda)$ in Eq. (19), we first slice the time-like parameter $\Lambda$ into $N$ sub-intervals, such that, $\Lambda/N = \varepsilon_j = \lambda_j - \lambda_{j-1}$, where $j = 1, 2, ..., N$. Letting $\vec{r}_j = \vec{r}\,(\lambda_j)$, $\vec{r}' = \vec{r}_0$, and, $\vec{r}\," = \vec{r}_N$, the time-sliced form of the particle propagator for winding number $n = 0$ is given by,

$$K_{n=0}(\vec{r}\,",\vec{r}';\Lambda) = \lim_{N\to\infty} \int \prod_{j=1}^{N} \exp(iS_j^0) \; \Theta(\bar{\eta} - \eta_j) \left(\frac{\mu}{2\pi i \varepsilon_j}\right)^{3/2} \prod_{j=1}^{N-1} (d^3 \, \vec{r}_j) \tag{21}$$

where the short-time action, from Eq. (20), can be written as $(g = e\Phi_0/2\pi)$,

$$S_j^0 = (\mu/2\varepsilon_j)(\Delta \, \vec{r}_j)^2 + g(\Delta\xi_j) + (k^2/2\mu)\varepsilon_j \quad . \tag{22}$$

Exponentiating the action, Eq. (22), we obtain the expression,

$$\prod_{j=1}^{N} \exp(iS_j^0) = \prod_{j=1}^{N} \exp\left\{ i\left[ (\mu/2\varepsilon_j)(\Delta \, \vec{r}_j)^2 + g(\Delta\xi_j) + (k^2/2\mu)\varepsilon_j \right] \right\}$$

$$= \exp\left\{ i\left[ g(\xi_N - \xi_0) + \left(\frac{k^2\Lambda}{2\mu}\right) \right] \right\} \prod_{j=1}^{N} \exp\left[ \left(\frac{i\mu}{2\varepsilon_j}\right)(\Delta \, \vec{r}_j)^2 \right], \tag{23}$$

where, $\Lambda = \sum_{j=1}^{N} \varepsilon_j$, $\xi_N = \xi"$, and, $\xi_0 = \xi'$. With Eq. (23), we can write Eq. (21) as,

$$K_{n=0}(\vec{r}\,",\vec{r}\,';\Lambda) = \exp\left\{i\left[g(\xi" - \xi') + (k^2\Lambda/2\mu)\right]\right\} K^f(\vec{r}\,",\vec{r}\,';\Lambda) \qquad (24)$$

where,

$$K^f(\vec{r}\,",\vec{r}\,';\Lambda) = \lim_{N\to\infty} \int \prod_{j=1}^{N} \exp\left[(i\mu/2\varepsilon_j)(\Delta\,\vec{r}_j)^2\right] \Theta(\bar\eta - \eta_j)$$

$$\times \left(\frac{\mu}{2\pi i\varepsilon_j}\right)^{3/2} \prod_{j=1}^{N-1}(d^3\,\vec{r}_j). \qquad (25)$$

Eq. (25) can be explicitly written in terms of toroidal coordinates with the help of the transformation equations from Cartesian coordinates given by,

$$x = \frac{a\sinh\eta\cos\phi}{\cosh\eta - \cos\xi}\ ;\quad y = \frac{a\sinh\eta\sin\phi}{\cosh\eta - \cos\xi}\ ;\quad z = \frac{a\sin\xi}{\cosh\eta - \cos\xi}\ . \qquad (26)$$

In particular, the kinetic term in Eq. (25) becomes,

$$\begin{aligned}
(\mu/2\varepsilon_j)(\Delta\,\vec{r}_j)^2 &= (\mu/2\varepsilon_j)\left[(\Delta x_j)^2 + (\Delta y_j)^2 + (\Delta z_j)^2\right] \\
&= \left[\mu a^2\,/\,(\cosh\eta_j - \cos\xi_j)(\cosh\eta_{j-1} - \cos\xi_{j-1})\varepsilon_j\right] \\
&\quad \times \left[\cosh\eta_j \cosh\eta_{j-1} - \sinh\eta_j \sinh\eta_{j-1}\cos(\Delta\phi_j)\right. \\
&\quad \left. - \cos(\Delta\xi_j)\,\right]
\end{aligned} \qquad (27)$$

and the volume element appears as,

$$d^3\,\vec{r}_j = \frac{a^3 \sinh\eta_j\,d\eta_j\,d\xi_j\,d\phi_j}{(\cosh\eta_j - \cos\xi_j)^3}\ . \qquad (28)$$

The path integration of Eq. (25), with Eqs. (27) and (28), is similar to the non-relativistic case[8]. This yields the result,

$$\begin{aligned}
K^f(\vec{r}\,",\vec{r}\,';\Lambda) &= \left(1/4\pi a^2\right)\left[(\cosh\eta" - \cos\xi")(\cosh\eta' - \cos\xi')\right]^{3/2} \\
&\quad \times \sum_{m=-\infty}^{+\infty} \exp\left[im(\phi" - \phi')\right] \\
&\quad \times \int_0^\infty d\rho\,2\rho\,\tanh(\pi\rho)\ d^{-\frac{1}{2}+i\rho,0}_{m\ 0}(\eta")\quad d^{-\frac{1}{2}+i\rho,0*}_{m\ 0}(\eta') \\
&\quad \times \left(\frac{\mu}{2\pi i\sigma}\right)^{1/2} \exp\left[\frac{i\mu a^2}{2\sigma}(\xi" - \xi')^2\right]\exp\left[(i/2\mu a^2)\left(\rho^2 + \tfrac{1}{16}\right)\sigma\right]
\end{aligned} \qquad (29)$$

where, $\sigma = \Lambda(\cosh\eta'' - \cos\xi'')(\cosh\eta' - \cos\xi')$, and, $d_{a\,b}^{l,\lambda}(\eta)$ are the Bargmann functions[12]. Note that in carrying out the integration for the $\eta$-variable, the ring which encloses a non-vanishing flux is taken to be infinitesimally thin, i.e., $\bar{\eta} \to \infty$ in Eq. (18). From Eqs. (24) and (29), we can obtain the partial propagator $K_n(\vec{r}\,'', \vec{r}\,'; \Lambda)$ corresponding to a non-zero winding number $n$ with the substitution of $(\xi'' - \xi')$ by $(\xi'' - \xi' + 2\pi n)$. Employing the relation,

$$(\mu/2\pi i\sigma)^{1/2} \exp\left[\frac{i\mu a^2}{2\sigma}(\xi'' - \xi' + 2\pi n)^2\right] = \left(\frac{1}{2\pi a}\right) \int_{-\infty}^{+\infty} \exp[i\gamma(\xi'' - \xi' + 2\pi n) - (i\gamma^2\sigma/2\mu a^2)]\,d\gamma, \tag{30}$$

we can write Eq. (19), with the help of Eqs. (24) and (29) as,

$$\left\langle \vec{r}\,'' \left|\exp[-iH\Lambda]\right| \vec{r}\,'\right\rangle = \sum_{n=-\infty}^{+\infty} \left([(\cosh\eta'' - \cos\xi'')(\cosh\eta' - \cos\xi')]^{3/2} / 8\pi^2 a^3\right)$$

$$\times \sum_{m=-\infty}^{+\infty} \exp[im(\phi'' - \phi')] \int_0^\infty d\rho\, 2\rho\, \tanh(\pi\rho)$$

$$\times d_{m\,0}^{-\frac{1}{2}+i\rho,0}(\eta'') \; d_{m\,0}^{-\frac{1}{2}+i\rho,0*}(\eta')$$

$$\times \int_{-\infty}^{+\infty} d\gamma \, \exp[i(\gamma + g)(\xi'' - \xi' + 2\pi n)]$$

$$\times \exp\left[ik^2\sigma / 2\mu(\cosh\eta'' - \cos\xi'')(\cosh\eta' - \cos\xi')\right]$$

$$\times \exp\left[i\left(\rho^2 - \gamma^2 + \tfrac{1}{16}\right)\sigma / 2\mu a^2\right] \tag{31}$$

where, $k^2 = E^2 - \mu^2$. Eq. (31) is a winding number representation of the propagator. Using Poisson's sum formula,

$$\sum_{n=-\infty}^{+\infty} \exp[i2\pi n(\gamma + g)] = \sum_{s=-\infty}^{+\infty} \delta(\gamma + g - s). \tag{32}$$

we can rewrite Eq. (31) as,

$$\left\langle \vec{r}\,'' \left|\exp[-iH\Lambda]\right| \vec{r}\,'\right\rangle = \sum_{s=-\infty}^{+\infty} \left([(\cosh\eta'' - \cos\xi'')(\cosh\eta' - \cos\xi')]^{3/2} / 8\pi^2 a^3\right)$$

$$\times \sum_{m=-\infty}^{+\infty} \exp[im(\phi'' - \phi')] \int_0^\infty d\rho\, 2\rho\, \tanh(\pi\rho)$$

$$\times d_{m\,0}^{-\frac{1}{2}+i\rho,0}(\eta'') \; d_{m\,0}^{-\frac{1}{2}+i\rho,0*}(\eta')$$

$$\times \int_{-\infty}^{+\infty} d\gamma \; \delta\,(\gamma + g - s) \exp\left[i(\gamma + g)(\xi'' - \xi')\right]$$
$$\times \; \exp\left[ik^2\sigma \,/\, 2\mu(\cosh\eta'' - \cos\xi'')(\cosh\eta' - \cos\xi')\right]$$
$$\times \; \exp\left[i\left(\rho^2 - \gamma^2 + \tfrac{1}{16}\right)\sigma \,/\, 2\mu a^2\right] \quad . \tag{33}$$

The $\delta$-function facilitates the integration over $\gamma$, and we obtain,

$$\left\langle \vec{r}\,'' \,|\exp[-iH\Lambda]|\,\vec{r}\,'\right\rangle = \left([(\cosh\eta'' - \cos\xi'')(\cosh\eta' - \cos\xi')]^{3/2} \,/\, 8\pi^2 a^3\right)$$
$$\times \sum_{s=-\infty}^{+\infty} \sum_{m=-\infty}^{+\infty} \exp\left[im(\phi'' - \phi')\right]\, \exp\left[is(\xi'' - \xi')\right]$$
$$\times \int_0^\infty d\rho\, 2\rho\, \tanh(\pi\rho)$$
$$\times \; d^{-\frac{1}{2}+i\rho,\,0}_{\;\;\;m\,0}(\eta'') \; d^{-\frac{1}{2}+i\rho,\,0*}_{\;\;\;m\,0}(\eta')$$
$$\times \; \exp\left[ik^2\sigma \,/\, 2\mu(\cosh\eta'' - \cos\xi'')(\cosh\eta' - \cos\xi')\right]$$
$$\times \; \exp\left[i\left(\rho^2 - (g-s)^2 + \tfrac{1}{16}\right)\sigma \,/\, 2\mu a^2\right] \quad . \tag{34}$$

Using Eq. (34) in Eq. (8), we now write the Green function for the iterated Dirac equation as (note, $d\sigma = d\Lambda(\cosh\eta'' - \cos\xi'')(\cosh\eta' - \cos\xi')$ ),

$$g(\vec{r}\,'',\vec{r}\,') = (i/2\mu)\int_0^\infty \left\langle \vec{r}\,'' \,|\exp[-iH\Lambda]|\,\vec{r}\,'\right\rangle \,(d\Lambda \,/\, d\sigma)\, d\sigma,$$
$$= \left(i\,[(\cosh\eta'' - \cos\xi'')(\cosh\eta' - \cos\xi')]^{1/2} \,/\, 16\mu\pi^2 a^3\right)$$
$$\times \sum_{s=-\infty}^{+\infty} \sum_{m=-\infty}^{+\infty} \exp\left[im(\phi'' - \phi')\right]\, \exp\left[is(\xi'' - \xi')\right]$$
$$\times \int_0^\infty d\rho\, 2\rho\, \tanh(\pi\rho)\; d^{-\frac{1}{2}+i\rho,\,0}_{\;\;\;m\,0}(\eta'') \; d^{-\frac{1}{2}+i\rho,\,0*}_{\;\;\;m\,0}(\eta')$$
$$\times \int_0^\infty \exp\left\{ i[\, k^2 \,/\, 2\mu(\cosh\eta'' - \cos\xi'')(\cosh\eta' - \cos\xi')\right.$$
$$\left. + \left(\rho^2 - (g-s)^2 + \tfrac{1}{16}\right) / 2\mu a^2]\sigma \right\} d\sigma \; . \tag{35}$$

Integration over $\sigma$ yields the result,

$$g(\vec{r}\,'',\vec{r}\,') = \left(\frac{-1}{8\pi^2 a^3}\right) \sum_{m=-\infty}^{+\infty} \sum_{s=-\infty}^{+\infty} [(\cosh\eta'' - \cos\xi'')(\cosh\eta' - \cos\xi')]^{3/2}$$
$$\times \; \exp\left[im(\phi'' - \phi')\right]\, \exp\left[is(\xi'' - \xi')\right]$$

$$\times \int_0^\infty d\rho \, 2\rho \, \tanh(\pi\rho) \; d^{-\frac{1}{2}+i\rho,0}_{m\ 0}(\eta") \; d^{-\frac{1}{2}+i\rho,0*}_{m\ 0}(\eta')$$
$$\times \left\{ (1/a^2) \left[ \rho^2 - (g-s)^2 + \tfrac{1}{16} \right] (\cosh\eta" - \cos\xi")(\cosh\eta' - \cos\xi') \right.$$
$$\left. + E^2 - \mu^2 + i\varepsilon \right\}^{-1} \tag{36}$$

where, $\varepsilon \to 0$ limit is taken. Eq. (36) gives us the the solution for the iterated Dirac equation (5). The Green function for the linear Dirac equation may then be obtained from Eq. (36) using Eq. (4).

## 5.2   Spin 0 Particle:

We now consider a spin 0 particle in the presence of a ring-shaped topological defect as exemplified by the toroidal Aharonov-Bohm set-up[9]. With the space being multiply-connected, we write Eq. (16) as,

$$\left\langle x" \left| \exp[-i\widetilde{H}\Lambda] \right| x' \right\rangle = \sum_{n=-\infty}^{+\infty} K_n(x", x'; \Lambda) , \tag{37}$$

where $n$ is the winding number. Again, slicing the parameter $\Lambda$ into $N$ sub-intervals, such that, $\Lambda/N = \varepsilon_j = \lambda_j - \lambda_{j-1}$, where $j = 1, 2, ..., N$, we have from Eqs. (17) and (37), the sum over paths for the case with zero winding, i.e., $n = 0$, given by,

$$K_{n=0}(x", x'; \Lambda) = \lim_{N \to \infty} \int \prod_{j=1}^{N} \exp(iS_j^0) \, \Theta(\bar{\eta} - \eta_j) \left( \frac{1}{2\pi i\varepsilon_j} \right)^{4/2}$$
$$\times \prod_{j=1}^{N-1} (idt_j \, d\vec{r}_j) , \tag{38}$$

where the short-time action in the presence of the vector potential, $\vec{A} = (\Phi_0/2\pi) \, \vec{\nabla} \xi$, appears as,

$$S_j^0 = \int_{\lambda_{j-1}}^{\lambda_j} \left[ \frac{1}{2} \left( -\dot{t}^2 + \dot{\vec{r}}^2 \right) + e\vec{A} \cdot \vec{r} - (\mu^2/2) \right] d\lambda$$
$$= -(\Delta t_j)^2/2\varepsilon_j + (\Delta\vec{r}_j)^2/2\varepsilon_j + g(\Delta\xi_j) - (\mu^2/2)\varepsilon_j . \tag{39}$$

This action, in fact, allows us to write Eq. (38) as,

$$K_{n=0}(x", x'; \Lambda) = \exp\left\{ i \left[ g(\xi" - \xi') - (\mu^2\Lambda/2) \right] \right\}$$
$$\times K(t", t'; \Lambda) \, K^0(\vec{r}", \vec{r}'; \Lambda) , \tag{40}$$

where the propagator along the time and space coordinates are given by,

$$K(t",t';\Lambda) = \lim_{N\to\infty} \int \prod_{j=1}^{N} \exp\left[-i(\Delta t_j)^2/2\varepsilon_j\right] \left(\frac{1}{2\pi i\varepsilon_j}\right)^{1/2} \prod_{j=1}^{N-1}(idt_j) \qquad (41)$$

and,

$$K^0(\vec{r}",\vec{r}';\Lambda) = \lim_{N\to\infty} \int \prod_{j=1}^{N} \exp\left[i(\Delta \vec{r}_j)^2/2\varepsilon_j\right]$$

$$\times \Theta(\bar{\eta} - \eta_j) \left(\frac{1}{2\pi i\varepsilon_j}\right)^{3/2} \prod_{j=1}^{N-1}(d\vec{r}_j) \ . \qquad (42)$$

The path integration for Eq. (41) is similar to that of a free particle propagator along the time coordinate which yields the result,

$$K(t",t';\Lambda) = (1/2\pi i) \int_{-\infty}^{+\infty} \exp\left[iE(t" - t') + i(E^2\Lambda/2)\right] \, dE \ . \qquad (43)$$

Eq. (42), on the other hand, is the particle propagator along the space part and its evaluation follows a similar procedure as the one encountered for the nonrelativistic case[8] (see also, Eq. (25) above ). Putting together the evaluated form of Eq. (42) with Eq. (43), we write Eq. (40) as,

$$K_{n=0}(x",x';\Lambda) = (-i/16\pi^3 a^3)\left[(\cosh\eta" - \cos\xi")(\cosh\eta' - \cos\xi')\right]^{3/2}$$

$$\times \sum_{m=-\infty}^{+\infty} \int_{-\infty}^{+\infty} dE \ \exp\left[im(\phi" - \phi')\right]\exp\left[iE(t" - t')\right]$$

$$\times \int_0^\infty d\rho \ 2\rho \ \tanh(\pi\rho) \ d_{m\ 0}^{-\frac{1}{2}+i\rho,0}(\eta") \ d_{m\ 0}^{-\frac{1}{2}+i\rho,0*}(\eta')$$

$$\times \int_{-\infty}^{+\infty} d\gamma \ \exp\left[i(\gamma + g)(\xi" - \xi')\right]$$

$$\times \ \exp\left\{i\left[(E^2 - \mu^2)/2(\cosh\eta" - \cos\xi")(\cosh\eta' - \cos\xi')\right]\sigma \right.$$
$$+ \ (i/2a^2)\left(\rho^2 - \gamma^2 + \tfrac{1}{16}\right)\sigma\right\} \qquad (44)$$

where, $\sigma = \Lambda(\cosh\eta" - \cos\xi")(\cosh\eta' - \cos\xi')$. The propagator for $n \neq 0$ can be obtained from Eq. (44) by substituting $(\xi" - \xi')$ by $(\xi" - \xi' + 2\pi n)$. Summing $K_n(x",x';\Lambda)$ over all $n$ as in Eq. (37), and using Eq. (32) such that the $\delta$-function facilitates the integration over $\gamma$, we obtain,

$$\left\langle x" \left| \exp[-i\widetilde{H}\Lambda] \right| x' \right\rangle = (-i/16\pi^3 a^3)\left[(\cosh\eta" - \cos\xi")(\cosh\eta' - \cos\xi')\right]^{3/2}$$

$$\times \sum_{s=-\infty}^{+\infty} \sum_{m=-\infty}^{+\infty} \exp\left[im(\phi" - \phi')\right] \exp\left[is(\xi" - \xi')\right]$$

$$\times \int_{-\infty}^{+\infty} dE \, \exp\left[iE(t" - t')\right] \quad \int_0^\infty d\rho \, 2\rho \, \tanh(\pi\rho)$$

$$\times d_{m\ 0}^{-\frac{1}{2}+i\rho, 0}(\eta") \; d_{m\ 0}^{-\frac{1}{2}+i\rho, 0*}(\eta')$$

$$\times \exp\left\{ i\left[(E^2 - \mu^2)/2\,(\cosh\eta" - \cos\xi")(\cosh\eta' - \cos\xi')\right]\sigma \right.$$
$$\left. + (i/2a^2)\left(\rho^2 - (s-g)^2 + \tfrac{1}{16}\right)\sigma \right\} \tag{45}$$

With this, the Green function for the Klein-Gordon particle, Eq. (15) becomes,

$$G(x", x') = (i/2)\int_0^\infty \left\langle x" \left| \exp[-i\widetilde{H}\Lambda] \right| x' \right\rangle \; (d\Lambda/d\sigma)\, d\sigma$$

$$= (1/32\pi^3 a^3)\left[(\cosh\eta" - \cos\xi")(\cosh\eta' - \cos\xi')\right]^{3/2} \sum_{m=-\infty}^{+\infty} \sum_{s=-\infty}^{+\infty}$$

$$\times \int_{-\infty}^{+\infty} dE \, \exp\left[iE(t" - t')\right] \exp\left[im(\phi" - \phi')\right] \exp\left[is(\xi" - \xi')\right]$$

$$\times \int_0^\infty d\rho \, 2\rho \, \tanh(\pi\rho) \; d_{m\ 0}^{-\frac{1}{2}+i\rho, 0}(\eta") \; d_{m\ 0}^{-\frac{1}{2}+i\rho, 0*}(\eta')$$

$$\times \int_0^\infty \exp\left\{ i\left[(E^2 - \mu^2)/2(\cosh\eta" - \cos\xi")(\cosh\eta' - \cos\xi')\right]\sigma \right.$$
$$\left. + (i/2a^2)\left(\rho^2 - (s-g)^2 + \tfrac{1}{16}\right)\sigma \right\} d\sigma , \tag{46}$$

and integration over $\sigma$ gives the result,

$$G(x", x') = (i/32\pi^3 a^3)\left[(\cosh\eta" - \cos\xi")(\cosh\eta' - \cos\xi')\right]^{5/2} \sum_{m=-\infty}^{+\infty} \sum_{s=-\infty}^{+\infty}$$

$$\times \int_{-\infty}^{+\infty} dE \, \exp\left[iE(t" - t')\right] \exp\left[im(\phi" - \phi')\right] \exp\left[is(\xi" - \xi')\right]$$

$$\times \int_0^\infty d\rho \, 2\rho \, \tanh(\pi\rho) \; d_{m\ 0}^{-\frac{1}{2}+i\rho, 0}(\eta") \; d_{m\ 0}^{-\frac{1}{2}+i\rho, 0*}(\eta')$$

$$\times \left\{ \left[(E^2 - \mu^2)/2\right] + \left[(1/2a^2)\left(\rho^2 - (s-g)^2 + \tfrac{1}{16}\right)\right] \right.$$
$$\left. \times (\cosh\eta" - \cos\xi")(\cosh\eta' - \cos\xi')\right] + i\varepsilon \right\}^{-1}, \tag{47}$$

where the limit, $\varepsilon \to 0$, is taken.

# 6 Conclusion

In this paper, a method for solving the relativistic Dirac and Klein-Gordon equations via the path integral approach was discussed. The method, in fact, can take advantage of results obtained from nonrelativistic quantum mechanics to facilitate the path integration for the relativistic case. We considered, in particular, a relativistic spin $\frac{1}{2}$ and spin 0 particle in the presence of a ring-shaped defect taking the toroidal Aharonov-Bohm experiment as an example where an explicit form for the Green function was obtained.

## ACKNOWLEDGEMENTS

Prof. Hiroshi Ezawa is gratefully acknowledged for his stimulating comments which prompted the author to investigate the relativistic toroidal Aharonov-Bohm set-up. This work is supported by a research grant from the National Research Council of the Philippines.

# REFERENCES

1. S. F. Edwards, *Proc. Phys. Soc.* **91** (1967) 513.
2. M. G. Laidlaw and C. M. DeWitt, *Phys. Rev.* **D3** (1971) 1375.
3. L. S. Schulman, *Techniques and Applications of Path Integration* (Wiley, New York, 1981).
4. F. W. Wiegel, *Introduction to Path Integral Methods in Physics and Polymer Science* (World Scientific, Singapore, 1986).
5. C. C. Gerry and V. A. Singh, *Phys. Rev.* **D20** (1979) 2550.
6. C. C. Bernido and A. Inomata, *J. Math. Phys.* **22** (1981) 715.
7. C. C. Bernido, M. V. Carpio-Bernido and A. Inomata, *Phys. Lett.* **A136** (1989) 259.
8. R. C. Ramos, Jr., C. C. Bernido, and M. V. Carpio-Bernido, *J. Phys. A: Math. Gen.* **27** (1994) 8251.
9. See, e.g., M. Peshkin and A. Tonomura. *The Aharonov-Bohm Effect*, Lecture Notes in Physics, vol. **340** (Springer Verlag, Heidelberg, 1989).
10. C. C. Bernido and M. V. Carpio-Bernido, in *Topics in Theoretical Physics*, ed. Y. M. Cho (World Scientific, Singapore, 1997) 200.
11. See, e.g., B. S. DeWitt in *Relativity, Groups and Topology*, C. DeWitt and B. DeWitt (Gordon and Breach, New York, 1964).
12. V. Bargmann, *Ann. Math.* **48** (1947) 568.

# SEMICLASSICAL QUANTUM MECHANICS ON A RIEMANN SURFACE

FREDERIK W. WIEGEL

Center for Theoretical Physics, University of Twente,
P.O. Box 217, 7500 AE Enschede, The Netherlands

Abstract. We study the simplest possible example of a topological quantum phenomenon: the movement of a free quantum mechanical particle on a Riemann surface with infinitely many sheets. The propagator is given, in the exact form as well as in a semiclassical approximation. It is shown that the semiclassical treatment of this topological model differs slightly from the usual WKB treatment which applies to situations in which topology plays no role; the difference is caused by the presence of the Edwards-Gulyaev term in the path integral in polar coordinates. Finally, we suggest that the Edwards-Gulyaev term sheds some light on the remarkable accuracy of the Geometric Theory of Diffraction.

## 1. INTRODUCTION

Semiclassical quantum mechanics is a separate part of physics, appropriate for dynamical processes the classical action of which is large as compared to Planck's constant. For quantum phenomena which have a topological aspect the semiclassical theory is not very well developed. As these phenomena are of some current interest (cf. the monographs [1,2]) we try to improve that state of affairs, using a particular model system for which both the exact quantum theory, and the proper form of the classical trajectories can be solved explicitly. The model system consists of a free particle which moves on a Riemann surface with infinitely many sheets (this is also the topology relevant to the Aharonov-Bohm effect).

In this early stage we would like to stress the difference between the real, physical space and the universal covering space for the topology under consideration. Real space is the Cartesian space of pairs $(x,y)$ with $-\infty<x<+\infty$, $-\infty<y<+\infty$. The points of real space tell you where the particle is, but not how it came to this position. On the other hand, the universal covering space is the set of polar coordinates $(r,\varphi)$ with $0<r<\infty$, $-\infty<\varphi<+\infty$. The points of this covering space correspond uniquely to the topology of the path of the particle, i.e. they tell you not only where the particle is, but also how often it had to wind around the origin of coordinates to come to its present position.

We now discuss the proper form of the Lagrangian. In real space it has the form

$$\mathcal{L}_0 = \frac{1}{2} m \left( \dot{x}^2 + \dot{y}^2 \right) \tag{1.1}$$

in an obvious notation. In the coordinates of the universal covering space the Lagrangian is found by substituting

$$x = r \cos \varphi, \tag{1.2a}$$

$$y = r \sin \varphi; \tag{1.2b}$$

this gives

$$\mathcal{L}_0 = \frac{1}{2} m \left( \dot{r}^2 + r^2 \dot{\varphi}^2 \right). \tag{1.3}$$

The classical action is the integral of the Lagrangian over time. It is perhaps appropriate to point out now already that (1.3) is *not* the correct form of the Lagrangian in the Feynman path integral. This was first realized by Edwards and Gulyaev, ref. [3].

The contents of this paper are as follows. In section 2 we give the exact form of the quantum mechanical propagator and the asymptotic form of the propagator in the semiclassical regime. Section 3 comments on the effective Lagrangian of a Feynman path integral in polar coordinates. In section 4 the classical trajectories which follow from the effective Lagrangian are studied. In section 5 we evaluate the classical action for the case in which the two points are on the same sheet of the Riemann surface and for the case where these points are situated on different sheets of the Riemann surface and we briefly comment on the geometric theory of diffraction.

## 2. THE QUANTUM PROPAGATOR

The propagator for a particle on the Riemann surface with infinitely many sheets follows from taking the limit $N \to \infty$ in the propagator $(G_N)$ for a particle on a Riemann surface with $N$ sheets. This Riemann surface has $0 \leq \varphi < 2\pi N$. Taking the limit $N \to \infty$ we find for the propagator of the particle on the Riemann surface with infinitely many sheets, in polar coordinates

$$G\left(r'', \varphi'', t | r', \varphi'\right) = \frac{m}{2\pi i \hbar t} \exp\left\{\frac{i m \left(r'^2 + r''^2\right)}{2\hbar t}\right\} \cdot$$

(2.1)

$$\int_{-\infty}^{\infty} i^{|\nu|} J_{|\nu|}\left(-\frac{m r' r''}{\hbar t}\right) \exp\left\{i\nu\left(\varphi''-\varphi'\right)\right\} d\nu.$$

The details of the calculations in this note will be published in a separate paper [5].

The exact expression (2.1) for the quantum propagator contains, as a factor, the integral

$$F(z) \equiv \int_{-\infty}^{\infty} i^{|\nu|} J_{|\nu|}(-z) \exp\left\{i\nu\left(\varphi''-\varphi'\right)\right\} d\nu,$$

(2.2)

which, to the best of my knowledge, cannot be evaluated in terms of "simple" functions.

The parameter

$$z = \frac{m r' r''}{\hbar t}$$

(2.3)

will be $\gg 1$ in the semiclassical region. We shall, therefore, now consider the question: what is the correct asymptotic form of the function $F(z)$ for $z \gg 1$? A straightforward, but somewhat lengthy analysis [5] gives the results

$$F(z) \cong \exp\left\{-iz \cos\left(\varphi''-\varphi'\right)\right\} +$$

$$\frac{1}{i\sqrt{i}} \sqrt{\frac{2\pi}{z}} \left\{\pi^2 - \left(\varphi''-\varphi'\right)^2\right\}^{-1} \exp(iz), \quad \left(\left|\varphi''-\varphi'\right| < \pi\right),$$

(2.4a)

$$F(z) \cong \frac{1}{i\sqrt{i}} \sqrt{\frac{2\pi}{z}} \left\{\pi^2 - \left(\varphi''-\varphi'\right)^2\right\}^{-1} \exp(iz), \quad \left(\left|\varphi''-\varphi'\right| > \pi\right).$$

(2.4b)

The total exponential factor in G is, in the case $\left|\varphi''-\varphi'\right| > \pi$ where the final point is on a different sheet of the Riemann surface than in the initial point, given by

$$\exp\left\{\frac{im}{2\hbar t}\left(r'^2 + r''^2\right) + iz\right\} = \exp\left\{\frac{im}{2\hbar t}\left(r' + r''\right)^2\right\}.$$

(2.5)

This shows that - in leading order of z - the "semi-classical" particle travels from r' to 0, then from 0 to r'', all in time t.

## 3. THE EFFECTIVE LAGRANGIAN

We now turn to the question how one could find the quantum propagator from the Feynman path integral instead of from the Schrödinger equation. That is, we ask how one should calculate the Feynman path integral, in polar coordinates, for this particle on the Riemann surface:

$$G\left(r'', \varphi'', t|r', \varphi'\right) = \int \exp\left(\frac{i}{\hbar} \int_0^t \mathcal{L}_{\text{eff}} \, dt'\right) d[r(t')] \, d[\phi(t')], \qquad (3.1)$$

where $\mathcal{L}_{\text{eff}}$ denotes the effective Lagrangian

$$\mathcal{L}_{\text{eff}} = \frac{1}{2} \, m \left(\dot{r}^2 + r^2 \dot{\varphi}^2\right) + \frac{\hbar^2}{8mr^2} \, . \qquad (3.2)$$

Note the Edwards-Gulyaev correction term $\hbar^2 / 8mr^2$, which is not due to an external potential $-\hbar^2 / 8mr^2$, but instead arises as a result of writing the path integral in terms of curvilinear coordinates (the polar coordinates r, $\varphi$ in this case). There is a fairly extensive literature which tries to justify this somewhat odd-looking term from first principles. This is a delicate task because the Feynman path integral does not yet have a fully satisfactory mathematical foundation. So I would like to give two more reasons to believe in the validity of the Edwards-Gulyaev correction term:

(i) One can evaluate the path integral (3.1), with the effective Lagrangian (3.2), rigorously and one finds the correct result (2.1) indeed. This calculation can be found in ref [4].

(ii) For the case of entangled macromolecules one finds Wiener path integrals for which there is a proper mathematical foundation. Here a similar term appears in the path integral; upon an analytic continuation which transforms the Wiener into the Feynman path integral this term is transformed into $\hbar^2 / 8mr^2$.

## 4. THE CLASSICAL TRAJECTORIES

In order to develop the semiclassical quantum mechanics one has to determine the paths for which the integrand of (3.1) is stationary, i.e. the solutions of the variational problem

$$\delta \int_0^t \mathcal{L}_{eff} \, dt' = 0, \tag{4.1}$$

which pass through $(r', \varphi')$ at t=0 and through $(r'', \varphi'')$ at time t. These solutions can be interpreted as the classical trajectories of a particle on the Riemann surface with infinitely many sheets. Because $\mathcal{L}_{eff}$ differs from the physical Lagrangian (1.3) these trajectories will differ from the classical trajectories of a free particle in a real plane; the latter are straight lines whereas the former will turn out to be curves which can wind many times around the origin in order to connect initial and final points situated on different sheets of the Riemann surface. It is in this sense that the semiclassical quantum mechanics for this topology differs slightly from the usual WKB result.

The analysis of the classical trajectories is straightforward but somewhat intricate; for its details we again refer to [5]. For most values of the parameters the shape of the trajectory is of the form

$$r(\varphi) = \frac{r_0}{\cos\left\{(\varphi - \varphi_0) \sqrt{1 - \frac{\hbar^2}{4m^2C^2}}\right\}}, \tag{4.2}$$

where the three constants $r_0$, $\varphi_0$ and C have to be determined by the two equations $r(\varphi') = r'$, $r(\varphi'') = r''$ and by the value of t. One finds that, for a given value of $\varphi'' - \varphi'$ one should first solve $\gamma$ from the transcendent equation

$$2z \frac{\sin \gamma}{\gamma} = \left\{(\varphi'' - \varphi')^2 - \gamma^2\right\}^{-1/2}, \tag{4.3}$$

after which the value of C can be solved from

$$\gamma = (\varphi'' - \varphi') \sqrt{1 - \frac{\hbar^2}{4m^2C^2}} . \tag{4.4}$$

The trajectories (4.2) look essentially as follows: from the initial point $(r', \varphi')$ the path will head for the branchpoint of the Riemann surface (i.e. the origin) along an almost straight line; it will encircle the origin several times, at a small distance, in a spiralling movement; finally it will move away from the vicinity of the origin towards the final point $(r'', \varphi'')$ along an almost straight line.

## 5. THE ACTION FOR THE CLASSICAL TRAJECTORIES.

It is again straightforward, but somewhat tedious, to calculate the action for the classical trajectories. One has to distinguish between the cases $|\varphi''-\varphi'| < \pi$, in which propagation happens between two points on the same sheet, and the case $|\varphi''-\varphi'| > \pi$, in which the two points are located on different sheets. The details will be published in ref. [5].

In the first case $|\varphi''-\varphi'| < \pi$ the action S is found to be given by

$$S = \frac{m}{2t} \left\{ r'^2 + r''^2 - 2r'r'' \cos\left(\varphi''-\varphi'\right) \right\}. \tag{5.1a}$$

In the second case $|\varphi''-\varphi'| > \pi$ one finds, in the semiclassical regime $z \gg 1$, the asymptotic result

$$S \cong \frac{m}{2t} \left( r'^2 + 2r'r'' + r''^2 \right). \tag{5.1b}$$

This means that the leading term in the exponent of the semiclassical propagator is found to have the following form

$$G\left(r'',\varphi'',t \mid r',\varphi'\right) \cong \exp\left\{ \frac{im}{2\hbar t} \left(r'^2 + r''^2\right) \right\} \exp\left\{ -iz\cos\left(\varphi''-\varphi'\right) \right\}, \tag{5.2a}$$
$$\left(|\varphi''-\varphi'| < \pi\right)$$

$$G\left(r'',\varphi'',t \mid r',\varphi'\right) \cong \exp\left\{ \frac{im}{2\hbar t} (r'+r'')^2 \right\}, \quad \left(|\varphi''-\varphi'| > \pi\right). \tag{5.2b}$$

Comparison with the exact results of section 2 shows that (5.2) actually reproduces the leading terms in the propagator, also in the case in which the endpoints are situated on different sheets. In this case, in which the de Broglie waves propagate between different sheets of the Riemann surface, the classical trajectory has - for $z \gg 1$ - essentially the form of the diffraction ray in the geometric theory of diffraction (c.f. ref. [6]): the trajectory moves from $(r',\varphi')$ practically in a straight line towards a small vicinity of the branchpoint $r=0$, winds around the origin several times to reach the proper sheet of the Riemann surface, and finally again moves towards $(r'',\varphi'')$ practically in a straight line. The calculations in this note therefore seem to suggest that the geometric theory of diffraction is "nothing but" the semiclassical theory for those cases in which a topological constraint is present. It is quite remarkable that this can be shown in a quantitative way only when the Edwards-Gulyaev correction term in the path integral in the universal covering space is taken into account. In this way it becomes clear that two somewhat mysterious aspects of Feynman's form of quantum mechanics are linked: the correction terms that appear for curvilinear coordinates, and the remarkable accuracy of the geometric theory of diffraction, which has fascinated many authors (c.f. for example, ref. [7]).

## ACKNOWLEDGMENTS

I am thankful to Prof. C.C. Bernido and to Dr. M.V. Carpio-Bernido for the organization of this meeting, and to Prof. H. Ezawa for this opportunity to meet him and several of his colleagues.

## REFERENCES

[1]  Yang, C.N. and Ge, M.L. (1989) *Braid Group, Knot Theory and Statistical Mechanics*. World Scientific, Singapore and London.
[2]  Wilczek, G.(1991) *Fractional Statistic in Action,* World Scientific, Singapore and London.
[3]  Edwards, S.F. and Gulyaev, Y.V. (1964) Proc. R. Soc. London A279, 229.
[4]  Khandekar, D.C., Bhagwat, K.V. and Wiegel, F.W. (1988). Phys. Lett. 127A, 379.
[5]  Wiegel, F.W. *Quantum Mechanics on a Riemann Surface*, Unpublished Report.
[6]  Hansen, R.C. Editor (1981) *Geometric Theory of Diffraction*, IEEE Press.
[7]  DeWitt-Morette, C., Low, S.G., Schulman, L.S, and Shiekh, A.Y. (1988) *Wedges I*. In: *Between Quantum and Cosmos*, Zurek, W., Merwe, A. van der and Miller, W.A., Eds., Princeton University Press, 459-497.

# ON THE PATH INTEGRAL EVALUATION OF THE S-MATRIX FOR NONCENTRAL POTENTIALS

M. V. CARPIO-BERNIDO[a], E. B. GRAVADOR[b], and C. C. BERNIDO[a,c]

[a]Research Center for Theoretical Physics, Central Visayan Institute
  6308 Jagna, Bohol, Philippines
[b]Department of Physics, MSU-Iligan Institute of Technology
  9200 Iligan City, Philippines
[c]National Institute of Physics, University of the Philippines
  1101 Diliman, Quezon City, Philippines

**Abstract**    Path integral representations of the S-matrix in polar coordinates for noncentral problems are discussed and applied to one of the classes of potentials with dynamical symmetries included in the work of Makarov, Smorodinsky, Valiev and Winternitz.

## 1  Introduction

A general framework for handling various types of collision problems and scattering by force fields, stationary or time-dependent, is provided by the scattering matrix or S-matrix. Since the S-matrix is closely related to the time-dependent Green function or propagator, we expect that we can express it as a Feynman path integral and thereby profit from all the advantages of this powerful and conceptually appealing approach. In 1975, several representations of the S-matrix were given in terms of path integrals, both global and time-sliced, over Cartesian coordinates and momenta by Campbell, Finkler, Jones and Misheloff (CFJM)[1]. A suitable form of the S-matrix from their work was then cast in spherical polar coordinates and applied to scattering from central potentials by Gerry and Singh[2]. A different approach would be to substitute the path integral for the propagator in the S-matrix which is already conveniently expressed in terms of the Green function. This was done by Ho[3] in 1982 to evaluate the S-matrix for the nonrelativistic Coulomb problem.

With later path integral treatments of various nonspherically symmetric potentials[4,5] such as the classes of Makarov, Smorodinsky, Valiev and Winternitz (MSVW)[6], it seemed natural to see whether one could also derive a suitable form of the S-matrix for such explicitly angle-dependent potentials. Apart from the mathematical interest in the functional integral formulation of formal quantum scattering theory for noncentral potentials, the MSVW potentials, for example, have dynamical symmetries and are interesting subjects for superintegrability studies, path integral evaluation in various curvilinear coordinate systems, and investigation of other group theoretical properties, among others[2,7]. On the other hand, physics-wise, the evaluation of the nonrelativistic quantum scattering problem for various exactly solvable nonspherically symmetric potentials could contribute in the improvement of theoretical results for scattering cross-sections arising from anisotropic parts in various interaction potentials. This becomes important, for example, at low temperatures (e.g. millidegrees Kelvin) where there is an increased sensitivity to the details of intermolecular potentials[8].

Using the approach followed by Ho for the Coulomb problem, we obtained the S-matrix for the relevant MSVW potentials and the mathematically similar dyonium and charge-dyon systems[9]. We also tried to adapt the approach of Gerry and Singh in treating the same problem[10]. In the following Sections, we show briefly how this is done and then discuss the attendant questions and open problems we encountered.

## 2   The S-matrix and the propagator

The S-matrix on the momentum basis is given by

$$S(\mathbf{k}", \mathbf{k}') = \langle \mathbf{k}" \mid S \mid \mathbf{k}' \rangle, \tag{1}$$

where the unitary scattering operator $S$ is defined by

$$S = \lim_{\substack{t" \to +\infty \\ t' \to -\infty}} U_I(t", t') \tag{2}$$

in terms of the interaction-picture time-development operator $U_I(t", t')$,

$$U_I(t", t') = \exp(iH_o t") \exp[-iH(t" - t')] \exp(-iH_o t'). \tag{3}$$

In Eq.(3), we take $H_o$ to be the free particle Hamiltonian and $H = H_o + V$. Inserting coordinate states in Eq.(1) and using the relations

$$\langle \mathbf{k}" \mid \exp(iH_o t") \mid \mathbf{x}" \rangle = (2\pi)^{-3/2} \exp[-i(\mathbf{k}" \cdot \mathbf{x}") + it" k"^2/2M] \tag{4}$$

$$\langle \mathbf{x}' \mid \exp(-iH_o t') \mid \mathbf{k}' \rangle = (2\pi)^{-3/2} \exp[i(\mathbf{k}' \cdot \mathbf{x}') - it' k'^2/2M] \tag{5}$$

the S-matrix can be written as

$$
\begin{aligned}
S(\mathbf{k}", \mathbf{k}') = \lim_{\substack{t" \to +\infty \\ t' \to -\infty}} (2\pi)^{-3} \int\!\!\int d^3x" \, d^3x' \; & \exp[-i(\mathbf{k}" \cdot \mathbf{x}") + it" k"^2/2M] \\
\times \, K(\mathbf{x}", \mathbf{x}'; t" - t') \; & \exp[i(\mathbf{k}' \cdot \mathbf{x}') - it' k'^2/2M],
\end{aligned} \tag{6}
$$

where $K(\mathbf{x}", \mathbf{x}'; t" - t') \equiv \langle \mathbf{x}" \mid \exp[-iH(t" - t')] \mid \mathbf{x}' \rangle$ is the quantum mechanical propagator.

For the scattering problem, the method of stationary phase may be used (see, e.g., Ref. 11) to evaluate the exponential terms in Eq.(6). With $\varphi(\mathbf{k}") = -(\mathbf{k}" \cdot \mathbf{x}") + t" k"^2/2M$, we consider the expansion, $\varphi(\mathbf{k}") = \varphi(\mathbf{k}_o) + \frac{1}{2}(\mathbf{k}" - \mathbf{k}_o)^2 \frac{d^2\varphi}{dk"^2}|_{\mathbf{k}_o}$ $+...$, about $\mathbf{k}_o = M \mathbf{x}"/t"$, to get

$$\exp[-i(\mathbf{k}" \cdot \mathbf{x}") + it" k"^2/2M] \underset{|t"| \to \infty}{\simeq} \exp(-iMx"^2/2t")$$
$$\times \exp[-M(\mathbf{x}" - t"\mathbf{k}"/M)^2/2it"], \qquad (7)$$

and similarly for

$$\exp[i(\mathbf{k}' \cdot \mathbf{x}') - it' k'^2/2M] \underset{|t'| \to -\infty}{\simeq} \exp(iMx'^2/2t')$$
$$\times \exp[-iM(\mathbf{x}' - t'\mathbf{k}'/M)^2/2t']. \qquad (8)$$

Furthermore, we use the definition

$$\delta(\mathbf{x} - \mathbf{x}_o) = \lim_{\in \to 0} (\pi \in^2)^{-3/2} \exp[-(\mathbf{x} - \mathbf{x}_o)^2/\in^2] \qquad (9)$$

to write the S-matrix as

$$S(\mathbf{k}", \mathbf{k}') = \lim_{\tau \to \infty} (i\tau/2M)^3 \int \int d^3x" \, d^3x' \, \exp[-(iM/\tau)(x"^2 + x'^2)$$
$$\times \delta(\mathbf{x}" - \tau\mathbf{k}"/2M) \, \delta(\mathbf{x}' + \tau\mathbf{k}'/2M) \, K(\mathbf{x}", \mathbf{x}'; \tau) \qquad (10)$$

where $\tau = t" - t'$. The integrations in Eq.(10) can then be performed yielding

$$S(\mathbf{k}", \mathbf{k}') = \lim_{\tau \to \infty} (i\tau/2M)^3 \exp[-(i\tau/4M)(k"^2 + k'^2)] K(\tau\mathbf{k}"/2M, -\tau\mathbf{k}'/2M; \tau).$$
$$(11)$$

Clearly, if we have a suitable form for the propagator, the S-matrix expressed as in Eq.(11) or Eq.(10) is convenient for use in treating a quantum scattering problem. The propagator may be obtained by evaluating the Feynman path integral

$$K(\mathbf{x}", \mathbf{x}'; \tau) = \int \exp(iS) \, \mathcal{D}[x], \qquad (12)$$

so that technically, we have a path integral representation of the S-matrix,

$$S(\mathbf{k}", \mathbf{k}') = \lim_{\tau \to \infty} (i\tau/2M)^3 \int \int d^3x" \, d^3x' \, \exp[-(iM/\tau)(x"^2 + x'^2)$$

$$\times \, \delta(\mathbf{x}" - \tau\mathbf{k}"/2M) \, \delta(\mathbf{x}' + \tau\mathbf{k}'/2M) \int \exp(iS) \, \mathcal{D}[x], \qquad (13)$$

or,

$$S(\mathbf{k}", \mathbf{k}') = \lim_{\tau \to \infty} (i\tau/2M)^3 \exp[-(i\tau/4M)(k"^2 + k'^2)] \int \exp(iS) \, \mathcal{D}[\tau k/2M]. \quad (14)$$

Furthermore, the S-matrix can be expressed in terms of the energy Green function, $G(\mathbf{x}", \mathbf{x}'; E) = i \int K(\mathbf{x}", \mathbf{x}'; \tau) \exp(iE\tau) \, d\tau$. This is applicable to the Coulomb potential, for which only the Green function is available in closed form, and not the propagator. This approach was followed by Ho[3] for the path integral evaluation of the Coulomb S-matrix.

## 3　Application to an MSVW potential

Consider the class of nonspherically symmetric potentials,

$$V_A(r, \vartheta, \phi) = \frac{\alpha + \gamma \cos \phi}{r^2 \sin^2 \vartheta \sin^2 \phi} + \frac{\beta}{r^2 \cos^2 \vartheta} \qquad , \qquad (15)$$

appearing among systems with dynamical symmetries discussed by MSVW[6]. The propagator for $V_A(r, \vartheta, \phi)$ has been evaluated by path integration in Ref. 4:

$$K(\mathbf{x}", \mathbf{x}'; \tau) = \sum_{\ell=0}^{\infty} \sum_{m=0}^{\infty} C_{\ell m} \, \Omega_{\ell m}(\vartheta", \phi") \, \Omega_{\ell m}(\vartheta', \phi') \exp\left[\frac{-i\pi}{2}(\ell' + \tfrac{1}{2})\right]$$

$$\times \left(\frac{M}{i\tau}\right) (r'r")^{-1/2} \exp\left[\frac{iM}{2\tau}(r"^2 + r'^2)\right] \, J_{\ell' + \frac{1}{2}}\left(\frac{Mr'r"}{\tau}\right) \quad (16)$$

where $J_{\ell' + \frac{1}{2}} (Mr'r"/\tau)$ is a Bessel function, $\ell' = \frac{1}{2}(\lambda_1 + \lambda_2) + \lambda_3 + m + 2\ell + 1$, $\lambda_1 = \frac{1}{2}[1 + 8(\alpha - \gamma)M]^{\frac{1}{2}}$, $\lambda_2 = \frac{1}{2}[1 + 8(\alpha + \gamma)M]^{\frac{1}{2}}$, and $\lambda_3 = \frac{1}{2}(1 + 8\beta M)^{\frac{1}{2}}$. The angular functions are given in terms of the Jacobi polynomials $P_\ell^{\mu, \nu}(\xi)$,

$$\Omega_{\ell m}(\vartheta, \phi) = (\sin \vartheta)^{\frac{1}{2}(\lambda_1 + \lambda_2) + m + \frac{1}{2}} (\cos \vartheta)^{\lambda_3 + \frac{1}{2}} [\sin(\phi/2)]^{\lambda_2 + \frac{1}{2}} [\cos(\phi/2)]^{\lambda_1 + \frac{1}{2}}$$

$$\times P_\ell^{\frac{1}{2}(\lambda_1 + \lambda_2) + m + \frac{1}{2}, \lambda_3}(\cos 2\vartheta) \, P_m^{\lambda_2, \lambda_1}(\cos \phi) \quad (17)$$

with normalization constants

$$C_{\ell m} = \left[ \frac{[\frac{1}{2}(\lambda_1 + \lambda_2) + \lambda_3 + m + 2\ell + \frac{3}{2}]\Gamma[\frac{1}{2}(\lambda_1 + \lambda_2) + \lambda_3 + m + \ell + \frac{3}{2}]}{\Gamma(\lambda_3 + \ell + 1) \ \Gamma(\frac{1}{2}\lambda_1 + \frac{1}{2}\lambda_2 + m + \ell + \frac{3}{2})} \right]$$

$$\times \left[ \frac{\ell! \ 2(\lambda_1 + \lambda_2 + 2m + 1) \ m! \ \Gamma(\lambda_1 + \lambda_2 + m + 1)}{\Gamma(\lambda_1 + m + 1) \ \Gamma(\lambda_2 + m + 1)} \right]. \tag{18}$$

With Eq.(16), the S-matrix of Eq.(10) can be expressed in polar coordinates as

$$
\begin{aligned}
S(\mathbf{k}", \mathbf{k}') = \lim_{\tau \to \infty} & \left( \frac{i\tau}{2M} \right)^3 \sum_{\ell=0}^{\infty} \sum_{m=0}^{\infty} C_{\ell m} \int \int [dr" \ d(\cos\vartheta") \ d\phi"] [dr' \ d(\cos\vartheta') \ d\phi'] \\
& \times \ \delta(r" - \tau k"/2M) \ \delta\left( \cos\vartheta" - \cos\widehat{\phi}" \right) \ \delta\left( \phi" - \widehat{\phi}" \right) \\
& \times \ \delta(r' - \tau k'/2M) \ \delta\left( \cos\vartheta' + \cos\widehat{\phi}' \right) \ \delta\left( \phi' - \widehat{\phi}' - \pi \right) \\
& \times \ (\sin\vartheta")^{\frac{1}{2}(\lambda_1+\lambda_2)+m+\frac{1}{2}} \ (\cos\vartheta")^{\lambda_3+\frac{1}{2}} \ P_\ell^{\frac{1}{2}(\lambda_1+\lambda_2)+m+\frac{1}{2},\lambda_3}(\cos 2\vartheta") \\
& \times \ [\sin(\phi"/2)]^{\lambda_2+\frac{1}{2}} \ [\cos(\phi"/2)]^{\lambda_1+\frac{1}{2}} \ P_m^{\lambda_2,\lambda_1}(\cos\phi") \\
& \times (\sin\vartheta')^{\frac{1}{2}(\lambda_1+\lambda_2)+m+\frac{1}{2}} \ (\cos\vartheta')^{\lambda_3+\frac{1}{2}} \ P_\ell^{\frac{1}{2}(\lambda_1+\lambda_2)+m+\frac{1}{2},\lambda_3}(\cos 2\vartheta') \\
& \times \ [\sin(\phi'/2)]^{\lambda_2+\frac{1}{2}} \ [\cos(\phi'/2)]^{\lambda_1+\frac{1}{2}} \ P_m^{\lambda_2,\lambda_1}(\cos\phi') \\
& \times \left( \frac{M}{i\tau} \right) \exp\left[ \frac{-i\pi}{2}(\ell + \tfrac{1}{2}) \right] \\
& \times \ (r'r")^{-1/2} \exp\left[ \frac{iM}{2\tau}(r"^2 + r'^2) \right] \ J_{\ell+\frac{1}{2}}\left( \frac{Mr'r"}{\tau} \right). \tag{19}
\end{aligned}
$$

The polar coordinates $\mathbf{x}" = (r", \vartheta", \phi")$, $\mathbf{x}' = (r', \vartheta', \phi')$, $\mathbf{k}" = (k", \widehat{\vartheta}", \widehat{\phi}")$, and $\mathbf{k}' = (k', \widehat{\vartheta}', \widehat{\phi}')$ describe the initial and final coordinate and momentum vectors, respectively.

For the scattering problem, we can make use of the large argument expansion of the Bessel function in Eq.(19):

$$J_{\ell+\frac{1}{2}}(z) \xrightarrow[z \to \infty]{} \sqrt{1/2\pi z} \ (-i) \ [\exp(iz)\exp(-i\pi\ell'/2) - \exp(-iz)\exp(i\pi\ell'/2)]. \tag{20}$$

We thus have

$$\lim_{\tau \to \infty} \left( \frac{i\tau}{2M} \right)^3 \left( \frac{M}{i\tau} \right) \int_0^\infty \int_0^\infty dr" \ dr' \ \delta(r" - \tau k"/2M) \ \delta(r' - \tau k'/2M)$$

$$\times \ (r'r")^{-1/2} \exp\left[\frac{iM}{2\tau}(r''^2 + r'^2)\right] J_{\ell'+\frac{1}{2}}\left(\frac{Mr'r"}{\tau}\right)$$

$$= \lim_{\tau\to\infty} \frac{i}{2} \ (2\pi M/\tau)^{-1/2}(k'k")^{-1}\{\exp[-(i\tau/8M)(k'^2 - 2k'k" + k"^2)]\exp(-i\pi\ell'/2)$$

$$- \exp[-(i\tau/8M)(k'^2 + 2k'k" + k"^2)]\exp(i\pi\ell'/2)\}$$

$$= \sqrt{i}(k'k")\{\exp(-i\pi\ell'/2)\delta(k" - k') - \exp(i\pi\ell'/2)\delta(k" + k')\}, \tag{21}$$

where, in the last line, we have used the relation Eq.(9).

Performing the angular integrations in Eq.(19), we get

$$S(\mathbf{k}",\mathbf{k}') = (k'k")^{-1}\delta(k" - k') \sum_{\ell=0}^{\infty}\sum_{m=0}^{\infty} e^{2i\Delta_{\ell m}}C_{\ell m}(\sin\widehat{\vartheta}")^{\frac{1}{2}(\lambda_1+\lambda_2)+m+\frac{1}{2}}$$

$$\times(\cos\widehat{\vartheta}")^{\lambda_3+\frac{1}{2}} P_{\ell}^{\frac{1}{2}(\lambda_1+\lambda_2)+m+\frac{1}{2},\lambda_3}(\cos 2\widehat{\vartheta}")$$

$$\times\left[\sin(\widehat{\phi}"/2)\right]^{\lambda_2+\frac{1}{2}}\left[\cos(\widehat{\phi}"/2)\right]^{\lambda_1+\frac{1}{2}} P_m^{\lambda_2,\lambda_1}(\cos\widehat{\phi}")$$

$$\times(\sin\widehat{\vartheta}')^{\frac{1}{2}(\lambda_1+\lambda_2)+m+\frac{1}{2}}(\cos\widehat{\vartheta}')^{\lambda_3+\frac{1}{2}} P_{\ell}^{\frac{1}{2}(\lambda_1+\lambda_2)+m+\frac{1}{2},\lambda_3}(\cos 2\widehat{\vartheta}')$$

$$\times\left[\sin(\widehat{\phi}'/2)\right]^{\lambda_2+\frac{1}{2}}\left[\cos(\widehat{\phi}'/2)\right]^{\lambda_1+\frac{1}{2}} P_m^{\lambda_2,\lambda_1}(\cos\widehat{\phi}') . \tag{22}$$

where $\Delta_{\ell m} = -\frac{1}{2}\pi[\frac{1}{2}(\lambda_1+\lambda_2)+2\lambda_3+2m+2\ell+\frac{3}{2}]$. In Eq.(22) we have not included the $\delta(k" + k')$-term in Eq.(21) as this makes no contribution.

To get the scattering amplitude $f(\mathbf{k}",\mathbf{k}')$, we compare Eq.(22) with[12]

$$S(\mathbf{k}",\mathbf{k}') = \delta(\mathbf{k}" - \mathbf{k}') + (i/2\pi k')\delta(k" - k')f(\mathbf{k}",\mathbf{k}'). \tag{23}$$

With the delta function expansion in Jacobi polynomials,

$$\delta(\cos\vartheta" - \cos\vartheta')\ \delta(\phi" - \phi') = \sum_{\ell=0}^{\infty}\ \sum_{m=0}^{\infty} C_{\ell m}\ (\sin\widehat{\vartheta}")^{\frac{1}{2}(\lambda_1+\lambda_2)+m+\frac{1}{2}}\ (\cos\widehat{\vartheta}")^{\lambda_3+\frac{1}{2}}$$

$$\times P_{\ell}^{\frac{1}{2}(\lambda_1+\lambda_2)+m+\frac{1}{2},\lambda_3}(\cos 2\widehat{\vartheta}")P_m^{\lambda_2,\lambda_1}(\cos\widehat{\phi}")$$

$$\times[\sin(\widehat{\phi}"/2)]^{\lambda_2+\frac{1}{2}}[\cos(\widehat{\phi}"/2)](\cos\widehat{\vartheta}')^{\lambda_3+\frac{1}{2}}$$

$$\times(\sin\widehat{\vartheta}')^{\frac{1}{2}(\lambda_1+\lambda_2)+m+\frac{1}{2}} P_{\ell}^{\frac{1}{2}(\lambda_1+\lambda_2)+m+\frac{1}{2},\lambda_3}(\cos 2\widehat{\vartheta}')$$

$$\times[\sin(\widehat{\phi}'/2)]^{\lambda_2+\frac{1}{2}} [\cos(\widehat{\phi}'/2)]^{\lambda_1+\frac{1}{2}} P_m^{\lambda_2,\lambda_1}(\cos\widehat{\phi}') \tag{24}$$

where $C_{\ell m}$ is given in Eq.(18), we obtain the scattering amplitude:

$$f(\mathbf{k}",\mathbf{k}') = (4\pi/k") \sum_{\ell=0}^{\infty}\ \sum_{m=0}^{\infty} C_{\ell m}e^{i\Delta_{\ell m}} \sin\Delta_{\ell m}(\sin\widehat{\vartheta}")^{\frac{1}{2}(\lambda_1+\lambda_2)+m+\frac{1}{2}}\ (\cos\widehat{\vartheta}")^{\lambda_3+\frac{1}{2}}$$

$$\times \, P_\ell^{\frac{1}{2}(\lambda_1+\lambda_2)+m+\frac{1}{2},\lambda_3}(\cos 2\widehat{\vartheta"})$$
$$\times [\sin(\widehat{\phi"}/2)]^{\lambda_2+\frac{1}{2}} \, [\cos(\widehat{\phi"}/2)]^{\lambda_1+\frac{1}{2}} \, P_m^{\lambda_2,\lambda_1}(\cos \widehat{\phi"})$$
$$\times (\sin \widehat{\vartheta'})^{\frac{1}{2}(\lambda_1+\lambda_2)+m+\frac{1}{2}} \, (\cos \widehat{\vartheta'})^{\lambda_3+\frac{1}{2}} \, P_\ell^{\frac{1}{2}(\lambda_1+\lambda_2)+m+\frac{1}{2},\lambda_3}(\cos 2\widehat{\vartheta'})$$
$$\times [\sin(\widehat{\phi'}/2)]^{\lambda_2+\frac{1}{2}} \, [\cos(\widehat{\phi'}/2)]^{\lambda_1+\frac{1}{2}} \, P_m^{\lambda_2,\lambda_1}(\cos \widehat{\phi'}). \tag{25}$$

Here, $\Delta_{\ell m} = -\frac{\pi}{2}[\frac{1}{2}(\lambda_1+\lambda_2)+2\lambda_3+2m+2\ell+\frac{3}{2}]$ is the phase shift in analogy to the central potential case. The differential scattering cross-section is then obtained from $d\sigma/d\Omega = |f(\mathbf{k"},\mathbf{k'})|^2$ .

## 4   Discussion

In the procedure given above, we simply substituted in Eq.(10) the propagator, obtained by whatever method, and expressed in asymptotic form for scattering, to get the corresponding S-matrix. We could very well ask whether the S-matrix can be expressed as a path integral at an earlier stage to give more weight to having the procedure called, *a path integral formulation*. In fact, this is what Campbell et al[1] did, followed by Gerry and Singh[2]. Of course, as expected, there is a stage in their procedure that eventually coincides with a result given in the previous Section. We show this briefly.

In the work of CFJM, they start with (note, we use their notation for convenience) the S-matrix given by

$$S(\mathbf{p"},\mathbf{p'}) = \lim_{\tau\to+\infty} \langle \mathbf{p"} \mid U_I(\tau/2,-\tau/2) \mid \mathbf{p'} \rangle, \tag{26}$$

where

$$U_I(t",t') = I - i \int_{t'}^{t"} dt H_I(t) U_I(t,t'), \tag{27}$$

$$H_I(t) = \exp(ip^2 t/2M) V(\mathbf{q}) \exp(-ip^2 t/2M) \tag{28}$$

in the interaction representation. For path integration, the time interval is broken up in segments,

$$\langle \mathbf{p"} \mid U_I(\tau/2,-\tau/2) \mid \mathbf{p'} \rangle = \int \cdots \int \prod_{j=0}^{N-1} \langle \mathbf{p}_{j+1} \mid U_I(t_{j+1},t_j) \mid \mathbf{p}_j \rangle \prod_{j=1}^{N} d^3 p_j \tag{29}$$

where $t_j = -\frac{\tau}{2} + \left(\frac{\tau}{N}\right)_j$ and $\mathbf{p}_N = \mathbf{p"}, \mathbf{p}_0 = \mathbf{p'}, j = 0,1,2,...N$. Also,

$$\langle \mathbf{p}_{j+1} \mid U_I(t_{j+1}, t_j) \mid \mathbf{p}_j \rangle \cong \langle \mathbf{p}_{j+1} \mid I - i(\tau/N)H_I(t_j) \mid \mathbf{p}_j \rangle$$
$$= \langle \mathbf{p}_{j+1} \mid \mathbf{p}_j \rangle - i(\tau/N)\exp[i(t_j/2M)(p_{j+1}^2 - p_j^2)]\langle \mathbf{p}_{j+1} \mid V \mid \mathbf{p}_j \rangle$$
$$= \int d^3 q_{j+1}\{(2\pi)^{-3}\exp[-i\mathbf{q}_{j+1} \cdot (\mathbf{p}_{j+1} - \mathbf{p}_j)]\} - \int d^3 q_{j+1}\{(2\pi)^{-3}$$
$$(\tau/N)V(\mathbf{q}_{j+1})\exp[-i\mathbf{q}_{j+1} \cdot (\mathbf{p}_{j+1} - \mathbf{p}_j) - i(t_j/2M)(p_{j+1}^2 - p_j^2)]\}. \quad (30)$$

Taking $\mathbf{q}_j \to \mathbf{q}_j - (t_{j-1}/2M)(\mathbf{p}_j + \mathbf{p}_{j-1})$ and substituting the transformed Eq.(30) in Eq.(26) gives the S-matrix as

$$S(\mathbf{p}'', \mathbf{p}') = \lim_{\tau \to +\infty} \exp[i(p''^2 + p'^2)\tau/4M] \lim_{N \to +\infty} \int \int \prod_{j=1}^{N-1} d^3 p_j \prod_{j=1}^{N} d^3 q_j$$

$$\times (2\pi)^{-3} \exp\{-i \sum_{j=1}^{N}[\mathbf{q}_j \cdot (\mathbf{p}_j - \mathbf{p}_{j-1})]\}$$

$$\times \exp\{-i \sum_{j=1}^{N}[(\tau/N)(p_j^2/2M) + (\tau/N)V(\mathbf{q}_j)]\}. \quad (31)$$

It is the representation of the S-matrix above, among others, that Gerry and Singh transformed to spherical polar coordinates and applied to central potentials[2]. With $\mathbf{q}_j \cdot (\mathbf{p}_j - \mathbf{p}_{j-1}) = r_j p_j \cos \Theta_j - r_j p_{j-1} \cos \Theta_j'$, where $\cos \Theta_j = \cos \theta_j \cos \theta_{p_j} + \sin \theta_j \sin \theta_{p_j} \cos(\phi_j - \phi_{p_j})$ and $\cos \Theta_j' = \cos \theta_j \cos \theta_{p_{j-1}} + \sin \theta_j \sin \theta_{p_{j-1}} \cos(\phi_j - \phi_{p_{j-1}})$, they used the expansion,

$$\exp(u \cos \Omega) = (\pi/2u)^{1/2} \sum_{\ell=0}^{\infty}(2\ell + 1)P_\ell(\cos \Omega)I_{\ell+\frac{1}{2}}(u), \quad (32)$$

where $P_\ell(\cos \Omega)$ is the Legendre polynomial and $I_{\ell+\frac{1}{2}}(u)$ is the modified Bessel function. Then, the addition theorem,

$$P_\ell(\cos \Omega) = \frac{4\pi}{2\ell + 1} \sum_{m=-\ell}^{\ell} Y_\ell^{m*}(\theta, \phi)Y_\ell^m(\theta_p, \phi_p) \quad (33)$$

for the spherical harmonics $Y_\ell^m(\theta, \phi)$ and the orthogonality relation,

$$\int \int d(\cos \theta_p)d\phi_p Y_\ell^{m*}(\theta_p, \phi_p)Y_{\ell'}^{m'}(\theta_p, \phi_p) = \delta_{\ell\ell'}\delta_{mm'}, \quad (34)$$

allow the straightforward integration over the momentum angles $(\theta_p, \phi_p)$.
The S-matrix can now be written as,

$$S(\mathbf{p}'', \mathbf{p}') = \lim_{\tau \to +\infty} \exp[i(p''^2 + p'^2)\tau/4M] \lim_{N \to +\infty} \int \int (2\pi)^{-3N}(4\pi)^{2N}$$

$$\times \sum_{\ell=0}^{\infty} R_\ell(\mathbf{r}_j; p_j, p_{j-1}) \sum_{m=-\ell}^{\ell} \prod_{j=1}^{N} Y_\ell^{m*}(\theta_j, \phi_j) Y_\ell^{m}(\theta_j, \phi_j)$$

$$\times Y_\ell^{m*}(\theta_p', \phi_p') Y_\ell^{m}(\theta_p'', \phi_p'') \prod_{j=1}^{N-1} p_j^2 dp_j \prod_{j=1}^{N} r_j^2 dr_j d(\cos\theta_j) d\phi_j, \quad (35)$$

where

$$R_\ell(\mathbf{r}_j; p_j, p_{j-1}) = (\pi/2r_j)(p_j p_{j-1})^{-\frac{1}{2}} I_{\ell+\frac{1}{2}}(-ir_j p_j) I_{\ell+\frac{1}{2}}(ir_j p_{j-1})$$

$$\times \exp\{-i(\tau/N)[p_j^2/2M + V(\mathbf{q}_j)]\}. \quad (36)$$

The radial momentum integrations may be done by using Weber's formula,

$$\int_0^{\infty} \exp(i\alpha z^2) I_\nu(-iaz) I_\nu(-ibz) dz = \frac{i}{2\alpha} \exp\left[\frac{-i(a^2 + b^2)}{4\alpha}\right] I_\nu(-iab/2\alpha). \quad (37)$$

This yields,

$$S(\mathbf{p}'', \mathbf{p}') = \lim_{\tau \to +\infty} \exp[i(p''^2 + p'^2)\tau/2M] \lim_{N \to +\infty} \int (2\pi)^{-3N}(4\pi)^{2N}(\pi/2)^N (M/i\epsilon)^{N-1}$$

$$\times \sum_{\ell=0}^{\infty} (p''p')^{-\frac{1}{2}} \exp[(iM/2\epsilon)(r_1^2 + r_N^2)] \exp(-i\epsilon p''^2/2M)$$

$$\times I_{\ell+\frac{1}{2}}(ir_1 p') I_{\ell+\frac{1}{2}}(-ir_N p'') \prod_{j=2}^{N} \exp[(iM/\epsilon)(r_j^2)] I_{\ell+\frac{1}{2}}(-ir_{j-1} r_j M/\epsilon)$$

$$\times \sum_{m=-\ell}^{\ell} \prod_{j=1}^{N} Y_\ell^{m*}(\theta_j, \phi_j) Y_\ell^{m}(\theta_j, \phi_j) Y_\ell^{m*}(\theta_p', \phi_p') Y_\ell^{m}(\theta_p'', \phi_p'')$$

$$\times \exp\{-i\epsilon V(\mathbf{r}_j)\} \prod_{j=1}^{N} r_j dr_j d(\cos\theta_j) d\phi_j, \quad (38)$$

where $\epsilon$ is the time-sliced interval, $t_j - t_{-1}$. For a spherically symmetric potential, $V(\mathbf{r}) = V(r)$, and the angular integrations can be done. This gives,

$$S(\mathbf{p}", \mathbf{p}') = \lim_{\tau \to +\infty} \exp[-i(p"^2 + p'^2)\tau/4M]\,(i\tau/2M)^3$$

$$\times \sum_{\ell=0}^{\infty} K_\ell \left(\frac{p"\tau}{2M}, \frac{p'\tau}{2M}; \tau\right) \sum_{m=-\ell}^{\ell} (-1)^\ell Y_\ell^{m*}(\theta'_p, \phi'_p) Y_\ell^m(\theta_p", \phi_p"),\quad (39)$$

where, for $p"\tau/2M \to r"$ and $p'\tau/2M \to r'$, the radial propagator for the $\ell$-wave is,

$$K_\ell(r", r'; \tau) = \lim_{N \to +\infty} (4\pi)^N (2\pi i\epsilon/M)^{-\frac{3N}{2}} \int (\pi i\epsilon/2M r_{j-1} r_j)^{\frac{1}{2}}$$

$$\times \exp[(iM/2\epsilon)(r_j^2 + r_{j-1}^2) - i\epsilon V(r_j)]$$

$$\times I_{\ell+\frac{1}{2}}(-ir_{j-1} r_j M/\epsilon) \prod_{j=1}^{N} r_j^2 dr_j. \qquad (40)$$

For the inverse-square potential, $V(r) = k/r^2$,

$$K_\ell(r", r'; \tau) = (r"r')^{-\frac{1}{2}} (M/i\tau) \exp[(iM/2\tau)(r"^2 + r'^2)] I_{\lambda+\frac{1}{2}}(-iMr"r'/\tau), \quad (41)$$

where $\lambda = [(\ell + \frac{1}{2})^2 + 2Mk]^{\frac{1}{2}} - \frac{1}{2}$. The S-matrix can then be simplified to the form,

$$S(\mathbf{p}", \mathbf{p}') = \frac{\delta(p" - p')}{p"p'} \sum_{\ell=0}^{\infty} e^{2i\delta_\ell} \sum_{m=-\ell}^{\ell} (-1)^\ell Y_\ell^{m*}(\theta'_p, \phi'_p) Y_\ell^m(\theta_p", \phi_p"), \qquad (42)$$

where $\delta_\ell = (\pi/2)(\ell - \lambda)$ is the phase shift.

We now note that the expression for the S-matrix, Eq.(39), obtained by Gerry and Singh corresponds to Eq.(11) if the propagator is expanded in spherical harmonics, as for the case of central potentials. Clearly, should we have an available expression for the propagator, or the related energy Green function, we can obtain the S-matrix as, in fact, we have shown in the preceding Section. Nevertheless, we thought it worthwhile to investigate the possibility of following the procedure of CFJM, transform to spherical polar coordinates, but expand not in terms of the spherical harmonics, but instead in terms of angular functions suitable to the noncentral potentials being considered. For example, we could consider expansion of the exponential factors in the path integral for the S-matrix, Eq.(31) in terms of Jacobi polynomials, hypergeometric functions, Wigner rotation functions, or, as in the case of the dyonium and charge-dyon systems, the monopole harmonics (which are,of course, related to Jacobi polynomials, hypergeometric functions, and Wigner rotation functions). In the course of our search for the appropriate expansion, we observed that, unlike the addition theorems for Legendre and Gegenbauer polynomials which belong to 19th century mathematics, an addition theorem for Jacobi polynomials was derived first by Koornwinder[13] only in 1973. Since then, Koornwinder and others have found other ways of deriving, and forms of, the addition theorem. These are now mentioned in the latest edition of Szego's classic work on Orthogonal Polynomials. However, at present, we have not come up with a form of the addition theorem suitable for the path integral expansion for the noncentral potentials we are considering. Or, perhaps, the addition theorem may not even be necessary. We leave these and other considerations (such as the observation that there may be other ways of handling the path integral without the time-slicing procedure[14]) for future work.

## ACKNOWLEDGMENTS

This work was supported in part by the National Research Council of the Philippines. Work on this project was also done while two of the authors, MVCB and CCB were at the Abdus Salam International Centre for Theoretical Physics on an Associateship Visit with financial support from the Government of Japan.

It is also a pleasure to thank Prof. C. DeWitt-Morette for discussions and special lectures she gave in Jagna. We hope these encounters will bear fruit in the near future.

## REFERENCES

1. W. B. Campbell, P. Finkler, C. E. Jones, and M. N. Misheloff, *Phys. Rev.* **D12** (1975) 2363.

2. C. C. Gerry and V. A. Singh, *Phys. Rev.* **D21** (1980) 2979.

3. R. Ho, Ph.D. Thesis (State Univ. of New York at Albany, 1982), unpublished.

4. M. V. Carpio-Bernido, *J. Math. Phys.* **32** (1991) 1799.

5. C. Grosche, G. S. Pogosyan and A. N. Sissakian, *Fort. der Physik* **43** (1995) 453; *Phys. Part. Nucl.* **27** (1996) 244.

6. A. A. Makarov, J. A. Smorodinsky, Kh. Valiev, and P. Winternitz, *Nuovo Cimento* **A52** (1967) 1061.

7. M. Kibler and C. Campigotto, *Phys. Lett.* **A181** (1993) 1.

8. R. Côté, A. Dalgarno and M. J. Jamieson, *Phys. Rev.* **A50** (1994) 399; R. Côté and A. Dalgarno, *Phys. Rev.* **A50** (1994) 4827.

9. E. B. Gravador, Masteral Thesis (Univ. of the Philippines, 1991), unpublished; E. B. Gravador, M. V. Carpio-Bernido and C. C. Bernido, *Vistas in Astron.* **37** (1993) 261.

10. E. B. Gravador, M. V. Carpio-Bernido and C. C. Bernido, in *Path Integrals in Physics*, eds. V. Sa-yakanit et al. (World Scientific, Singapore, 1994) 455; M. V. Carpio-Bernido, in *Proc. of the 1st Jagna Int'l. Workshop on Advances in Theoretical Physics* (RCTP-CVI, Jagna,1996) 162.

11. L. S. Schulman, *Techniques and Applications of Path Integration* (Wiley, New York, 1981).

12. Schiff, L., *Quantum Mechanics* (McGraw-Hill Kogakusha, Tokyo, 1968).

13. Koornwinder, T., *SIAM J. Appl. Math.* **25** (1973) 236.

14. P. Cartier, C. DeWitt-Morette, A. Wurm and D. Collins, in *Functional Integration: Basics and Applications*, NATO ASI Series, Eds. C. DeWitt-Morette, P. Cartier and A. Folacci (Plenum Press, New York, 1997) 1.

# SPIN SYSTEMS AND COHERENT STATES

# PATH INTEGRAL METHOD FOR TUNNELLING OF SPIN SYSTEMS AND THE MACROSCOPIC QUANTUM COHERENCE

JIUQING LIANG[1,2], YUNBO ZHANG[2,1], FU-CHO PU[3,2]

[1] Department of Physics, Shanxi University, Taiyuan, China, 030006
[2] Institute of Physics and Center for Condensed Matter Physics, Chinese Academy of Sciences, Beijing, China, 100080
[3] Department of Physics, Guangzhou Normal College, Guangzhou, China, 510400

**ABSTRACT** The quantum tunneling effect in a ferromagnetic particle which can be evaluated only for the ground state previously is extended to excited states within the framework of instanton method. The tunneling between $n$-th degenerate states of neighboring wells is dominated by a periodic pseudoparticle configuration with which a formula of level-splitting valid for entire region of energies is derived. The low-lying level-splitting is also obtained with the LSZ method in field theory in which the tunneling is viewed as the transition of $n$ bosons induced by the usual (vacuum) instanton. Two results coincide exactly in the low energy limit. The tunneling effect increases at excited states. The results should be useful in investigation of the macroscopic quantum coherence in ferromagnetic particles.

## 1. INTRODUCTION

Motivated by the searching for macroscopic quantum phenomena, in recent years the tunneling of quantum spins and the possibility of its observation in macroscopic ferro- or antiferromagnetic particles have attracted considerable interests as the rapidly expanding literature in the field demonstrates[1-6]. The principal idea is that in such particles of mesoscopic size the electronic spins can form an aligned magnetic state which can assume several directions so that quantum mechanics suggests the possibility to lift this degeneracy by tunneling from one direction to another. This tunneling effect has been explored in numerous investigations. In most of them the imaginary time path-integral or instanton method is used, however, is only restricted to the WKB leading order approximation. The instanton which is a classical solution of nonlinear field equation of imaginary time with finite Euclidean action satisfies manifestly the vacuum (ground state) boundary condition therefore is only suitable to evaluate the tunneling at ground state. These restrictions have become an obstacle to the further investigation of quantum tunneling of spin systems with the instanton method.

In the present paper we first demonstrate the imaginary path-integral method for quantum tunneling of spin systems including the one-loop correction which results in a prefactor of the WKB exponential and then the instanton method is extended to calculation of tunneling at excited states by means of periodic instanton configuration[7]. In section 2 the Hamiltonian of a spin system is mapped on to a particle problem in terms of spin coherent state path-integral. A formula of energy spectrum is derived showing the level splitting due to tunneling for the integer-spin and suppression of tunneling effect for half-integer-spin. The level splitting of ground state with the correction of quantum fluctuation is calculated with the help of the usual instanton method in section 3. In section 4 the level splitting of low-lying excited states is obtained with the LSZ method[12−15] in field theory in which the tunneling at $n$-th excited states is viewed as transition of $n$ bosons induced by the usual instanton instead. In section 5 we demonstrate the periodic instanton method[8−11] to obtain the level splitting of excited state. In the low energy region these two results coincide exactly.

## 2. SPIN COHERENT STATE PATH-INTEGRAL AND THE EFFECTIVE PARTICLE MODEL OF A SPIN SYSTEM

We begin with the Hamiltonian of a ferromagnetic particle

$$\hat{H} = K_1 \hat{S}_z^2 + K_2 \hat{S}_y^2 \tag{1}$$

which describes XOY easy plane anisotropy and an easy axis along the $x$ direction with the anisotropy constants $K_1 > K_2 > 0$. Here $\hat{S}_i$, $i = x, y, z$, are spin operators with the usual commutation relation $[\hat{S}_i, \hat{S}_j] = i\epsilon_{ijk}\hat{S}_k$ (using natural units throughout). The first step is the conversion of the discrete spin system eq. (1) into a continuous one. To this end we evaluate the matrix element of the evolution operator in the spin coherent state representation by means of the coherent state path-integral, i.e.

$$\left\langle \vec{n}_f \middle| e^{-2i\hat{H}T} \middle| \vec{n}_i \right\rangle = \int \{ \prod_{k=1}^{M-1} d\mu(\vec{n}_k) \} \{ \prod_{k=1}^{M} \left\langle \vec{n}_k \middle| e^{-i\epsilon\hat{H}} \middle| \vec{n}_{k-1} \right\rangle \}. \tag{2}$$

Here we define $\left| \vec{n}_M \right\rangle = \left| \vec{n}_f \right\rangle, \left| \vec{n}_0 \right\rangle = \left| \vec{n}_i \right\rangle$ and $t_f - t_i = 2T, \epsilon = \frac{2T}{M}$ respectively. Further $\left| \vec{n} \right\rangle$ denotes the spin coherent state generated from the reference spin eigenstate $|S, S\rangle$ such that

$$\left| \vec{n} \right\rangle = e^{\varsigma^* \hat{S}_- - \varsigma \hat{S}_+} |S, S\rangle \tag{3}$$

where $\varsigma = \frac{\theta}{2} e^{-i\phi}$ with polar angle $\theta$ and azimuthal angle $\phi$ of the unit vector $\vec{n} = (\sin\theta\cos\phi, \sin\theta\sin\phi, \cos\theta)$. The measure of integration is defined by[16]

$$d\mu(\vec{n}_k) = \frac{2S+1}{4\pi}, \qquad d\,\vec{n}_k = \sin\theta_k d\theta_k d\phi_k. \tag{4}$$

The infinitesimal evolution operator may be rewritten as

$$e^{-i\epsilon\hat{H}} = e^{-i\epsilon\frac{K_2}{2}\hat{S}^2} e^{i\epsilon\frac{K_2}{4}\hat{S}_-^2} e^{-i\epsilon(K_1-\frac{K_2}{2})\hat{S}_z^2} e^{i\epsilon\frac{K_2}{4}\hat{S}_+^2} \tag{5}$$

The matrix element in eq.(2) can be evaluated approximately by replace of the operators $\hat{S}^2, \hat{S}_z$ and $\hat{S}_+$ with $S(S+1)$, $S\cos\theta$ and $Se^{i\phi}\sin\theta$ respectively in the large $S$ limit with $\hat{S}_\pm = \hat{S}_z \pm i\hat{S}_y$. Then the matrix element of the evolution operator eq.(2) is written as the path integral in phase space, i.e.

$$e^{-iS(\phi_f - \phi_i)} \int \mathcal{D}\phi \mathcal{D}p e^{i\int_{t_i}^{t_f} \mathcal{L}(\phi, p)dt} \tag{6}$$

with canonical variables $\phi$ and $p = S\cos\theta$, where

$$\mathcal{L} = \dot{\phi}\, p - H(\phi, p) \tag{7}$$

is the phase space (or first order) Lagrangian, and

$$H = \frac{p^2}{2m(\phi)} + V(\phi) \tag{8}$$

is the classical Hamiltonian. The mass

$$m(\phi) = \frac{1}{2K_1(1 - \lambda\sin^2\phi)}$$

is position dependent (reflecting the fact that there is a velocity dependent force ) and $V(\phi) = S(S+1)K_2\sin^2\phi$ is the potential , and $\lambda = \frac{K_2}{K_1}$ is a dimensionless parameter. The position dependent mass also implies that the original space on which the continuous classical mechanics is defined for the spin system has an intrinsic curvature. The position dependent kinetic term corresponds to a velocity dependent potential, and this requires that one starts from the phase space path-integral as pointed long ago[17]. The phase of the prefactor $e^{-iS(\phi_f - \phi_i)}$ which originated from the inner product of coherent states. i. e. $\prod_{k=1}^{M}\left\langle \vec{n}_k \mid \vec{n}_{k-1}\right\rangle$ can be put into the Lagrangian, i. e. the angle difference $\phi_f - \phi_i$ is written as $\phi_f - \phi_i = \int_{t_i}^{t_f}\dot{\phi}\,dt$, and has been identified formally as a Weiss-Zumino term[18].

Integrating out the momentum in the path-integral we obtain the configuration space functional integral, i.e.

$$\left\langle \vec{n}_f \left| e^{-2i\hat{H}T} \right| \vec{n}_i \right\rangle = e^{-iS(\phi_f - \phi_i)}K(\phi_f, t_f = T; \phi_i, t_i = -T) \tag{9}$$

where

$$K(\phi_f, t_f; \phi_i, t_i) = \int \tilde{\mathcal{D}}\phi\, e^{i\int_{t_i}^{t_f} \mathcal{L}(\phi, \dot{\phi})dt}$$

The functional $K$ is the Feynman propagator in configuration space with the second order Lagrangian

$$\mathcal{L} = \frac{1}{2}m(\phi)\,\dot{\phi}^2 - V(\phi) \tag{10}$$

and the measure

$$\tilde{\mathcal{D}}\phi = \prod_{k=1}^{M-1}\sqrt{\frac{m(\phi_k)}{2\pi i\epsilon}}d\phi_k$$

The canonical momentum is defined as usual, i.e. $p = m(\phi)\,\dot{\phi}$. The potential is periodic with period $\pi$ and has an infinite number of degenerate minima by successive $2\pi$ extension at $\Phi_n = n\pi, n$ being an integer. Considering the periodic potential as a super-lattice with lattice constant $\pi$ , we can derive the energy spectrum in the tight-binding approximation. The energy spectrum is seen to be given by[19]

$$E_m = \epsilon_m - 2\triangle\epsilon_m \cos(S + \xi)\pi \tag{11}$$

where

$$\triangle\epsilon_m = -\int u_m^*(\phi, \Phi_n)\hat{H}u_m(\phi, \Phi_{n+1})d\phi \tag{12}$$

is the usual overlap integral or simply the level shift due to tunneling through any one of the barriers. Here $u_m(\phi-\Phi_n)$ denotes the eigenfunction of the harmonic oscillator-approximated Hamiltonian $\hat{H}_0$ in the $n$-th well, i.e.

$$H_0 = \frac{p^2}{2m_0} + \frac{1}{2}m_0\omega_0{}^2(\phi - \Phi_n)^2 \tag{13}$$

with $m_0 = \frac{1}{2K_1}$ and $\omega_0{}^2 = 4K_1K_2s(s+1)$. $\xi$ is an integer and here can takes only either of the two values "0" and "1". The level splitting is obtained from the energy spectrum eq. (11) by the absolute difference between $\xi = 0$ and 1, namely

$$\triangle E_m = 2\triangle\epsilon_m \left|\cos(s+1)\pi - \cos s\pi\right| = \begin{cases} 4\triangle\epsilon_m & for\ integer\ spin\ s \\ 0 & for\ half\ integer\ spin\ s \end{cases} \tag{14}$$

For half integer spin $s$ the spectrum eq. (11) is quenched to a single degenerate level with degeneracy two. The quenching is seen to be a consequence of Kramer's theorem which says that for half integer spin $s$ the degeneracy cannot be removed in the absence of the crystal field[19].

## 3. CALCULATION OF THE TUNNELLING PARAMETER $\triangle\epsilon_0$ OF GROUND STATE AND THE CONVENTIONAL INSTANTON METHOD

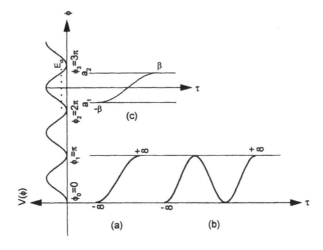

Fig. 1 The periodic potential and the instanton trajectories:
(a) For one instanton (i.e. vacuum instanton),
(b) For one instanton plus one instanton-anti-instanton pair,
(c) One half of the periodic instanton.

Above we derived the level splitting of the energy spectrum of a giant spin particle in the large spin limit. The only parameter left unknown is the overlap integral $\triangle\epsilon_m$ which can be evaluated with the help of the instanton method. Instantons in field theory of $1+0$ dimensions are viewed as pseudoparticles with trajectories existing in barriers, are therefore responsible for tunneling. Since instanton trajectories interpolate between degenerate vacua (See Fig. 1. (a)) and satisfy vacuum boundary conditions, the instanton method is only suitable for the calculation of tunneling parameter $\Delta\epsilon_0$ between neighboring vacua. In the following we first consider tunneling at the vacuum level (i. e. $m = 0$), which leads to the level splitting of the ground state energy, i.e. $4\Delta\epsilon_0$. Passing to imaginary time by Wick rotation $\tau = it, \beta = iT$, the amplitude for tunneling from the initial well, say that with $n = 0$ (and $\phi_i = 0$), to the neighboring well with $n = 1$ (and $\phi_f = \pi$), and considering large $\beta$, the amplitude for the transition between the corresponding coherent states can be shown to be

$$
\begin{aligned}
< \mathbf{n}(\pi)|e^{-2\beta\hat{H}}|\mathbf{n}(0) > &= < \mathbf{n}(\pi)|0, \Phi_1 >< 0, \Phi_0|\mathbf{n}(0) > e^{-2\beta\epsilon_0} \sinh(2\beta\Delta\epsilon_0) \\
&= e^{-i\pi s} K_E(\phi_f = \pi, \beta; \phi_i = 0, -\beta)
\end{aligned} \tag{15}
$$

where

$$
K_E = \int \mathcal{D}\phi e^{-S_E}
$$

is the Euclidean propagator with Euclidean action defined by

$$
S_E = \int_{-\beta}^{\beta} \mathcal{L}_E d\tau, \quad \mathcal{L}_E = \frac{1}{2}m(\phi)\dot{\phi}^2 + V(\phi) \tag{16}
$$

In eq. (15) $\epsilon_0$ is the ground state energy of $\hat{H}_0$. From now on $\dot{\phi} = \frac{d\phi}{d\tau}$ denotes the imaginary time derivative.

In the following the Euclidean propagator $K_E$ is evaluated with the instanton method. After evaluation we compare the result with eq. (15) to find the tunneling parameter $\Delta\epsilon_0$. The instanton configuration which minimizes the Euclidean action $S_E$ satisfies the equation of motion

$$
\frac{1}{2}m(\phi)\left(\frac{d\phi}{d\tau}\right)^2 - V(\phi) = 0 \tag{17}
$$

can be found to be

$$
\phi_c = \arcsin[\cosh^2 \omega_0(\tau - \tau_0) - \lambda \sinh^2 \omega_0(\tau - \tau_0)]^{-\frac{1}{2}} \tag{18}
$$

with position $\tau_0$ which is the integration constant. The instanton trajectory eq.(18) is shown in Fig. 1 (a) with $\tau_0 = 0$. The Euclidean action evaluated for the instanton trajectory eq.(18), sometimes called the instanton mass, is

$$
S_c = \int_{-\infty}^{\infty} m(\phi_c)\dot{\phi}_c^2 d\tau = \sqrt{s(s+1)} \ln \frac{1+\sqrt{\lambda}}{1-\sqrt{\lambda}} \approx (s + \frac{1}{2}) \ln \frac{1+\sqrt{\lambda}}{1-\sqrt{\lambda}} \tag{19}
$$

The functional integral $K_E$ can be evaluated with the stationary phase method by expanding $\phi$ about the instanton trajectory $\phi_c$ such that $\phi = \phi_c + \eta$, where $\eta$ is the small fluctuation with boundary conditions $\eta(\beta) = \eta(-\beta) = 0$. Up to the one-loop approximation we have

$$
K_E = e^{-S_c} I \tag{20}
$$

where

$$
I = \int_{\eta(-\beta)=0}^{\eta(\beta)=0} \mathcal{D}\eta e^{-\delta S_E} \tag{21}
$$

is the fluctuation functional integral with the fluctuation action

$$
\delta S_E = \int_{-\beta}^{\beta} \eta \hat{M} \eta d\tau \tag{22}
$$

where

$$\hat{M} = -\frac{1}{2}\frac{d}{d\tau}m(\phi_c)\frac{d}{d\tau} + \tilde{V}(\phi_c) \tag{23}$$

with

$$\tilde{V}(\phi_c) = \frac{1}{2}[-m'(\phi_c)\ddot{\phi}_c - \frac{1}{2}m''(\phi_c)\dot{\phi}_c^2 + V''(\phi_c)] \tag{24}$$

Here $\hat{M}(\phi_c)$ is the operator of the second variation of the action and $m'(\phi_c) = \frac{\partial m(\phi)}{\partial \phi}|_{\phi=\phi_c}$. Let $\{\psi_n\}$ be the set of eigenmodes of the operator $\hat{M}$ such that $\hat{M}\psi_n = E_n\psi_n$, we expand the fluctuation variable in terms of the eigenmodes of $\hat{M}$, i.e.

$$\eta = \sum_n c_n\psi_n \tag{25}$$

The functional integral eq. (21) can be evaluated to be

$$I = \left|\frac{\partial\eta}{\partial c_n}\right|\sum_n\left[\frac{\pi}{E_n}\right]^{\frac{1}{2}} \tag{26}$$

where the first factor denotes the Jacobian of the transformation eq. (25). In view of the time translation symmetry of the equation of motion eq. (17), the functional integral $K_E$ is not well defined when expanded about the classical solution. The translational symmetry results in zero eigenmodes of the second variation operator $\hat{M}$ of the action. This problem can be cured by the Faddeev-Popov procedure[20] or in a more systematic way with the help of the BRST transformation[21]. Then the zero-mode functional integral is converted into an integral over the collective coordinate, i.e. the instanton position $\tau_0$, and leads to $2\beta$ and a Faddeev-Popov determinate $\sqrt{D}$ with

$$D = \int_{-\infty}^{\infty}\dot{\phi}_c^2 d\tau = \omega_0(1 + \frac{1-\lambda}{2\sqrt{\lambda}}\ln\frac{1+\sqrt{\lambda}}{1-\sqrt{\lambda}}) \tag{27}$$

The functional integral eq. (21) then becomes

$$I = \left|\frac{\partial\eta}{\partial c_n}\right|\sum_{n\neq 0}\left[\frac{\pi}{E_n}\right]^{\frac{1}{2}}\sqrt{D} \tag{28}$$

The complicated determinant of $\hat{M}$ and the related Jacobian are avoided by the direct integration involved in the transformation of the well-known shift method[8−11], i.e.

$$\eta = y + \dot{\phi}_c\int_{\tau_i}^{\tau}\frac{\ddot{\phi}_c(\tau')}{\dot{\phi}_c^2(\tau')}y(\tau')d\tau' \tag{29}$$

where the derivative of the instanton solution, i.e. $\dot{\phi}_c$, is simply the zero eigenmode of $\hat{M}$. The resulting fluctuation integral is

$$I = \frac{1}{\sqrt{2\pi}}[\dot{\phi}_c(\beta)\dot{\phi}_c(-\beta)\int_{-\beta}^{\beta}\frac{d\tau}{\dot{\phi}_c^2(\tau)m(\phi_c(\tau))}]^{-\frac{1}{2}} \tag{30}$$

Taking the large $\beta$ limit of the above eq. (30) and comparing with eqs. (26) and (28) we obtain

$$I = \frac{2\beta}{\sqrt{2\pi}}\lambda^{\frac{1}{4}}s^{\frac{1}{2}}(\frac{\mathcal{E}_0}{\pi})^{\frac{1}{2}}D^{\frac{1}{2}} \tag{31}$$

where $\mathcal{E}_0$ denotes the ground state energy of the operator $\hat{M}$ such that $\hat{M}\tilde{\psi}_0 = \mathcal{E}_0 \tilde{\psi}_0$ with the boundary condition $\tilde{\psi}_0(\beta) = \tilde{\psi}_0(-\beta) = 0$. $\mathcal{E}_0$ can be evaluated by so-called boundary perturbation method from the zero-mode $\phi_c$, i.e.,

$$
\mathcal{E}_0^{\frac{1}{2}} = \left[ \frac{m(\phi_c(\beta))\dot{\phi}_c(\beta)\,\ddot{\phi}_c(\beta) - m(\phi_c(-\beta))\dot{\phi}_c(-\beta)\,\ddot{\phi}_c(-\beta)}{\int_{-\beta}^{\beta} \dot{\phi}_c^2 d\tau} \right]^{\frac{1}{2}}
$$

$$
= \frac{4\sqrt{K_2}Se^{-\omega_0\beta}}{\left[(1-\lambda)(1 + \frac{1-\lambda}{2\lambda} \ln \frac{1+\sqrt{\lambda}}{1-\sqrt{\lambda}})\right]^{\frac{1}{2}}} \tag{32}
$$

Substituting eq. (32) into eq. (31) yields the one-instanton contribution to the propagator in the one-loop approximation,

$$
K_E^{(1)} = 2\beta \frac{4}{\pi} \left( \frac{\omega_0 K_2}{2(1-\lambda)} \right)^{\frac{1}{2}} \lambda^{\frac{1}{4}} s^{\frac{3}{2}} e^{-\omega_0\beta} e^{-S_c} \tag{33}
$$

The path-integral calculation requires a sum over all possible paths. In our case, the path fall into an infinite number of classes according to the one instanton plus the number of instanton and anti-instanton pairs for a given time interval, i.e.

$$
K_E = \sum_{n=0}^{\infty} K_E^{(2n+1)}(\phi_f, \beta; \phi_i, -\beta) \tag{34}
$$

where $K_E^{(2n+1)}$ denotes the propagator for one instanton plus $n$ pairs. With the help of the group property of the Feynman propagator, the kernel $K_E^{(2n+1)}$ is obtained as

$$
K_E^{(2n+1)} = \frac{(2\beta)^{2n+1}}{(2n+1)!} [\frac{4}{\pi} \{ \frac{\omega_0 K_2}{2(1-\lambda)} \}^{\frac{1}{2}} \lambda^{\frac{1}{4}} s^{\frac{3}{2}} e^{-S_c}]^{2n+1} e^{-\omega_0\beta} \times \left[ \frac{\pi}{\frac{\partial^2 S_c}{\partial \phi_f^2}} \right]^{2n} \tag{35}
$$

The last factor comes from the integrations of intermediate end points in the dilute instanton-gas approximation. Inserting the value

$$
\lim_{\beta \to \infty} \frac{\partial^2 S_c}{\partial \phi_f^2} = \frac{\omega_0}{2K_1} \tag{36}
$$

into eq .(35) the final result for the Feynman kernel eq .(34) is therefore

$$
K_E = \lambda^{\frac{1}{4}} (\frac{s}{\pi})^{\frac{1}{2}} e^{-\beta\omega_0} \sinh[2\beta.2^2 \{ \frac{K_1 K_2}{(1-\lambda)\pi} \}^{\frac{1}{2}} \lambda^{\frac{1}{4}} s^{\frac{3}{2}} e^{-S_c}] \tag{37}
$$

Comparing with eq.(15) the tunneling parameter of ground state is

$$
\Delta\epsilon_0 = 2^2 \{ \frac{K_1 K_2}{(1-\lambda)\pi} \}^{\frac{1}{2}} \lambda^{\frac{1}{4}} s^{\frac{3}{2}} e^{-S_c} \tag{38}
$$

which agrees with the results of ref. [22-27]. Since $s$ is a large number (500~5000 for the giant ferromagnetic particle) the tunneling effect is unobservably small except that $\lambda \ll 10^{-4}$. It is certainly interesting to investigate the tunneling at excited states which is not available in the framework of the previous instanton calculation. In the following section we first extend the tunneling effect to low-lying levels with the help of LSZ method on field theory.

## 4. LEVEL SPLITTING OF LOW-LYING LEVELS AND THE LSZ METHOD

The idea of a tunneling transition from one side of a potential barrier to the other has recently also been linked with the LSZ reduction mechanism of a transition from asymptotic in-state to asymptotic out-states. But although this idea is very attractive and worth pursuing it has not been studied in detail. In the following we therefore go beyond a cursory employment of the method and use the LSZ reduction procedure in a modified way in order to calculate the tunneling in the one-instanton sector for the effective potential of the spin system including the contribution of quantum fluctuations up to the one-loop approximation.

We recall first the case of a one-dimensional harmonic oscillator described by the Hamiltonian

$$H = \frac{1}{2}p^2 + \frac{1}{2}\omega^2 q^2 \tag{39}$$

for mass $m = 1$. Here $q$ and $p$ are dynamical observables which become operators when subjected to the Heisenberg algebra of ordinary canonical quantization. The solution of the Heisenberg equation of motion, $\ddot{q} + \omega^2 q = 0$, then becomes

$$q(t) = \frac{1}{\sqrt{2\omega}}[ae^{-i\omega t} + a^\dagger e^{i\omega t}] \tag{40}$$

where $a, a^\dagger$ are time-independent operators defined by the initial $(t = 0)$ values of $q$ and $p$, i.e.

$$q(0) = \frac{1}{\sqrt{2\omega}}[a + a^\dagger], p(0) = \frac{-i\omega}{\sqrt{2\omega}}[a - a^\dagger] \tag{41}$$

The operators $a, a^\dagger$ can be obtained from $q(t), p(t) = \dot{q}(t)$. Thus

$$
\begin{aligned}
a^\dagger &= -\frac{i}{\sqrt{2\omega}}e^{-i\omega t}[\dot{q}(t) + i\omega q(t)] \\
&\equiv -\frac{i}{\sqrt{2\omega}}e^{-i\omega t}\frac{\overleftrightarrow{\partial}}{\partial t}\, q(t)
\end{aligned}
\tag{42}
$$

and $a$ follows with complex conjugation. One should note the extra minus sign in the definition of the symbol $\frac{\overleftrightarrow{\partial}}{\partial t}$ when acting to the left. Operators of this type are well-known in the literature.

We now consider the $(1 + 0)$ dimensional theory defined by the Euclidean Lagrangian eq.(16). Crucial aspects of the LSZ procedure are its asymptotic conditions which require the theory to have an interpretation in terms of observables for stationary states in coming and outgoing states. We can simulate such a situation here artificially by imagining the central barrier of the potential to be extremely high and the neighboring wells " − " and " + " on either side as extremely far apart. We therefore construct appropriate functions $\phi_\pm(\tau)$ which become

$$\phi_+ = \pi - \phi_c, \qquad \phi_- = \phi_c \tag{43}$$

such that the interaction fields vanish in their respective asymptotic regions, i.e.

$$\lim_{\tau \to +\infty} \phi_+(\tau) = 0, \qquad \lim_{\tau \to -\infty} \phi_-(\tau) = 0 \tag{44}$$

The subscripts " − " and " + " here denote the wells with minima at $\Phi_0 = 0$ and $\Phi_1 = \pi$ respectively. The Euclidean creation and annihilation operators $\hat{a}_\pm^\dagger$ and $\hat{a}_\pm$ which create and annihilate an effective boson in wells " + " and " − " respectively are related to the interaction field operators $\phi_\pm$ by

$$\hat{a}_\pm^\dagger(\tau) := \sqrt{\frac{2m_0}{\omega_0}} e^{-\omega_0 \tau} \overset{\leftrightarrow}{\frac{\partial}{\partial \tau}} \phi_\pm(\tau),$$

$$\hat{a}_\pm(\tau) := -\sqrt{\frac{2m_0}{\omega_0}} e^{\omega_0 \tau} \overset{\leftrightarrow}{\frac{\partial}{\partial \tau}} \phi_\pm(\tau) \tag{45}$$

From the viewpoint of the LSZ method the transition amplitude between $m$-th degenerate eigenstates in any two neighboring wells (here for $n = 0, 1$) is viewed as the transition of $m$ bosons induced by the instanton of eq.(18) and is related with the tunneling parameter $\Delta \epsilon_m$ by

$$A_{f,i}^m = \; <m, \Phi_1|e^{-2\beta \hat{H}}|m, \Phi_0> \; = e^{-2\beta \epsilon_m} \sinh 2\beta \Delta \epsilon_m \tag{46}$$

The transition amplitude as well as the $S$-matrix can be related to the Green's function through the procedure known as the LSZ reduction technique. To this end we rewrite the transition amplitude as

$$A_{f,i}^m = S_{f,i}^m e^{-2\beta \omega_0} \tag{47}$$

with $S$-matrix element

$$S_{f,i}^m = \lim_{\substack{\tau^i \to -\infty \\ \tau^f \to \infty}} \frac{1}{m!} \langle 0|\hat{a}_+(\tau_m^f) \ldots \hat{a}_+(\tau_1^f) \hat{a}_-^\dagger(\tau_1^i) \ldots \hat{a}_-^\dagger(\tau_m^i)|0\rangle \tag{48}$$

The $S$-matrix element can be evaluated in terms of the Green's function $G$ which arises in its evaluation. Thus

$$\begin{aligned}
S_{f,i}^m &= \lim_{\substack{\tau^i \to -\infty \\ \tau^f \to \infty}} \frac{1}{m!} \prod_{l=1}^m \left( -\sqrt{\frac{2m_0}{\omega_0}} e^{\omega_0 \tau_l^f} \overset{\leftrightarrow}{\frac{\partial}{\partial \tau_l^f}} \right) \left( \sqrt{\frac{2m_0}{\omega_0}} e^{-\omega_0 \tau_l^i} \overset{\leftrightarrow}{\frac{\partial}{\partial \tau_l^i}} \right) G \\
&= \frac{1}{m!} \prod_{l=1}^m \left( \frac{-2m_0}{\omega_0} \right) e^{\omega_0(\tau_l^f - \tau_l^i)} \left[ \frac{\partial^2 G}{\partial \tau_l^f \partial \tau_l^i} + \omega_0 \left( \frac{\partial G}{\partial \tau_l^f} - \frac{\partial G}{\partial \tau_l^i} - \omega_0 G \right) \right]
\end{aligned} \tag{49}$$

where the $2m$-point Green's function is defined as usual, i.e.

$$G = \langle 0|\hat{\phi}_+(\tau_m^f) \ldots \hat{\phi}_+(\tau_1^f) \hat{\phi}_-(\tau_1^i) \ldots \hat{\phi}_-(\tau_m^i)|0\rangle \tag{50}$$

We evaluate $G$ by inserting complete sets of states of final and initial field configurations $\phi_f, \phi_i$. Thus

$$G(\tau^f, \tau^i) = \phi_+(\tau^f) \phi_-(\tau^i) A_{f,i}^0 \tag{51}$$

which vanishes in the limit $\tau^i \to -\infty, \tau^f \to \infty$, due to eq. (44). Thus in eq. (49) the only nonvanishing contribution in these limits results from the second derivative, then

$$\frac{\partial^2 G}{\partial \tau^f \partial \tau^i} = \frac{\partial \phi_+(\tau^f)}{\partial \tau^f} \frac{\partial \phi_-(\tau^i)}{\partial \tau^i} A_{f,i}^0 \tag{52}$$

the $S$-matrix element for the transition of $m$ bosons is thus

$$S_{f,i}^m = \frac{1}{m!} \prod_{l=1}^m \left\{ \left( -\frac{2m_0}{\omega_0} \right) \left[ \frac{d\phi_+(\tau_l^f)}{d\tau_l^f} \frac{d\phi_-(\tau_l^i)}{d\tau_l^i} \right] \right\} A_{f,i}^0 \tag{53}$$

The transition amplitude between degenerate ground state can be calculated from the definition eq. (46) in terms of the tunneling parameter $\Delta\epsilon_0$, i.e.

$$A_{f,i}^0 = 2\beta\Delta\epsilon_0 e^{-\beta\omega_0} \tag{54}$$

After performing the imaginary time derivatives, we finally obtain the transition amplitude by observing that each pair of vertices in eq. (53) contributes a factor $-4\omega_0^2$. Then

$$A_{f,i}^m = \frac{1}{m!}\left(\frac{2^3 m_0\omega_0}{1-\lambda}\right)^m e^{-\omega_0 m 2\beta} A_{f,i}^0 \tag{55}$$

The funneling parameter at $m$-th excited state is seen to be

$$\Delta\epsilon_m = \frac{1}{m!}\left(\frac{2^3\lambda^{\frac{1}{2}}s}{1-\lambda}\right)^m \Delta\epsilon_0 \tag{56}$$

## 5. PERIODIC INSTANTON METHOD AND THE GENERALIZED FORMULA FOR THE LEVEL SPLITTING

The periodic instanton method has become a powerful tool for the evaluation of quantum tunneling over the entire region of energy. The model at hand can be looked at as one for tunneling at the level of excited states of a sine-Gordon-type potential with a position-dependent mass, which has not been reported in the literature. The periodic instanton configuration $\phi_p$ which minimizes the Euclidean action eq. (16) is seen to satisfy the equation of motion

$$\frac{m(\phi_p)}{2}\left(\frac{d\phi_p}{d\tau}\right)^2 - V(\phi_p) = -E_{cl} \tag{57}$$

where $E_{cl} > 0$, which is a constant of integration, may be viewed as the classical energy of the pseudoparticle configuration. Through the usual procedure of derivation of the periodic instanton solution, after a laborious but straightforward integration of eq. (57), we obtain the periodic instanton configuration[27]

$$\phi_p = \arcsin\left[\frac{1 - k^2\operatorname{sn}^2(\omega\tau|k)}{1 - \lambda k^2\operatorname{sn}^2(\omega\tau|k)}\right]^{\frac{1}{2}} \tag{58}$$

where $\operatorname{sn}(\omega\tau|k)$ denotes a Jacobian elliptic function of modulus $k$,

$$k = \sqrt{\frac{n_1^2 - 1}{n_1^2 - \lambda}} \tag{59}$$

with

$$n_1^2 = \frac{K_2 s(s+1)}{E_{cl}}, \qquad \omega = \omega_0\sqrt{1 - \frac{\lambda E_{cl}}{K_2 s(s+1)}} \tag{60}$$

One can check with $\operatorname{cn}(u|1) = \tanh u$ that for $k^2 = 1$ (corresponding to $E_{cl} \to 0$) the configuration (58) reduces to the instanton of eq.(18). The Jacobian elliptic function cn has real periods $4nK(k)$, $n$ being an integer and $K(k)$ the quarter period given by the usual elliptic integral of the first kind. The parameter values (59), (60) ensure the periodicity of $\phi_p$ at $\tau = -2\beta, +2\beta$ with a crossover from negative to positive values at $\tau = 0$ and $\phi = \frac{\pi}{2}\operatorname{mod}2\pi$. Thus the one-way transition from a turning point $a_1$ to the other turning point $a_2$ is mediated by the instanton–like part or one half of the periodic instanton extending from $\tau = -\beta$ to $\tau = +\beta$ (the periodic instanton itself returning to its original position after time $4\beta$). The trajectory of the periodic instanton eq.(58) is shown schematically in Fig. 1(c) where the trajectory is shifted by an amount $2\pi$. The instanton–like part starts at time $-\beta$ from turning point $\phi = a_1 = \arcsin\sqrt{\frac{E_{cl}}{K_{2s}(s+1)}}$ and reaches the other turning point $\phi = a_2 = \pi - \arcsin\sqrt{\frac{E_{cl}}{K_{2s}(s+1)}}$ at time $\beta$. The Euclidean action of the periodic instanton configuration eq.(58) over the domain from $\tau = -\beta$ to $\tau = +\beta$ can be found to be

$$S_p = \int_{-\beta}^{\beta} \left[ m(\phi_p)\dot{\phi}_p^2 + E_{cl} \right] d\tau = W + 2E_{cl}\beta \tag{61}$$

where

$$W = \frac{\omega}{\lambda K_1} \left[ K(k) - (1 - \lambda k^2)\Pi(\lambda k^2, k) \right] \tag{62}$$

and $\Pi(\lambda k^2, k)$ is the complete elliptic integral of the third kind.

It is now necessary to calculate the amplitude $A^{(1)}$ for the transition mediated by one pseudoparticle configuration – in the present case the instanton–part of the periodic instanton

$$A^{(1)} = \int \psi_{E_{cl}}^*(\phi_f)\psi_{E_{cl}}(\phi_i)K_E(\phi_f, \tau_f; \phi_i, \tau_i)d\phi_f d\phi_i \tag{63}$$

Evaluating the Feynman Kernel with help of shift method as shown in Section 3 and using the WKB wave functions in the barrier i. e.

$$\psi_{E_{cl}}(\phi_f) = \frac{C\exp\left(-\int_{\phi_f}^{a_2} m(\phi)\,\dot{\phi}\,d\phi\right)}{\sqrt{\dot{\phi}_f}} \tag{64}$$

$$\psi_{E_{cl}}(\phi_i) = \frac{C\exp\left(-\int_{a_1}^{\phi_i} m(\phi)\,\dot{\phi}\,d\phi\right)}{\sqrt{\dot{\phi}_i}}$$

with the renormalization constant evaluated as

$$C = \left[\frac{\omega}{4K(k')}\right]^{\frac{1}{2}} \tag{65}$$

we obtain the transition amplitude $A^{(1)}$ as

$$A^{(1)} = 2\beta\frac{\omega}{4K(k')}e^{-W}e^{-2E_{cl}\beta} \tag{66}$$

The contribution of one instanton plus $m$ instanton-anti-instanton pairs is given by

$$A^{(2m+1)} = \int_{-\beta}^{\beta} d\tau_1 \int_{-\beta}^{\tau_1} d\tau_2 \cdots \int_{-\beta}^{\tau_{2m}} d\tau \left[\frac{\omega}{4K(k')}\right]^{2m+1} e^{-(2m+1)W}e^{-2E_{cl}\beta} \tag{67}$$

The total amplitude is given by the sum

$$A = \sum_{m=0}^{\infty} A^{(2m+1)} = e^{-2E_{cl}\beta}\sinh\left\{\frac{2\beta\omega}{4K(k')}e^{-W}\right\} \tag{68}$$

The tunneling parameter is then

$$\Delta E_{cl} = \frac{\omega}{4K(k')} e^{-W} \tag{69}$$

which is valid over the whole range of $\lambda$.

We now consider the low energy limit where $E_{cl}$ is much less than the sphaleron energy (or barrier height $K_2 s(s + 1)$ of the potential), i.e. $k \to 1, k' \to 0$. Expanding $W$ as power series of $k'$ up to quadratic order as in refs. [8-11] and making use of the oscillator approximation of the periodic potential around one of the minima with the quantization replacement $E_{cl} \to \epsilon_m = \left(m + \frac{1}{2}\right) \omega_0$ (as in refs. [8-11]), we then have

$$W = \left(s + \frac{1}{2}\right) \ln \frac{1 + \sqrt{\lambda}}{1 - \sqrt{\lambda}} + \left(m + \frac{1}{2}\right) \ln \left[\frac{1 - \lambda}{8\sqrt{\lambda}s} \left(m + \frac{1}{2}\right)\right] - \left(m + \frac{1}{2}\right) \tag{70}$$

which leads exactly to the tunneling parameter $\Delta \epsilon_m$ eq. (56) obtained with the help of LSZ method. the new observation made with our new result eq. (69) as well as eq. (56) is that the tunneling effect increases at excited states.

## 6. CONCLUSION

We have shown that the periodic instanton method is useful for the calculation of tunneling effects at exited states of a spin system at finite temperature. A formula valid for entire region of energy is obtained and reduces to the result of LSZ method in the low energy limit. With the simple model of eq. (1) for a ferromagnetic particle we conclude that the tunneling effect at level of excited states increases.

# References

[1] A. J. Leggett, S. Chakravarty, A. T. Dorsey, M. P. A. Fisher, A. Garg and W. Zwerger, Rev. Mod. Phys. **59** (1987) 1; *Quantum Tunneling of Magnetization*, ed. L. Gunther and B. Barbara, Kluwer, Dordrecht(1995).

[2] A. Garg, Phys. Rev. Lett. **71** (1993) 4249

[3] D. D. Awschalom, J. F. Smyth, G. Grinstein, D. P. DiVincenzo and D. Loss, Phys. Rev. Lett. **68** (1992) 3092.

[4] D. D. Awschalom, D. P. DiVincenzo, G. Grinstein and D. Loss, Phys. Rev. Lett. **71** (1993) 4276.

[5] D. D. Awschalom, M.A. McCord and G. Grinstein, Phys. Rev. Lett. **65** (**1990**) 783.

[6] L. Gunther, Phys. World **3**, No.12 (1990) 28.

[7] N. S. Manton and T. S. Samols, Phys. Lett. **B207** (1988) 179; J.-Q. Liang, H. J. W. Müller–Kirsten and D. H. Tchrakian, Phys. Lett. **B282** (1992) 105.

[8] J.-Q. Liang and H. J. W. Müller-Kirsten, Phys. Rev. **D46** (1992) 4685.

[9] J.-Q. Liang and H.J.W. Müller–Kirsten, Phys. Rev. **D50** (1994) 6519 .

[10] J.-Q. Liang and H.J. W. Müller–Kirsten, Phys. Rev. **D51** (1995) 718.

[11] For a review see J.-Q. Liang and H. J. W. Müller–Kirsten, *Topics in Quantum Field Theory - Modern Methods in Fundamental Physics*, ed. D. H. Tchrakian, World Scientific(1996), pp. 54 –68.

[12] J.-Q. Liang and H. J. W. Müller–Kirsten, Phys. Lett. **B332** (1994) 129.

[13] J.-G. Zhou, J.-Q. Liang, J. Burzlaff, H.J.W. Müller–Kirsten, Phys. Lett. **A223** (1996) 142.

[14] For the LSZ procedure see for instance S. S. Schweber, *An Introduction to Relativistic Quantum Field Theory*, Harper and Row, New York(1961), p. 691.

[15] J.-Q. Liang, H. J. W. Müller–Kirsten, J.-G. Zhou, F.-C. Pu, Phys. Lett. **A228** (1997) 97;in eq. (25) of this paper '8' has to be replaced by '4' and in eq. (27) $m_0$ by $\frac{m_0}{1-\lambda}$.

[16] A. Perelomov, *Generalized Coherent States and Their Applications*, Springer, Heidelberg(1986).

[17] E. S. Abers and B. W. Lee, Phys. Rep. **9** (1973) 1.

[18] E. Fradkin, *Field Theories of Condensed Matter Systems*, Addison-Wesley, New York(1991).

[19] J.-Q. Liang, H. J. W. Müller–Kirsten, J.-G. Zhou, Z. Physik **B102** (1997) 525; note that $\sqrt{2}$ in formula (27) must be removed, the factors '3' in (30) and (32) are misprints, factors '8' in (28), (29) must be '4' and $2\triangle\epsilon$ in (30), (32) must be $4\triangle\epsilon$.

[20] L. D. Faddeev and V. N. Popov, Phys. Lett. **B25** (1967) 29.

[21] J.-G. Zhou, F. Zimmerschied, J.-Q. Liang and H. J. W. Müller–Kirsten, Phys. Lett. **B365** (1996) 163.

[22] M. Enz and R. Schilling, J. Phys. C, Solid State Phys. **19** (1986) 1765 .

[23] M. Enz and R. Schilling, J. Phys. C, Solid State Phys. **19** (1986) L711 .

[24] E. M. Chudnovsky and L. Gunther, Phys. Rev. Lett. **60** (1988) 661.

[25] D. A. Garanin, J. Phys. A: Math. Gen. **24** (1991) L61.

[26] J.-Q. Liang, H.J.W. Müller–Kirsten and J.M.S. Rana, Phys. Lett.**A** (1997), to appear.

[27] J.-Q. Liang, H.J.W. Müller–Kirsten, A.V. Shurgaia and F. Zimmerschied, Kaiserslautern report KL–TH/97/5, unpublished.

[28] Y. B. Zhang and J.-Q. Liang (unpublished).

# CLASSICAL PROPERTIES OF GENERALIZED COHERENT STATES: FROM PHASE-SPACE DYNAMICS TO BELL'S INEQUALITY

C. BRIF, A. MANN, and M. REVZEN

Department of Physics, Technion—Israel Institute of Technology, Haifa 32000, Israel

**Abstract**    We review classical properties of harmonic-oscillator coherent states. Then we discuss which of these classical properties are preserved under the group-theoretic generalization of coherent states. We prove that the generalized coherent states of quantum systems with Lie-group symmetries are the unique Bell states, i.e., the pure quantum states preserving the fundamental classical property of satisfying Bell's inequality upon splitting.

## 1. INTRODUCTION

A few years ago, a joint paper with Prof. Ezawa[1] started: "Some 25 years ago Aharonov, Falkoff, Lerner, and Pendleton characterized the quantum state of the radiation field known as coherent state by a classical attribute it possesses." In the present work dedicated to Prof. Ezawa upon his 65th birthday (Dear Hiroshi: Tanjobi omedeto gozaimas!) we would like to discuss a few more aspects of the quantum-classical relationship.

## 2. HARMONIC-OSCILLATOR COHERENT STATES

Coherent states (CS) for a quantum harmonic oscillator were first discovered in 1926 by Schrödinger[2] who looked for wave packets with minimum possible

215

dispersions that will preserve their form while moving along a classical trajectory. An enormous interest in the CS was stimulated by pioneering works of
Glauber[3] who introduced them in the context of quantum optics and invented
their name. Glauber proposed two criteria for the CS: (i) quantum states of
the radiation field produced by a classically prescribed current; (ii) states pos-
sessing the property of maximal coherence (which is the origin of the name):
normal-ordered correlation functions of all orders factorize.

There are two (equivalent) formal definitions of the CS: (i) eigenstates of
the boson annihilation operator:

$$a|\alpha\rangle = \alpha|\alpha\rangle, \qquad \alpha \in \mathbb{C} \tag{1}$$

(this property actually implies another possible definition of $|\alpha\rangle$ as minimum-
uncertainty states); (ii) states produced by the action of the displacement
operator $D(\alpha)$ on the vacuum,

$$|\alpha\rangle = D(\alpha)|0\rangle = \exp(\alpha a^\dagger - \alpha^* a)|0\rangle. \tag{2}$$

There exists an intimate relation between the CS and the concept of phase
space in quantum mechanics. The scaled position and momentum of a quantum harmonic oscillator (which is the mathematical model of a single mode of
the quantized radiation field) are given by

$$q = \frac{1}{\sqrt{2}}(a + a^\dagger), \qquad p = \frac{1}{i\sqrt{2}}(a - a^\dagger). \tag{3}$$

For the CS $|\alpha\rangle$ one obtains:

$$\langle q\rangle = \sqrt{2}\,\mathrm{Re}\,\alpha, \qquad \langle p\rangle = \sqrt{2}\,\mathrm{Im}\,\alpha, \qquad \Delta q = \Delta p = 1/\sqrt{2}. \tag{4}$$

Therefore the phase space $(q, p)$ is just the complex $\alpha$ plane.

The coherent-state wave packets not only have minimal possible dispersions, but also preserve this property during the dynamical evolution. The
evolution of a single-mode quantized radiation field interacting with a classical
external source is governed by the Hamiltonian

$$H = \omega a^\dagger a + \lambda(t)a^\dagger + \lambda^*(t)a, \tag{5}$$

which is linear in the generators of the oscillator group $H_4$. If the initial state
$|\psi(0)\rangle$ is the vacuum, the solution of the Schrödinger equation is[4]

$$\begin{aligned}
|\psi(t)\rangle &= \mathcal{T}\exp\left(-i\int H(t)dt\right)|0\rangle \\
&= e^{i\eta(t)}\exp\left[\alpha(t)a^\dagger - \alpha^*(t)a\right]|0\rangle = e^{i\eta(t)}|\alpha(t)\rangle,
\end{aligned} \tag{6}$$

where

$$\alpha(t) = -ie^{-i\omega t} \int_0^t \lambda^*(\tau)e^{i\omega \tau}d\tau,$$

$$\eta(t) = -\frac{\omega t}{2} - \int_0^t \text{Re}[\lambda(\tau)\alpha(\tau)]d\tau.$$

This result means that the system will remain for all times in a coherent state. Furthermore, if the initial state is a coherent state (including the vacuum) and the Hamiltonian is linear in the generators of $H_4$, then the state will evolve into a coherent state, i.e., "*once a coherent state, always a coherent state*". The CS will evolve along a classical trajectory

$$\alpha(t) = [q(t) + ip(t)]/\sqrt{2} \tag{7}$$

in the phase space. In the next section we will consider how this important property of the harmonic-oscillator CS can be generalized for other quantum systems.

## 3. THE GENERALIZED COHERENT STATES

The Gilmore-Perelomov generalization[4,5] of the CS consists of the following components: $G$ is a Lie group (the dynamical symmetry group of a quantum system), $\Gamma_\Lambda$ is unitary irreducible representation (irrep) of $G$ acting on the Hilbert space $\mathcal{H}_\Lambda$, $|\Psi_0\rangle$ is a fixed normalized reference state in $\mathcal{H}_\Lambda$. The generalized CS $|\Psi_g\rangle$ are produced by the action of group elements on the reference state:

$$|\Psi_g\rangle = \Gamma_\Lambda(g)|\Psi_0\rangle, \qquad g \in G. \tag{8}$$

Elements of the isotropy subgroup $H \subset G$ leave the reference state invariant up to a phase factor:

$$\Gamma_\Lambda(h)|\Psi_0\rangle = e^{i\phi(h)}|\Psi_0\rangle, \qquad h \in H. \tag{9}$$

For every element $g \in G$, there is a decomposition:

$$g = \Omega h, \qquad g \in G, \; h \in H, \; \Omega \in G/H \tag{10}$$

where $G/H$ is the coset space. Group elements $g$ and $g'$ with different $h$ and $h'$ but with the same $\Omega$ produce CS which differ only by a phase factor: $|\Psi_g\rangle = e^{i\delta}|\Psi_{g'}\rangle$, where $\delta = \phi(h) - \phi(h')$. Therefore a coherent state $|\Omega\rangle \equiv |\Psi_\Omega\rangle$ is determined by a point $\Omega = \Omega(g)$ in the coset space.

The standard (maximum-symmetry) system of the CS is obtained when the reference state is an 'extreme' state of the Hilbert space (e.g., the vacuum state of an oscillator or the lowest/highest spin state). Then $G/H$ will be a homogeneous Kählerian manifold. *Then the coset space is the phase space of the system.* Examples of phase spaces: (1) The Glauber CS of the Heisenberg-Weyl group $H_3$ are defined on the complex plane $\mathbb{C} = H_3/U(1)$. (2) The spin CS are defined on the unit sphere $\mathbb{S}^2 = SU(2)/U(1)$.

Under the action of the group elements the generalized CS transform among themselves. If the Hamiltonian is linear in the group generators

$$H = \sum_i \beta_i T_i, \tag{11}$$

the evolution operator will be an element of the group. *Then the generalized CS will evolve along a classical trajectory $\Omega(t)$ in the phase space.*

## 4. ANOTHER CRITERION OF CLASSICALITY

In 1966 Aharonov, Falkoff, Lerner and Pendleton (AFLP) proposed[6] a characterization of a quantum state of the radiation field by a classical attribute it possesses. They used the following arguments: In classical physics, two observers $B$ and $C$ cannot ascertain by any local measurements (including correlating their observations) whether two radiation beams emanated from one source that was split—or they came from two independent sources (see Figure 1). In quantum physics, quantum correlations created between the beams upon splitting will, in general, enable the observers to distinguish between the two situations.

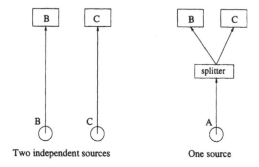

Figure 1: A sketch of the AFLP scheme.

However, the Glauber CS are the only quantum states of the radiation field which possess the classical attribute: The beams $B$ and $C$ that were split from one source $A$ can be simulated by two beams from independent sources. The reason is that no quantum correlations are created for the coherent state upon splitting. This means that the field coherent state will factorize upon splitting and the beams $B$ and $C$ will be disentangled.

We present here a brief description of the proof given by AFLP[6]. A state of the radiation field in the mode $A$ can be written as

$$|\psi_A\rangle_A = f_A(a_A^\dagger)|0\rangle, \tag{12}$$

and analogously for the modes $B$ and $C$. Splitting of the light beam can be realized by a half-silvered mirror (with the vacuum in the second input port). The corresponding transformation is

$$a_A^\dagger = \mu a_B^\dagger + \nu a_C^\dagger, \qquad |\mu|^2 + |\nu|^2 = 1. \tag{13}$$

One looks for the state that factorizes upon splitting,

$$|\psi_A\rangle_A = |\psi_B\rangle_B \otimes |\psi_C\rangle_C. \tag{14}$$

This leads to the functional equation:

$$f_A(\mu a_B^\dagger + \nu a_C^\dagger) = f_B(a_B^\dagger)f_C(a_C^\dagger). \tag{15}$$

Following AFLP, the unique solution of this equation is given by the functions (including normalization):

$$f_A(a_A^\dagger) = D_A(\alpha) = \exp(\alpha a_A^\dagger - \alpha^* a_A),$$
$$f_B(a_B^\dagger) = D_B(\alpha_B) = \exp(\alpha_B a_B^\dagger - \alpha_B^* a_B),$$
$$f_C(a_C^\dagger) = D_C(\alpha_C) = \exp(\alpha_C a_C^\dagger - \alpha_C^* a_C),$$

where $\alpha_B = \mu \alpha_A$, $\alpha_C = \nu \alpha_A$. These operators are exactly the displacement operators producing the Glauber CS. Therefore, the Glauber CS are the unique field states which factorize upon splitting:

$$|\alpha\rangle_A = |\mu\alpha\rangle_B \otimes |\nu\alpha\rangle_C, \qquad |\mu|^2 + |\nu|^2 = 1. \tag{16}$$

## 5. BELL'S INEQUALITY AND BELL STATES

A state of a quantum system, consisting of two subsystems, is called *entangled*, if it cannot be represented as a direct product of states of the subsystems. This situation is in contradiction with classical physics as expressed mathematically by the violation of Bell's inequality[7].

Let $B$ and $C$ be the two subsystems. The operator $\hat{C}(\rho)$ acts on the subsystem $C$ and has eigenvalues $\pm 1$, depending on the parameter $\rho$. The operator $\hat{B}(\sigma)$ acts in the same way on the subsystem $B$. Then Bell's inequality (in the version of Clauser, Horne, Shimony, and Holt[8]) reads

$$\left| \langle \psi | \hat{C}(\rho) \hat{B}(\sigma) + \hat{C}(\rho) \hat{B}(\sigma') + \hat{C}(\rho') \hat{B}(\sigma) - \hat{C}(\rho') \hat{B}(\sigma') | \psi \rangle \right| \leq 2. \tag{17}$$

It is clear that the product state

$$|\psi\rangle = |\psi_B\rangle_B \otimes |\psi_C\rangle_C$$

satisfies Bell's inequality (17). Furthermore, it has been proven[9–11] that any entangled (i.e., non-product) pure state will always violate the inequality. In other words, for pure states there is equivalence between entanglement and the violation of Bell's inequality.

Let us now consider a quantum system $A$ which is split or decays into two subsystems $B$ and $C$. A pure quantum state that upon splitting will not violate Bell's inequality is called the *Bell state*[11]. This definition means that the Bell state of the system $A$ must factorize upon splitting into the direct product of states of the subsystems $B$ and $C$. Then the AFLP result means that the Glauber CS are the unique Bell states of the quantized radiation field.

Let us address the following question: What are the Bell states for quantum systems other than the single-mode radiation field? Does this classical attribute exist for a quantum system of general symmetry?

## 6. COHERENT STATES FOR SEMISIMPLE LIE GROUPS

We will consider quantum systems, whose dynamical symmetry groups are semisimple Lie groups. In particular, we will consider SU(2) which is the dynamical symmetry group of a spin-$j$ particle. We use the following notation: $G$ is a semisimple Lie group, $\mathfrak{G}$ is the corresponding Lie algebra, $\mathfrak{H}$ is the Cartan subalgebra, $\Delta$ is the set of non-zero roots, $\{H_i, E_\alpha\}$ is the Cartan-Weyl basis $(H_i \in \mathfrak{H}; \alpha, \beta \in \Delta)$:

$$[H_i, H_j] = 0,$$
$$[H_i, E_\alpha] = \alpha(H_i)E_\alpha,$$
$$[E_\alpha, E_\beta] = \begin{cases} 0, & \text{if } \alpha + \beta \neq 0 \text{ and } \alpha + \beta \notin \Delta, \\ H_\alpha, & \text{if } \alpha + \beta = 0, \\ N_{\alpha,\beta} E_{\alpha+\beta}, & \text{if } \alpha + \beta \in \Delta. \end{cases}$$

The generators $H_i$ are diagonal in any irrep $\Gamma_\Lambda$, while $E_\alpha$ are the "shift operators". For a Hermitian irrep, one has $H_i^\dagger = H_i$ and $E_\alpha^\dagger = E_{-\alpha}$.

A physically sensible choice of the reference state $|\Psi_0\rangle$ is the ground state of the system, i.e., an "extremal state" in the Hilbert space $\mathcal{H}_\Lambda$. *This choice of the reference state determines the classical properties of the CS.* The extremal state is the lowest-weight state $|\Lambda, -\Lambda\rangle$:

$$E_\alpha|\Lambda, -\Lambda\rangle = 0, \qquad \alpha < 0, \tag{18}$$

$$H_i|\Lambda, -\Lambda\rangle = \Lambda_i|\Lambda, -\Lambda\rangle, \qquad H_i \in \mathfrak{H}. \tag{19}$$

Thus the Lie algebra of the isotropy subgroup $H$ is just the Cartan subalgebra $\mathfrak{H}$. Therefore, elements $\Omega$ of the coset space $G/H$ are

$$\Omega = \exp\left[\sum_{\alpha>0}(\eta_\alpha E_\alpha - \eta_\alpha^* E_{-\alpha})\right], \tag{20}$$

where $\eta_\alpha$ are complex numbers. Using the Baker-Campbell-Hausdorff formula, one obtains

$$\exp\left[\sum_{\alpha>0}(\eta_\alpha E_\alpha - \eta_\alpha^* E_{-\alpha})\right] = \exp\left[\sum_{\alpha>0}\tau_\alpha E_\alpha\right]\exp\left[\sum_i \gamma_i H_i\right]\exp\left[-\sum_{\alpha>0}\tau_\alpha^* E_{-\alpha}\right].$$

The relation between $\tau_\alpha$, $\gamma_i$ and $\eta_\alpha$ can be derived[4] from the matrix representation of $G$. The generalized CS $|\Lambda, \Omega\rangle$ are given by

$$|\Lambda, \Omega\rangle = \Omega|\Lambda, -\Lambda\rangle = \mathcal{N}\exp\left[\sum_{\alpha>0}\tau_\alpha E_\alpha\right]|\Lambda, -\Lambda\rangle, \tag{21}$$

where the normalization factor is

$$\mathcal{N} = \exp\left[\sum_i \gamma_i \Lambda_i\right]. \tag{22}$$

As an example we consider the SU(2) group for a spin-$j$ particle. The su(2) algebra is spanned by $\{J_0, J_+, J_-\}$,

$$[J_0, J_\pm] = \pm J_\pm, \qquad [J_+, J_-] = 2J_0. \tag{23}$$

The Casimir operator $J^2 = J_0^2 + (J_+J_- + J_-J_+)/2$ for any irrep $\Gamma_j$ is the identity operator times a number: $J^2 = j(j+1)I$ (where $j = 0, 1/2, 1, 3/2, \ldots$). The lowest-weight state is $|j, -j\rangle$ (annihilated by $J_-$). The isotropy subgroup $H$=U(1) consists of all group elements $h$ of the form $h = \exp(i\delta J_3)$. The coset space is SU(2)/U(1) (the sphere), and the coherent state is specified by a unit vector

$$\mathbf{n} = (\sin\theta\cos\varphi, \sin\theta\sin\varphi, \cos\theta). \tag{24}$$

Then an element $\Omega$ of the coset space is

$$\Omega = \exp(\xi J_+ - \xi^* J_-), \tag{25}$$

where $\xi = -(\theta/2)e^{-i\varphi}$. The SU(2) CS are given by

$$|j, \varsigma\rangle = \Omega|j, -j\rangle = (1 + |\varsigma|^2)^{-j} \exp(\varsigma J_+)|j, -j\rangle, \tag{26}$$

where $\varsigma = (\xi/|\xi|) \tan|\xi| = -\tan(\theta/2)e^{-i\varphi}$.

## 7. FACTORIZATION UPON SPLITTING OR DECAY

Consider a quantum system $A$ (whose dynamical symmetry group is $G$), that is split or decays into two subsystems $B$ and $C$ *of the same symmetry*. The commutation relations are satisfied by operators describing each of the three systems if and only if

$$E_{A\alpha} = E_{B\alpha} \otimes I_C + I_B \otimes E_{C\alpha}, \tag{27}$$

$$H_{Ai} = H_{Bi} \otimes I_C + I_B \otimes H_{Ci}, \tag{28}$$

where $\alpha \in \Delta$ and $H_{Xi} \in \mathfrak{H}$, $X = A, B, C$. For SU(2), this condition is equivalent to the rule for the addition of angular momenta,

$$\mathbf{J}_A = \mathbf{J}_B \otimes I_C + I_B \otimes \mathbf{J}_C. \tag{29}$$

We look for quantum states factorizing upon splitting. First, we require that the lowest-weight state factorize:

$$|\Lambda_A, -\Lambda_A\rangle_A = |\Lambda_B, -\Lambda_B\rangle_B \otimes |\Lambda_C, -\Lambda_C\rangle_C, \tag{30}$$

giving the necessary condition for the factorization:

$$\Lambda_{Ai} = \Lambda_{Bi} + \Lambda_{Ci}. \tag{31}$$

In particular, for SU(2) this condition reads

$$j_A = j_B + j_C. \tag{32}$$

If the representations of the subsystems $B$ and $C$ satisfy the condition (31), then *any generalized coherent state factorizes upon splitting*. Indeed, using Equations (27) and (30), we obtain

$$|\Lambda_A, \Omega(\eta)\rangle_A = \exp\left[\sum_{\alpha>0}(\eta_\alpha E_{A\alpha} - \eta_\alpha^* E_{A\alpha}^\dagger)\right]|\Lambda_A, -\Lambda_A\rangle_A$$

$$= \exp\left[\sum_{\alpha>0}(\eta_\alpha E_{B\alpha} - \eta_\alpha^* E_{B\alpha}^\dagger)\right]|\Lambda_B, -\Lambda_B\rangle_B$$

$$\otimes \exp\left[\sum_{\alpha>0}(\eta_\alpha E_{C\alpha} - \eta_\alpha^* E_{C\alpha}^\dagger)\right]|\Lambda_C, -\Lambda_C\rangle_C,$$

which gives

$$|\Lambda_A, \Omega(\eta)\rangle_A = |\Lambda_B, \Omega(\eta)\rangle_B \otimes |\Lambda_C, \Omega(\eta)\rangle_C. \tag{33}$$

The amplitudes $\eta_\alpha$ are the same for the system $A$ and for the subsystems $B$ and $C$. In particular, we find for spin

$$|j_A, \zeta\rangle_A = |j_B, \zeta\rangle_B \otimes |j_C, \zeta\rangle_C, \tag{34}$$

i.e., $\zeta_A = \zeta_B = \zeta_C$. (The lowest and highest spin states $|j, -j\rangle$ and $|j, j\rangle$ are particular cases of the SU(2) CS.) The situation is somewhat different for the Glauber CS $|\alpha\rangle$ of the radiation field, where the coherent-state amplitude is split according to Equation (16). The reason for this distinction lies in the differing structures of the nilpotent group $H_3$ and a semisimple group $G$.

As a simple example we consider a spin-one particle (the system $A$) decaying into two spin-half particles (the subsystems $B$ and $C$). In this case

$$|1, 1\rangle_A = |\tfrac{1}{2}, \tfrac{1}{2}\rangle_B \otimes |\tfrac{1}{2}, \tfrac{1}{2}\rangle_C,$$

$$|1, 0\rangle_A = \frac{1}{\sqrt{2}} \left( |\tfrac{1}{2}, \tfrac{1}{2}\rangle_B \otimes |\tfrac{1}{2}, -\tfrac{1}{2}\rangle_C + |\tfrac{1}{2}, -\tfrac{1}{2}\rangle_B \otimes |\tfrac{1}{2}, \tfrac{1}{2}\rangle_C \right),$$

$$|1, -1\rangle_A = |\tfrac{1}{2}, -\tfrac{1}{2}\rangle_B \otimes |\tfrac{1}{2}, -\tfrac{1}{2}\rangle_C.$$

The explicit form of the SU(2) CS is

$$|\tfrac{1}{2}, \zeta\rangle = \frac{|\tfrac{1}{2}, -\tfrac{1}{2}\rangle + \zeta|\tfrac{1}{2}, \tfrac{1}{2}\rangle}{\sqrt{1 + |\zeta|^2}} \tag{35}$$

for $j = 1/2$ and

$$|1, \zeta\rangle = \frac{|1, -1\rangle + \sqrt{2}\zeta|1, 0\rangle + \zeta^2|1, 1\rangle}{1 + |\zeta|^2} \tag{36}$$

for $j = 1$. Assume that the spin-one particle $A$ was prepared before the decay in the coherent state. Then we obtain

$$\begin{aligned}
|1, \zeta\rangle_A &= \frac{|\tfrac{1}{2}, -\tfrac{1}{2}\rangle_B + \zeta|\tfrac{1}{2}, \tfrac{1}{2}\rangle_B}{\sqrt{1 + |\zeta|^2}} \otimes \frac{|\tfrac{1}{2}, -\tfrac{1}{2}\rangle_C + \zeta|\tfrac{1}{2}, \tfrac{1}{2}\rangle_C}{\sqrt{1 + |\zeta|^2}} \\
&= |\tfrac{1}{2}, \zeta\rangle_B \otimes |\tfrac{1}{2}, \zeta\rangle_C.
\end{aligned} \tag{37}$$

The direct products $|1, 1\rangle_A$ and $|1, -1\rangle_A$ are obtained as particular cases of (37) for $\zeta \to \infty$ and $\zeta = 0$, respectively.

We can prove[12] that the generalized CS are the *only* states which factorize upon splitting. We first note that the states of the orthonormal basis are obtained by applying the raising operators to the lowest-weight state one or more times:

$$(E_\alpha)^p|\Lambda, -\Lambda\rangle = |\Lambda, -\Lambda + p\alpha\rangle \times \text{factor}, \tag{38}$$

where $\alpha > 0$ and $p \in \mathbb{N}$. Therefore, any state $|\Phi\rangle \in \mathcal{H}_\Lambda$ can be obtained by applying a function of the raising operators to the lowest-weight state:

$$|\Phi\rangle = f(\{E_\alpha\})|\Lambda, -\Lambda\rangle, \qquad \alpha > 0. \tag{39}$$

For example, for the SU(2) group one has

$$|j, m\rangle = \left( \begin{array}{c} 2j \\ j+m \end{array} \right)^{1/2} \frac{(J_+)^{j+m}}{(j+m)!} |j, -j\rangle, \tag{40}$$

and $|\Phi\rangle = f(J_+)|j, -j\rangle$.

If the state $|\Phi_A\rangle_A$ factorizes upon splitting,

$$|\Phi_A\rangle_A = |\Phi_B\rangle_B \otimes |\Phi_C\rangle_C, \tag{41}$$

then the following functional equation must be satisfied,

$$f_A(\{E_{A\alpha}\}) = f_B(\{E_{B\alpha}\}) f_C(\{E_{C\alpha}\}) \tag{42}$$

(here and in what follows $\alpha > 0$). Since $E_{A\alpha} = E_{B\alpha} + E_{C\alpha}$ (for the sake of simplicity, we omit the identity operators), we obtain

$$f_A(\{E_{B\alpha} + E_{C\alpha}\}) = f_B(\{E_{B\alpha}\}) f_C(\{E_{C\alpha}\}). \tag{43}$$

Using the same method as AFLP[6], we can easily prove that Equation (43) has the unique solution—the three functions $f_X$ $(X = A, B, C)$ must be exponentials:

$$f_X(\{E_{X\alpha}\}) = f_X(0) \exp\left[ \sum_{\alpha > 0} \tau_\alpha E_{X\alpha} \right], \tag{44}$$

where $f_X(0)$ is a normalization factor. Here $\tau_\alpha$ are complex parameters which are the same for the three systems $A$, $B$, and $C$. For example, in the case of the SU(2) group, we find

$$f_X(J_{X+}) = (1 + |\zeta|^2)^{-jx} \exp(\zeta J_{X+}), \tag{45}$$

where $X = A, B, C$ and $\zeta \in \mathbb{C}$. The operator-valued function $f_X(\{E_{X\alpha}\})$ for each of the three systems $(A, B, \text{and } C)$ is precisely the function that produces the generalized CS. (The factor $f_X(0)$ is recognized as the normalization factor $\mathcal{N}$). This completes the proof of uniqueness.

## 8. CONCLUSIONS

As was shown, the generalized CS keep two most important classical attributes of the Glauber CS. (1) The known fact: each point $\Omega$ of the phase space corresponds to the coherent state $|\Omega\rangle$. If the Hamiltonian is linear in group generators, the generalized CS evolve along a classical trajectory $\Omega(t)$ in the phase space. (2) The new fact: For the semisimple Lie groups, the generalized CS are the unique Bell states, i.e., the pure quantum states preserving the fundamental classical property of satisfying Bell's inequality upon splitting.

## ACKNOWLEDGEMENTS

This work was supported by the Fund for Promotion of Research at the Technion and by the Technion VPR Fund—Promotion of Sponsored Research.

## REFERENCES

1. H. Ezawa, A. Mann, K. Nakamura, and M. Revzen, *Ann. Phys. (N.Y.)* **209**, 216 (1991).
2. E. Schrödinger, *Naturwissenschaften* **14**, 664 (1926).
3. R. J. Glauber, *Phys. Rev.* **131**, 2766 (1963).
4. W.-M. Zhang, D. H. Feng, and R. Gilmore, *Rev. Mod. Phys.* **62**, 867 (1990).
5. A. M. Perelomov, *Generalized Coherent States and Their Applications* (Springer, Berlin, 1986).
6. Y. Aharonov, D. Falkoff, E. Lerner, and H. Pendleton, *Ann. Phys. (N.Y.)* **39**, 498 (1966).
7. J. S. Bell, *Physics* **1**, 195 (1964-1965).
8. J. F. Clauser, M. A. Horne, A. Shimony, and R. A. Holt, *Phys. Rev. Lett.* **23**, 880 (1969).
9. A. Peres, *Am. J. Phys.* **46**, 745 (1978).
10. N. Gisin, *Phys. Lett.* A **154**, 201 (1991).
11. A. Mann, M. Revzen, and W. Schleich, *Phys. Rev.* A **46**, 5363 (1992).
12. C. Brif, A. Mann, and M. Revzen, *Phys. Rev.* A **57**, 742 (1998).

# GAUGE FIELD THEORY
# AND
# SUPERSYMMETRY

# ASPECTS OF GAUGE QUANTUM MECHANICS ON NON-SIMPLY CONNECTED SPACES

ASAO ARAI

Department of Mathematics, Hokkaido University
Sapporo 060, Japan

**Abstract** A class of representations of the canonical commutation relations is constructed in a quantum theory of a particle moving in an external (non-Abelian) gauge field on a non-simply connected space in three space dimensions. A coupling to a Bose field of the particle is also discussed.

**Key Words** gauge theory, Aharonov-Bohm effect, canonical commutation relations, rotation algebra, quantum group, quantum field

## 1   INTRODUCTION

In a series of papers (Arai, 1992, 1993, 1995a, 1995b, 1996, 1998), the author studied representation-theoretic aspects of a (non-Abelian) gauge theory of a quantum particle moving on a non-simply connected space of the form $\mathbf{R}^2 \setminus \mathbf{D}$ with $\mathbf{D}$ a finite discrete subset (Arai, 1992, 1993, 1995a, 1995b) or an infinite discrete subset of $\mathbf{R}^2$ (Arai, 1996, 1998). In such a gauge theory, a representation of the canonical commutation relations (CCR) with two degrees of freedom is realized by the position and the physical momentum operators of the particle if the gauge field under consideration is flat. This representation is not necessarily equivalent to the Schrödinger representation. A complete characterization for the representation to be equivalent to the Schrödinger one is given in terms of "Wilson loops". An interesting fact is that the case where the representation is inequivalent to the Schrödinger one corresponds to the (non-Abelian) Aharonov-Bohm effect (Aharonov and Bohm, 1959). Moreover, connections to representations of the rotation algebra and the quantum group $U_q(sl_2)$ with $|q| = 1$ were discovered (Arai, 1996, 1998).

In this paper we further continue to study gauge quantum mechanics on non-simply connected spaces. We consider in Section 2 a quantum system of a particle in an external (non-Abelian) gauge field on a non-simply connected space in $\mathbf{R}^3$. This part is a straightforward extension of the framework and some results of the work (Arai, 1998) to a three-dimensional non-Abelian case. In the last section we construct a representation of CCR in a coupled system of the particle and a Bose field.

# 2 REPRESENTATION OF CCR IN GAU-GE QUANTUM MECHANICS

## 2.1 Physical Momentum Operators

Let $\{\mathbf{a}_n = (a_{n1}, a_{n2}) \in \mathbf{R}^2 | n \in \mathbf{N}\}$ be a discrete subset of $\mathbf{R}^2$ such that the sequences $\{a_{nj}\}_{n \in \mathbf{N}}$, $j = 1, 2$, have no accumulation points in $\mathbf{R}$ and let

$$\mathbf{D} := \{(\mathbf{a}_n, x_3) \in \mathbf{R}^3 | n \in \mathbf{N}, x_3 \in \mathbf{R}\}. \tag{2.1}$$

Then the set

$$\mathbf{M} := \mathbf{R}^3 \setminus \mathbf{D} = \{\mathbf{x} = (x_1, x_2, x_3) \in \mathbf{R}^3 | \mathbf{x} \notin \mathbf{D}\}, \tag{2.2}$$

is non-simply connected. We consider a system of a quantum particle moving in $\mathbf{M}$ under the influence of an external gauge field. We take the gauge group of the gauge field to be the $N$-dimensional unitary group $U(N)$ ($N \geq 1$). We denote its Lie algebra by $\mathsf{u}(N)$, which is the set of $N \times N$ anti-Hermitian matrices. Then the gauge field is given by a $\mathsf{u}(N)$-valued 1-form on $\mathbf{M}$

$$A := A_1 dx_1 + A_2 dx_2 + A_3 dx_3. \tag{2.3}$$

We assume that, for some $r \geq 0$, $A_j \in C^r(\mathbf{M}; \mathsf{u}(N))$ (the set of $r$ times continuously differentiable $\mathsf{u}(N)$-valued functions on $\mathbf{M}$), $j = 1, 2, 3$ (but $A_j$ may be singular on a subset of $\mathbf{D}$). The Hilbert space of state vectors of the quantum system under consideration is taken to be $L^2(\mathbf{M}; \mathbf{C}^N)$, the Hilbert space of $\mathbf{C}^N$-valued square Lebesgue-integrable functions on $\mathbf{M}$, which can be naturally identified with $L^2(\mathbf{R}^3; \mathbf{C}^N)$, since the Lebesgue measure $|\mathbf{D}|$ of $\mathbf{D}$ is zero.

The *physical momentum operator* $\mathbf{p}(A) := (p_1(A), p_2(A), p_3(A))$ of the particle is defined by

$$p_j(A) = -i(D_j + A_j), \quad j = 1, 2, 3, \tag{2.4}$$

acting in $L^2(\mathbf{R}^3; \mathbf{C}^N)$, where $D_j$ is the generalized partial differential operator in the variable $x_j$. It is easy to see that each $p_j(A)$ is a symmetric operator with $D(p_j(A)) \supset C_0^\infty(\mathbf{M}; \mathbf{C}^N)$ ( the set of $\mathbf{C}^N$-valued, infinitely differentiable functions on $\mathbf{M}$ with bounded support in $\mathbf{M}$). We denote the closure of $p_j(A)$ by $\bar{p}_j(A)$.

For a continuous, piecewise differentiable curve $C$ in $\mathbf{M}$ with parametrization $\gamma(\tau) = (\gamma_1(\tau), \gamma_2(\tau), \gamma_3(\tau))$, $\tau \in [a, b]$ ($a < b, a, b \in \mathbf{R}$), we define

$$W_A[C] := \prod_a^b \exp\left(-\sum_{j=1}^3 A_j(\gamma(\tau))\dot{\gamma}_j(\tau)d\tau\right) \tag{2.5}$$

the product integral for the $\mathbf{u}(N)$-valued function: $\tau \to -\sum_{j=1}^3 A_j(\gamma(\tau))\dot{\gamma}_j(\tau)$ on $[a, b]$ ($\dot{\gamma}_j(\tau) := d\gamma_j(\tau)/d\tau$) [§1.1 in Dollard and Friedman (1979)].[1]. Geometrically $W_A[C]$ is called the parallel transporter along the curve $C$ (Göckeler and Schücker, 1987). It follows that $W_A[C] \in U(N)$.

We introduce two subsets of $\mathbf{M}$:

$$\mathbf{M}_1 := \{\mathbf{x} \in \mathbf{M} | x_2 \neq a_{n2}, n \in \mathbf{N}\}, \ \mathbf{M}_2 := \{\mathbf{x} \in \mathbf{M} | x_1 \neq a_{n1}, n \in \mathbf{N}\}. \tag{2.6}$$

By the assumption for $\{\mathbf{a}_n\}_{n\in\mathbf{N}}$, each $\mathbf{M}_j$ is an open set of $\mathbf{R}^3$. For notational convenience, we set

$$\mathbf{M}_3 := \mathbf{M}. \tag{2.7}$$

For $\mathbf{x}, \mathbf{y} \in \mathbf{R}^3$, we denote by $L_{\mathbf{x};\mathbf{y}}$ the straight line from $\mathbf{x}$ to $\mathbf{y}$.

Let $\{\mathbf{e}_j\}_{j=1}^3$ be the standard basis of $\mathbf{R}^3$:

$$\mathbf{e}_1 = (1, 0, 0), \quad \mathbf{e}_2 = (0, 1, 0), \quad \mathbf{e}_3 = (0, 0, 1). \tag{2.8}$$

If $\mathbf{x} \in \mathbf{M}_j$, then $\tau x_j \mathbf{e}_j + \sum_{k\neq j} x_k \mathbf{e}_k \in \mathbf{M}_j$ for all $\tau \in \mathbf{R}$. Hence we can define a $U(N)$-valued function $U_j$ on $\mathbf{M}_j$ by

$$U_j(\mathbf{x}) := W_A[L_{\sum_{k\neq j} x_k \mathbf{e}_k; \mathbf{x}}], \quad \mathbf{x} \in \mathbf{M}_j. \tag{2.9}$$

Since $|\mathbf{M} \setminus \mathbf{M}_j| = 0$, the function $U_j$ defines uniquely a matrix-multiplication unitary operator on $L^2(\mathbf{R}^3; \mathbf{C}^N)$. We denote it by the same symbol $U_j$.

**Theorem 2.1** *For $j = 1, 2, 3$, the operator $p_j(A)$ is essentially self-adjoint on $C_0^\tau(\mathbf{M}_j; \mathbf{C}^N)$ and the operator equation*

$$\bar{p}_j(A) = U_j(-iD_j)U_j^{-1} \tag{2.10}$$

*holds.*

---

[1]In the usual physics notation, $W_A[C] = P e^{-\int_C A}$, the path-ordered exponential of $-A$. It is sometimes called a *Wilson loop* or a *Wilson operator*.

## 2.2 The Unitary Groups Generated by the Physical Momentum Operators

**Theorem 2.2** *For all $u \in L^2(\mathbf{R}^3; \mathbf{C}^N)$ and $t \in \mathbf{R}$,*

$$(e^{it\bar{p}_j(A)}u)(\mathbf{x}) = W_A\left[L_{\mathbf{x}+te_j;\mathbf{x}}\right]u(\mathbf{x}+te_j), \quad \text{a.e. } \mathbf{x} \in \mathbf{R}^3. \tag{2.11}$$

*Proof.* Similar to the proof of Theorem 2.2 in Arai(1995a). An alternative proof is made by using (2.10) and the associative law of product integral. ∎

For $t, s \in \mathbf{R}$ and $j, k = 1, 2, 3, j \neq k$, we denote by $C_{\mathbf{x}}(j, k; s, t)$ the rectangular closed curve starting at $\mathbf{x}$ and going around as $\mathbf{x} \rightarrow \mathbf{x} + se_j \rightarrow \mathbf{x} + se_j + te_k \rightarrow \mathbf{x} + te_k \rightarrow \mathbf{x}$:

$$\begin{aligned} C_{\mathbf{x}}(j, k; s, t): \ &= \ \{\mathbf{x} + \lambda e_j | 0 \leq \lambda \leq s\} \circ \{\mathbf{x} + se_j + \mu e_k | 0 \leq \mu \leq t\} \\ &\circ \{\mathbf{x} + (s-\lambda)e_j + te_k | 0 \leq \lambda \leq s\} \\ &\circ \{\mathbf{x} + (t-\mu)e_k | 0 \leq \mu \leq t\}. \end{aligned} \tag{2.12}$$

Let

$$L_{n,j} := \{(\mathbf{a}_n, x_3) + \mu e_j | x_3, \mu \in \mathbf{R}\}, \quad n \in \mathbf{N}, \quad j = 1, 2, 3. \tag{2.13}$$

and

$$\mathbf{M}(j, k; s, t) := \mathbf{R}^3 \setminus \cup_{n=1}^{\infty}\left[L_{n,j} \cup (L_{n,j} - te_k) \cup L_{n,k} \cup (L_{n,k} - se_j)\right]. \tag{2.14}$$

Note that, if $\mathbf{x} \in \mathbf{M}(j, k; s, t)$, then $C_{\mathbf{x}}(j, k; s, t)$ does not intersect $\mathbf{D}$. Hence we can define a $U(N)$-valued function $W_{j,k;s,t}^A$ on $\mathbf{M}(j, k; s, t)$ by

$$W_{j,k;s,t}^A(\mathbf{x}) = W_A\left[C_{\mathbf{x}}(j, k; s, t)\right], \quad \mathbf{x} \in \mathbf{M}(j, k; s, t).$$

We set $W_{j,j;s,t}^A(\mathbf{x}) \equiv 1$, $\mathbf{x} \in \mathbf{M}$. Since $|\mathbf{R}^3 \setminus \mathbf{M}(j, k; s, t)| = 0$, the function $W_{j,k;s,t}^A$ defines a unique matrix-multiplicaiton unitary operator on $L^2(\mathbf{R}^3; \mathbf{C}^N)$. We denote by $q_j$ the multiplication operator by $x_j$ on $L^2(\mathbf{R}^3; \mathbf{C}^N)$.

**Theorem 2.3** *For all $s, t \in \mathbf{R}$ and $j, k = 1, 2, 3$,*

$$e^{is\bar{p}_j(A)}e^{it\bar{p}_k(A)} = \left(W_{j,k;s,t}^A\right)^{-1}e^{it\bar{p}_k(A)}e^{is\bar{p}_j(A)}, \tag{2.15}$$

$$e^{isq_j}e^{it\bar{p}_k(A)} = e^{-ist\delta_{jk}}e^{it\bar{p}_k A}e^{isq_j}. \tag{2.16}$$

*Proof.* Similar to the proof of Theorem 2.2 in Arai(1995a). ∎

## 2.3   Representation of CCR

As for the terminology of representation theory of CCR, we follow that given in §1 in Arai(1996).

Let $r \geq 1$. Then we can define the gauge field strength

$$F(A) := dA + A \wedge A = \sum_{j<k} F_{jk}^A dx_j \wedge dx_k \qquad (2.17)$$

on $\mathbf{M}$, where

$$F_{jk}^A := D_j A_k - D_k A_j + [A_j, A_k]. \qquad (2.18)$$

The gauge field $A$ is said to be *flat* on $\mathbf{M}$ if $F(A) = 0$ on $\mathbf{M}$. We denote by $\mathcal{G}_0(\mathbf{M})$ the set of flat gauge fields on $\mathbf{M}$.

It is easy to see that, for all $j, k = 1, 2, 3$,

$$[q_j, p_k(A)] = i\delta_{jk}, \quad [p_j(A), p_k(A)] = -F_{jk}^A \qquad (2.19)$$

on $C_0^2(\mathbf{M}; \mathbf{C}^N)$. These commutation relations imply the following.

**Proposition 2.4** *The triple*

$$\pi_A := \{L^2(\mathbf{R}^3; \mathbf{C}^N), C_0^2(\mathbf{M}; \mathbf{C}^N), \{q_j, \bar{p}_j(A)\}_{j=1}^3\} \qquad (2.20)$$

*is a representation of the CCR with three degrees of freedom if and only if* $A \in \mathcal{G}_0(\mathbf{M})$.

A characterization for the representation $\pi_A$ of CCR with $A \in \mathcal{G}_0(\mathbf{M})$ to be a Schrödinger 3-system is given by the following theorem. We introduce a subset of $\mathcal{G}_0(\mathbf{M})$:

$$\mathcal{G}_0^Q(\mathbf{M}) := \{A \in \mathcal{G}_0(\mathbf{M}) | W_{j,k;s,t}^A = I \text{ for all } s, t \in \mathbf{R} \text{ and } j, k = 1, 2, 3\}. \quad (2.21)$$

[The letter "Q" refers to "quantization" of the gauge field strength in a sense; cf. the case $N = 1$ (Arai, 1992, 1996, 1998).]

**Theorem 2.5** *Let* $A \in \mathcal{G}_0(\mathbf{M})$. *Then* $\pi_A$ *is a Schrödinger 3-system if and only if* $A \in \mathcal{G}_0^Q(\mathbf{M})$.

*Proof.* This follows from Theorem 2.3 and the von Neumann uniqueness theorem on the Weyl representation of CCR (Von Neumann, 1931; Putnum, 1967). ∎

**Remark 2.1** (i) The set $\mathcal{G}_0(\mathbf{M}) \setminus \mathcal{G}_0^{\mathrm{Q}}(\mathbf{M})$ is an infinite one (cf. Arai, 1995a, 1995b, 1996, 1998).

(ii) Theorem 2.3 implies that $\bar{p}_j(A)$ strongly commutes with $\bar{p}_k(A)$ for all $j, k = 1, 2, 3$ if and only if $A \in \mathcal{G}_0^{\mathrm{Q}}(\mathbf{M})$.[2] If $A$ is in $\mathcal{G}_0(\mathbf{M}) \setminus \mathcal{G}_0^{\mathrm{Q}}(\mathbf{M})$, then $\{\bar{p}_j(A)\}_{j=1}^3$ is not strongly commuting by (2.15), although it is commuting on $C_0^2(\mathbf{M})$. This is an operator-theoretical reason why it is possible for the representation $\pi_A$ of CCR not to be a Schrödinger 3-system.

(iii) In the case where $A$ is flat on $\mathbf{M}$ (hence the gauge field strength is concentrated on $\mathbf{D}$ in the distribution sense), but not in $\mathcal{G}_0^{\mathrm{Q}}(\mathbf{M})$, formula (2.15) gives a mathematical form of the (non-commutative) Aharonov-Bohm effect. Thus, as in the previous cases (Arai, 1995a, 1995b, 1996), there is a correspondence between the Aharonov-Bohm effect and the case where the representation $\pi_A$ of CCR is inequivalent to the Schrödinger representation.

# 3   Coupling to a Bose Field

## 3.1   Perturbation of the particle physical momentum operator

Let $\mathcal{H}$ be a Hilbert space. Then the boson Fock space over $\mathcal{H}$ is defined by

$$\mathcal{F}_{\mathrm{b}}(\mathcal{H}) := \oplus_{n=0}^{\infty} \otimes_{\mathrm{s}}^n \mathcal{H}, \tag{3.1}$$

where $\otimes_{\mathrm{s}}^n \mathcal{H}$ is the $n$-fold symmetric tensor product of $\mathcal{H}$ ($\otimes_{\mathrm{s}}^0 \mathcal{H} := \mathbf{C}$) [§II.4, Example 2 in Reed and Simon (1972)]. We denote by $a(f)$, $f \in \mathcal{H}$, the annihilation operators on $\mathcal{F}_{\mathrm{b}}(\mathcal{H})$ [§X.7 in Reed and Simon(1975)]. Then the Segal field operator

$$\Phi_{\mathrm{S}}(f) := \frac{1}{\sqrt{2}} \{a(f)^* + a(f)\} \tag{3.2}$$

with $f \in \mathcal{H}$ (an abstract Bose field smeared out in $f$) is essentially self-adjoint on the space of finite particle vectors

$$\mathcal{F}_0(\mathcal{H}) := \Big\{ \psi = \{\psi^{(n)}\}_{n=0}^{\infty} \in \mathcal{F}_{\mathrm{b}}(\mathcal{H}) | \psi^{(n)} = 0 \text{ for all but}$$
$$\text{finitely many } n\text{'s} \Big\} \tag{3.3}$$

[§X.7 in Reed and Simon(1975)]. We denote the closure of $\Phi_{\mathrm{S}}(f)$ by $\overline{\Phi_{\mathrm{S}}(f)}$.

---

[2]Two self-adjoint operators on a Hilbert space are said to *strongly commute* if their spectral measures commute. A characterization for strong commutativity is given in Theorem VIII.13 in Reed and Simon(1972).

The Hilbert space of a coupled system of the particle and the Bose field is taken to be the tensor product

$$\mathcal{F} := L^2(\mathbf{R}^3; \mathbf{C}^N) \otimes \mathcal{F}_b(\mathcal{H}) \tag{3.4}$$

of $L^2(\mathbf{R}^3; \mathbf{C}^N)$ and $\mathcal{F}_b(\mathcal{H})$. We denote by $\oplus^N \mathcal{F}_b(\mathcal{H})$ the $N$ direct sum of $\mathcal{F}_b(\mathcal{H})$ and introduce

$$L^2(\mathbf{R}^3; \oplus^N \mathcal{F}_b(\mathcal{H}))$$
$$:= \left\{ \Psi : \mathbf{R}^3 \to \oplus^N \mathcal{F}_b(\mathcal{H}) \Big| \int_{\mathbf{R}^3} \|\Psi(\mathbf{x})\|^2_{\oplus^N \mathcal{F}_b(\mathcal{H})} d\mathbf{x} < \infty \right\}, \tag{3.5}$$

the Hilbert space of $\oplus^N \mathcal{F}_b(\mathcal{H})$-valued, square integrable measurable functions on $\mathbf{R}^3$ [Chapter II, Example 6 in Reed and Simon (1975)]. For $u = (u_1, \cdots, u_N) \in L^2(\mathbf{R}^3; \mathbf{C}^N)$ and $\psi \in \mathcal{F}_b(\mathcal{H})$, we define $u\psi \in L^2(\mathbf{R}^3; \oplus^N \mathcal{F}_b(\mathcal{H}))$ by

$$(u\psi)(\mathbf{x}) := (u_1(\mathbf{x})\psi, \cdots, u_N(\mathbf{x})\psi), \quad \mathbf{x} \in \mathbf{R}^3. \tag{3.6}$$

As in Theroem II.10(b) in Reed and Simon (1972), there exists a unitary operator $U$ from $\mathcal{F}$ to $L^2(\mathbf{R}^3; \oplus^N \mathcal{F}_b(\mathcal{H}))$ such that

$$U(u \otimes \psi) = u\psi \in L^2(\mathbf{R}^3; \oplus^N \mathcal{F}_b(\mathcal{H})), \quad u \in L^2(\mathbf{R}^3; \mathbf{C}^N), \psi \in \mathcal{F}_b(\mathcal{H}). \tag{3.7}$$

Let $f : \mathbf{R}^3 \to \mathcal{H}$ be an $\mathcal{H}$-valued strongly continuous function on $\mathbf{R}^3$. Then one can define a self-adjoint, decomposable operator

$$\tilde{\Phi}[f] := \int_{\mathbf{R}^3}^{\oplus} \overline{\Phi_S(f(\mathbf{x}))} d\mathbf{x} \tag{3.8}$$

acting in $L^2(\mathbf{R}^3; \oplus^N \mathcal{F}_b(\mathcal{H})) = \int_{\mathbf{R}^3}^{\oplus} \oplus^N \mathcal{F}_b(\mathcal{H}) d\mathbf{x}$ [§XIII.16, pp.280–285, in Reed and Simon(1978)]. With this operator, we define

$$\Phi[f] := U^{-1}\tilde{\Phi}[f]U, \tag{3.9}$$

acting in $\mathcal{F}$.

Let $h$ be a nonnegative self-adjoint operator on $\mathcal{H}$ such that $\ker h = \{0\}$. If $f(\mathbf{x})$ is in $D(h^r)$ for all $\mathbf{x} \in \mathbf{R}^3$ $(r \in \mathbf{R})$, then we define a nonnegative number $\|f\|_{r,\infty}$ by

$$\|f\|_{r,\infty} := \sup_{\mathbf{x} \in \mathbf{R}^3} \|h^r f(\mathbf{x})\|_{\mathcal{H}} \tag{3.10}$$

(which may be infinite).

To describe an interaction between the particle and the Bose field, we suppose that there exist $\mathcal{H}$-valued functions $g_j^{(l)} : \mathbf{R}^3 \to \mathcal{H}$ $(j = 1, 2, 3, l = 1, \cdots, L)$ on $\mathbf{R}^3$ $(L \in \mathbf{N})$ which have the following properties:

**(g.1)** For all $\mathbf{x} \in \mathbf{R}^3$, each $g_j^{(l)}(\mathbf{x}) \in D(h^{-1/2}) \cap D(h)$ and, for $r = -1/2, 1$, $h^r g_j^{(l)}$ is strongly continuous on $\mathbf{R}^3$ with $\|g_j^{(l)}\|_{r,\infty} < \infty$, $j = 1, 2, 3$, $l = 1, \cdots, L$.

**(g.2)** For $j, k = 1, 2, 3$, $l = 1, \cdots, L$, (i) $g_j^{(l)}$ and $h^{-1/2} g_j^{(l)}$ are strongly differentiable on $\mathbf{R}^3$ with $\partial_k g_j^{(l)} \in D(h^{-1/2})$ and $h^{-1/2} \partial_k g_j^{(l)}(\mathbf{x}) = \partial_k h^{-1/2} g_j^{(l)}(\mathbf{x})$, $\mathbf{x} \in \mathbf{R}^3$; (ii) $\partial_k g_j^{(l)}$ and $\partial_k h^{-1/2} g_j^{(l)}$ are strongly continuous on $\mathbf{R}^3$ with $\|\partial_k g_j^{(l)}\|_{r,\infty} < \infty$, $r = -1/2, 0$.

For a subspace $D$ of $L^2(\mathbf{R}^3; \mathbf{C}^N)$ and a subspace $\mathcal{W}$ of $\mathcal{F}_b(\mathcal{H})$, we define

$$D \otimes_{\mathbf{alg}} \mathcal{W} := \mathcal{L}\{u \otimes \psi | u \in D, \ \psi \in \mathcal{W}\}, \qquad (3.11)$$

where $\mathcal{L}\{\cdots\}$ denotes the linear span of the set $\{\cdots\}$. We denote by $\Omega_0 := \{1, 0, 0, \cdots\}$ the vacuum in $\mathcal{F}_b(\mathcal{H})$ [§X.7 in Reed and Simon (1975)] and, for a subspace $\mathcal{D}$ of $\mathcal{H}$, we set and

$$\mathcal{F}_{\mathrm{fin}}(\mathcal{D}) := \mathcal{L}\{a(f_1)^* \cdots a(f_n)^* \Omega_0, \ \Omega_0 | n \geq 1, \ f_j \in \mathcal{D}, \ j = 1, \cdots, n\}. \qquad (3.12)$$

Let $V_j^{(l)}(\mathbf{x}) \, (j = 1, 2, 3, \ l = 1, \cdots, L)$ be an $N \times N$ Hermitian matrix-valued, continuously differentiable function on $\mathbf{M}$ such that

**(V.1)** $[A_k(\mathbf{x}), V_j^{(l)}(\mathbf{x})] = 0$, $\mathbf{x} \in \mathbf{M}$ and $\sup_{\mathbf{x} \in \mathbf{M}} \|V_j^{(l)}(\mathbf{x})\| < \infty$, $\sup_{\mathbf{x} \in \mathbf{M}} \|\partial_k V_j^{(l)}(\mathbf{x})\| < \infty$ for $j, k = 1, 2, 3$, $l = 1, \cdots, L$.

Then the operators

$$W_j := \sum_{l=1}^{L} (V_j^{(l)} \otimes I) \Phi[g_j^{(l)}], \quad j = 1, 2, 3, \qquad (3.13)$$

acting in $\mathcal{F}$ are symmetric.

Let $\{h_j\}_{j=1}^3$ be a family of strongly commuting self-adjoint operators on $\mathcal{H}$ such that each $h_j$ strongly commutes with $h$ and $\pm h_j \leq h$, $j = 1, 2, 3$.

We are concerned with the operators

$$P_j(A) := p_j(A) \otimes I + I \otimes d\Gamma(h_j) - W_j, \quad j = 1, 2, 3, \qquad (3.14)$$

acting in $\mathcal{F}$, where, for a self-adjoint operator $S$ on $\mathcal{H}$, we denote by $d\Gamma(S)$ the second quantization of $S$ [§VIII.10, Example 2 in Reed and Simon (1972)]. The operator $P_j(A)$ may be regarded as an extension of the particle physical momentum operator to the case where the particle interacts with a Bose field too.

Let

$$C^\infty(h) := \cap_{n=1}^\infty D(h^n). \qquad (3.15)$$

Then we can prove the following theorem.

**Theorem 3.1** *Assume that $A_j \in C^r(\mathbf{M}; \mathfrak{u}(N))$, $j = 1, 2, 3$, for some $r \geq 1$. Then each $P_j(A)$ is essentially self-adjoint on $C_0^r(\mathbf{M}_j; \mathbf{C}^N) \otimes_{\mathrm{alg}} \mathcal{F}_{\mathrm{fin}}(C^\infty(h))$.*

*Proof* (Outline). We first show that $P_j(0)$ is essentially self-adjoint on $C_0^\infty(\mathbf{M}_j; \mathbf{C}^N) \otimes_{\mathrm{alg}} \mathcal{F}_{\mathrm{fin}}(C^\infty(h))$. This can be done by applying the Nelson commutator theorem [§X.5, Theorem X.37 in Reed and Simon (1975)] to the pair $\langle P_j(0), H_j \rangle$ of operators with $H_j := p_j^2 \otimes I + I \otimes d\Gamma(h) + 1$. Then, using the relation $(U_j \otimes I)P_j(0)(U_j^{-1} \otimes I) = P_j(A)$ on $C_0^r(\mathbf{M}_j; \mathbf{C}^N) \otimes_{\mathrm{alg}} \mathcal{F}_{\mathrm{fin}}(C^\infty(h))$, we obtain the desired result. ∎

## 3.2  Representation of CCR

For the functions $g_j^{(l)}$ and $V_j^{(l)}$, we assume in addition to (g.1), (g.2) and (V.1) the following:

**(g.3)** For $j, k = 1, 2, 3$, $l = 1, \cdots, L$, $\partial_k g_j^{(l)} = -ih_k g_j^{(l)}$.

**(V.2)** Each $V_j^{(l)}$ is constant and mutually commuting: $[V_j^{(l)}, V_k^{(m)}] = 0$, $j, k = 1, 2, 3$, $l, m = 1, \cdots, L$.

Condition (g.3) implies that, for all $j, k = 1, 2, 3$, $l, m = 1, \cdots, L$, the **C**-valued function: $\mathbf{x} \to (g_j^{(l)}(\mathbf{x}), g_k^{(m)}(\mathbf{x}))_{\mathcal{H}}$ is constant. We denote this constant by $g_{jk}^{(lm)}$:

$$g_{jk}^{(lm)} := (g_j^{(l)}(\mathbf{x}), g_k^{(m)}(\mathbf{x}))_{\mathcal{H}}. \tag{3.16}$$

In what follows we assume that $r \geq 1$.

**Theorem 3.2** *For all $s, t \in \mathbf{R}$ and $j, k = 1, 2, 3$,*

$$e^{is\bar{P}_j(A)} e^{it\bar{P}_k(A)} = \left( \left( W_{j,k;s,t}^A \right)^{-1} e^{-ist \sum_{l,m=1}^L V_j^{(l)} V_k^{(m)} \Im g_{jk}^{(lm)}} \otimes I \right)$$
$$\times e^{it\bar{P}_k(A)} e^{is\bar{P}_j(A)}, \tag{3.17}$$
$$e^{isq_j \otimes I} e^{it\bar{P}_k(A)} = e^{-ist\delta_{jk}} e^{it\bar{P}_k(A)} e^{isq_j \otimes I}. \tag{3.18}$$

**Lemma 3.3** *The triple*

$$\Pi_A := \left\{ \mathcal{F}, C_0^2(\mathbf{M}; \mathbf{C}^N) \otimes_{\mathrm{alg}} \mathcal{F}_{\mathrm{fin}}(C^\infty(h)), \{q_j \otimes I, \bar{P}_j(A)\}_{j=1}^3 \right\} \tag{3.19}$$

*is a representation of the CCR with three degrees of freedom if and only if*

$$F_{jk}^A(\mathbf{x}) = i \sum_{l,m=1}^L V_j^{(l)} V_k^{(m)} \Im g_{jk}^{(lm)}, \quad j, k = 1, 2, 3, \ \mathbf{x} \in \mathbf{M}. \tag{3.20}$$

By Theorem 3.2, we obtain the following characterization of the representation $\Pi_A$ of CCR given in Lemma 3.3.

**Corollary 3.4** *Suppose that (3.20) holds. Then the representation $\Pi_A$ of CCR is a Schrödinger 3-system if and only if*

$$W^A_{j,k;s,t} = e^{-ist \sum_{l,m=1}^{L} V_j^{(l)} V_k^{(m)} \Im g_{jk}^{(lm)}}$$

*for all $s, t \in \mathbf{R}$ and $j, k = 1, 2, 3$.*

## 3.3  Structure of the Representation $\Pi_A$

Suppose that (3.20) holds, so that $\Pi_A$ is a representation of CCR. As we shall see below, this representation has a simple structure.

It is easy to see that $\{q_j \otimes d\Gamma(h_j)\}_{j=1}^3$ is a set of strongly commuting self-adjoint operators. Hence the operator

$$T := \sum_{j=1}^{3} q_j \otimes d\Gamma(h_j) \tag{3.21}$$

is essentially self-adjoint. We can show that

$$S := \sum_{l=1}^{L} q_j V_j^{(l)} \otimes \overline{\Phi_{\mathrm{s}}(g_j^{(l)}(0))} \tag{3.22}$$

is essentially self-adjoint. Hence we can define a unitary operator

$$Y := e^{-iS} e^{iT}. \tag{3.23}$$

**Theorem 3.5** *Let*

$$\tilde{A}_j(\mathbf{x}) := A_j(\mathbf{x}) - \frac{i}{2} \sum_{k=1}^{3} \sum_{l,m=1}^{L} x_k V_k^{(l)} V_j^{(m)} \Im g_{kj}^{(lm)}, \quad j = 1, 2, 3, \tag{3.24}$$

*and $\tilde{A} := \sum_{j=1}^{3} \tilde{A}_j dx_j$. Then the following operator equations hold:*

$$Y \bar{P}_j(A) Y^{-1} = \bar{p}_j(\tilde{A}) \otimes I, \quad Y(q_j \otimes I) Y^{-1} = q_j \otimes I, \quad j = 1, 2, 3. \tag{3.25}$$

**Remark 3.1** By the present assumption (3.20), $\tilde{A}$ is flat on **M**.

Theorem 3.5 implies that the representation $\Pi_A$ of CCR is unitarily equivalent to an inifinite direct sum of the representation $\{q_j, \bar{p}_j(\tilde{A})\}_{j=1}^3$ of CCR. Hence the essential change in passing to the representation $\Pi_A$ is a modification (3.24) of the gauge field which comes from the interaction of the particle with the Bose field.

**Example 3.1** Let $s \in \mathbf{N}$ and consider the case where $\mathcal{H} = L^2(\mathbf{R}_\mathbf{k}^3; \mathbf{C}^s)$ ($\mathbf{R}_\mathbf{k}^3 := \{\mathbf{k} = (k_1, k_2, k_3), |k_j \in \mathbf{R}, j = 1, 2, 3\}$). Let $\omega$ be a nonnegative Borel measurable function on $\mathbf{R}_\mathbf{k}^3$, a.e. finite with respect to the Lebesgue measure on $\mathbf{R}_\mathbf{k}^3$ such that $|k_j| \leq \omega(\mathbf{k})$, a.e.$\mathbf{k}$, $j = 1, 2, 3$. For $j = 1, 2, 3$ and $L = 1, \cdots, l$, let $G_j^{(l)}$ be a $\mathbf{C}^s$-valued Borel measurable function on $\mathbf{R}_\mathbf{k}^3$ such that $\omega G_j^{(l)}, \omega^{-1/2} G_j^{(l)} \in L^2(\mathbf{R}_\mathbf{k}^3; \mathbf{C}^s)$ and define $g_j^{(l)} : \mathbf{R}^3 \to L^2(\mathbf{R}_\mathbf{k}^3; \mathbf{C}^s)$ by $g_j^{(l)}(\mathbf{x})(\mathbf{k}) := G_j^{(l)}(\mathbf{k})e^{-i\mathbf{k}\mathbf{x}}$. Then it is easy to check that conditions (g.1)-(g.3) hold with $h = \omega$ and $h_j(\mathbf{k}) = k_j$. In the present example, we have

$$g_{jj'}^{(ll')} = (G_j^{(l)}, G_{j'}^{(l')})_{L^2(\mathbf{R}_\mathbf{k}^3; \mathbf{C}^s)}.$$

# ACKNOWLEDGEMENT

This work was supported by the Grant-In-Aid No.08454021 for science research from the Ministry of Education, Japan.

# REFERENCES

Aharonov, Y. and Bohm, D. (1959) Significance of electromagnetic potentials in the quantum theory, *Phys. Rev.*, **115**, 485–491

Arai, A. (1992) Momentum operators with gauge potentials, local quantization of magnetic flux, and representation of canonical commutation relations. *J. Math. Phys.*, **33**, 3374–3378

Arai, A. (1993) Properties of the Dirac-Weyl operator with a strongly singular gauge potential. *J. Math. Phys.*, **34**, 915–935

Arai, A. (1995a) Gauge theory on a non-simply conneted domain and representations of canonical commutation relations. *J. Math. Phys.*, **36**, 2569–2580

Arai, A. (1995b) Representation of canonical commutation relations in a gauge theory, the Aharonov-Bohm effect, and the Dirac-Weyl operator. *J. Nonlinear Math. Phys.*, **2**, 247–262

Arai, A. (1996) Canonical commutation relations in a gauge theory, the Weierstrass Zeta function, and infinite dimensional Hilbert space representations of the quantum group $U_q(sl_2)$. *J. Math. Phys.*, **37**, 4203–4218

Arai, A. (1998) Representation-theoretic aspects of two-dimensional quantum systems in singular vector potentials: canonical commutation relations, quantum algebras, and reduction to lattice quantum systems. *J. Math. Phys.*, **39**, to be published

Dollard, J. D. and Friedman, C. N. (1979) Product Integration. In *Encyclopedia of Mathematics and Its Applicaiton Vol. 10: Analysis.* Reading, Massachusetts: Addison-Wesley

Göckeler, M. and Schücker, T. (1987) *Differential Geometry, Gauge Theories, and Gravity.* Cambridge: Cambridge University Press

Putnam, C. R. (1967) *Commutation Properties of Hilbert Space Operators.* Berlin: Springer

Reed, M. and Simon, B. (1972) *Methods of Modern Mathematical Physics Vol.I.* New York: Academic Press.

Reed, M. and Simon, B. (1975) *Methods of Modern Mathematical Physics Vol.II.* New York: Academic Press.

Reed, M. and Simon, B. (1978) *Methods of Modern Mathematical Physics Vol.IV.* New York: Academic Press.

Von Neumann, J. (1931) Die Eindeutigkeit der Schrödingerschen Operatoren. *Math.Ann.*, **104**, 570–578

# YANG-MILLS THEORY IN A NON-LINEAR GAUGE

JOSE A. MAGPANTAY

National Institute of Physics, University of the Philippines
Diliman, Quezon City 1101, PHILIPPINES

Abstract The Coulomb gauge is first shown to be an incomplete gauge-fixing condition for non-Abelian theories. A non-linear gauge condition, which is quadratic in $A_\mu$ and includes the Coulomb gauge is presented and its consequences explored. The effective dynamics of the new degrees of freedom strongly hint of non-perturbative effects. Finally, we show that the non-linear gauge fixing condition can no longer be generalized to include higher derivatives and higher order in $A$.

## I. INTRODUCTION

The man we are honoring in this workshop is a friend of the Filipino physics community. In his visits to the country he has not only talked about his latest research but also about physics education and the experience of Japan in building its scientific tradition. Thus, when Chris Bernido invited me to join the workshop and present a talk, I readily agreed. I hope the ideas I present here will measure up to the prestige of our honoree, Professor Hiroshi Ezawa.

In quantizing theories with gauge symmetries, it is important that every gauge-equivalent fields are counted once and only once. It is also equally important that each orbit of gauge-equivalent fields is represented in the count for otherwise some important field configurations may be left out in the path-integral and physical effects associated with these fields brushed aside. This means that when we specify how to do the counting, i.e., when we specify the gauge condition, it better be globally unique and realizable. Uniqueness of the gauge condition implies that the orbit intersects the gauge-fixing surface only once. Realizability means that starting from a field configuration not on the gauge-fixing surface, we can follow the orbit until it reaches the surface.

241

Up to this time, only linear gauge conditions have been used to remove the excess degrees of freedom. For the Abelian case, linear gauge conditions are always realizable because gauge transformation is only a "translation". Also, the incomplete removal of all the gauge degrees of freedom, like in the axial and Lorentz gauges, can easily be compensated for by imposing subsidiary conditions.

The ideal gauge-fixing condition for the Abelian theory is the Coulomb gauge because it is unique and always realizable. The physical degrees of freedom of the electromagnetic field, the transverse photon, is also transparent in this gauge. Because of this it was naturally assumed that the same thing is true for the non-Abelian case, until Gribov pointed out in 1977[1] that gauge copies exist for certain field configurations satisfying the Coulomb gauge. This means that the gauge condition is not unique.

Since Gribov's pioneering paper, various aspects of the Coulomb gauge formulation of non-abelian gauge theories have been discussed. Peccei[2] had shown that the existence of copies does not affect the perturbative calculations. Fujikawa,[3] Hirschfield[4] showed the possibility that the Gribov copies may not necessarily affect the functional integral. Bender, et. al.[5] looked for confinement by investigating the effect of the Gribov phenomena on the continuum Hamiltonian. Cutkosky[6] had a similar idea and searched for a mass gap in the Hamiltonian using the lattice approach.

On the more technical aspect of the Gribov problem, Maskawa and Nakajima[7] made a geometrical formulation of the Coulomb gauge by considering the functional space of $L^2$ fields and showed that the Coulomb gauge follows from minimizing $\left|A^{\Omega}\right|^2 = \int d^3x \left[\Omega \vec{A}\Omega^{-1} + (\vec{\partial}\Omega)\Omega^{-1}\right]^2$. Zwanziger and Dell Antonio[8] (i) verified the ideas of Semionov-Tian-Shansky and Franke[9] that there are Gribov copies inside the horizon, (ii) claimed that all gauge orbits pass through the central Gribov region, and (iii) proved that the central Gribov region has ellipsoidal bounds and forms a convex set.

In a 1994 paper,[10] Magpantay looked at the Coulomb gauge again and showed that there is no need to modify the Fadeev-Popov determinant $\det(\vec{\partial} \cdot \vec{D})$ even in cases where gauge copies exist. More precisely, the Fadeev-Popov determinant was shown to be exactly zero on the horizon, thus effectively showing that the horizon does not contribute to the path-integral.

In the same paper, it was also shown that there are field configurations that cannot be gauge transformed to the Coulomb gauge, thus disproving the claim of Zwanziger and Dell-Antonio that all orbits pass inside the horizon. These are the field configurations that satisfy

$$(\vec{\partial} \cdot \vec{D})(\vec{\partial} \cdot \vec{A}) = (\vec{D} \cdot \vec{\partial})(\vec{\partial} \cdot \vec{A}) = 0 , \tag{1a}$$

$$\vec{\partial} \cdot \vec{A} = f(\vec{x}) \neq 0 . \tag{1b}$$

An explicit example of a class of field configurations that satisfy (1;a,b) was constructed by Magpantay and Cuansing[11]. These field configurations are shown to be not transformable to the Coulomb gauge if we require the square-integrability ($L^2$) of the vector fields.

The above results show that non-perturbative physics, like confinement, is not connected with the Gribov horizon. Furthermore, these results suggest that the Coulomb gauge has to be generalized and that the appropriate generalization is defined by equation (1a). This gauge condition has two parts – the usual Coulomb gauge $\vec{\partial} \cdot \vec{A} = 0$ and a quadratic "surface" in configuration space.

This paper presents some new results in the use of the non-linear gauge condition in SU(2). In particular, we will derive the independent degrees of freedom and their effective dynamics. We will see that the effective action of the independent degrees of freedom clearly hints of non-perturbative effects. But before we present these results, we discuss first in Section II some heuristic arguments for the non-linear gauge condition. Section III will show in a very simple way that field configurations on the Coulomb surface and those that satisfy equations (1;a,b) do not mix, i.e., they are not connected by a gauge transformation. Section IV differentiates the Hamiltonian formulations of the Coulomb gauge and the non-linear gauge. The new degrees of freedom and their effective dynamics is the subject of Section V. Why the non-linear gauge condition can no longer be extended to gauge-fixing with higher derivatives and higher order than quadratic in $A_\mu$ is the topic of Section VI. The conclusion summarizes what had been accomplished and presents a conjecture.

## II. HEURISTIC ARGUMENTS

The non-linear gauge-fixing condition can be written as

$$\nabla^2 (\vec{\partial} \cdot \vec{A}^a) - g \in^{abc} (\vec{A}^c \cdot \vec{\partial})(\vec{\partial} \cdot \vec{A}^b) = 0 \qquad (2)$$

In the Abelian limit, this gives $\nabla^2 (\vec{\partial} \cdot \vec{A}) = 0$. Since $\nabla^2$ is a positive definite operator (for simply connected spaces), we get back the Coulomb gauge.

In small distance scales, the coupling constant goes to zero and we can effectively neglect the second term again giving back the Coulomb gauge. This is the situation well within the hadrons where the quarks are loosely bound interacting with transverse vector fields.

Since the first term is proportional to $p^3$ (p is the momenta) and the running coupling becomes stronger at small momenta (note the second term is proportional to $p^2 g(p)$) then even before we get to the distance scales where the perturbative running coupling is no longer valid, we see that the second term becomes important. Because of this the "gluon" is no longer sufficiently described by transverse degrees of freedom. And indeed in non-Abelian theory, because of the gluon self-interaction, the transverse degrees of freedom are not physical in large distance scales.

In this sense, the Coulomb gauge is not an appropriate gauge-fixing condition for Yang-Mills theories. The gauge-fixing condition defined by equation (2) is intuitively better because it is consistent with the asymptotic freedom and confinement phases of unbroken gauge theories.

Furthermore, as we will show in the next section, the non-Coulomb configurations of (2) cannot be gauge transformed to the Coulomb surface. And as will be shown also, these field configurations are horizons on surfaces defined by $\vec{\partial} \cdot \vec{A}^a = f^a$ ($f^a$ defines the arbitrary surface) and that fields on the horizon belong to an orbit.

## III. NO MIXING OF FIELD CONFIGURATIONS

In this section, we will supplement the results of references (10) and (11) by providing a simple proof, albeit using infinitesimal transformations, that field configurations on the Coulomb surface and those on the surface defined by (1;a,b) cannot be gauge transformed to each other.

Consider a vector field $\vec{A}$ that satisfies equations (1;a,b). Is there a gauge transformation that will take this field configuration to the Coulomb surface? Let

$$\vec{A}' = \vec{A} + \vec{D}\Lambda. \tag{3}$$

Imposing $\vec{A}'$ to be on the Coulomb surface yields

$$(\vec{\partial} \cdot \vec{D})\Lambda = -(\vec{\partial} \cdot \vec{A}). \tag{4}$$

But since the zero mode of $(\vec{\partial} \cdot \vec{D})$ is also $(\vec{\partial} \cdot \vec{A})$, then equation (4) will not have an $L^2$ solution. Thus, there is no gauge transformation that will take field configurations satisfying the non-linear condition to the Coulomb surface.

Conversely, suppose $\vec{\partial} \cdot \vec{A} = 0$. Is there an $\vec{A}' = \vec{A} + \vec{D}\Lambda$ such that $\vec{A}'$ satisfies equations (1;a,b)? This means $\Lambda$ must satisfy

$$(\vec{\partial} \cdot \vec{D})\Lambda = f(\vec{x}) \neq 0, \tag{5a}$$

$$(\vec{\partial} \cdot \vec{D})(\vec{\partial} \cdot \vec{D})\Lambda = 0. \tag{5b}$$

Equation (5b) says that $\vec{\partial} \cdot \vec{D}$ is a singular operator with $(\vec{\partial} \cdot \vec{D})\Lambda = f(\vec{x})$ as its zero mode. To solve for $\Lambda$ in equation (5a), for an $L^2$ solution to exist, $f(\vec{x})$ must be orthogonal to the zero mode of $(\vec{\partial} \cdot \vec{D})$. But the zero mode is also the source, thus no solution exists. This proves that field configurations on the Coulomb surface cannot be gauge transformed to configurations that satisfy equations (1;a,b).

Let us now establish two major points about the non-linear gauge condition. First the field configurations satisfying equations (1;a,b) are actually on the horizon of the surface $\vec{\partial} \cdot \vec{A}^a = f^a(\vec{x}) \neq 0$ because given a certain $f^a(\vec{x})$, the condition $(\vec{\partial} \cdot \vec{D}) f(\vec{x}) = 0$ says that the operator $(\vec{\partial} \cdot \vec{D})$ is singular. Since the Fadeev-Popov operator is linear in $\vec{A}$ then the horizon on the surface $\vec{\partial} \cdot \vec{A} = f(\vec{x})$ defined by $(\vec{\partial} \cdot \vec{D}) f = 0$ is also convex just like the horizon on the Coulomb surface.

Second, since the field configurations on the surface defined by equations (1;a,b) are not gauge transformable to the Coulomb gauge, it is natural to ask where the orbits go? The answer is that the orbits of these field configurations stay on the surface and on the horizon.

Proof of this is as follows. Let $\vec{A}$ satisfy equations (1;a,b). Let $\vec{A}' = \vec{A} + \vec{D}\Lambda$. Imposing that $\vec{A}'$ is also on the same surface as $\vec{A}$, i.e. $\vec{\partial} \cdot \vec{A} = f(\vec{x})$ then $(\vec{\partial} \cdot \vec{D})\Lambda = 0$. This means $\Lambda = c(\vec{\partial} \cdot \vec{A})$ where $c$ is an infinitesimal constant (dimension length squared so that $\Lambda$ is dimensionless). It also follows that $(\vec{\partial} \cdot \vec{D}(\vec{A}'))f = 0$, thus $\vec{A}'$ is not only on the same surface but also on the horizon. We can repeat the same procedure for $\vec{A}'$ and $\vec{A}''$ and so on with each step resulting in a gauge transformed field that is also on the surface and on the horizon. This complete the proof that the orbit of field configurations that satisfy equation (1;a,b) stay on the surface and on the horizon.

Another way to prove the second point is by explicitly verifying that any gauge transformation that attempts to get off the surface does not have an $L^2$ solution. Consider the field configuration (used in reference (11))

$$A_i^a = \delta_i^a k(r) + \frac{x^a x_i}{r^2} \ell(r), \tag{6a}$$

$$\partial_i A_i^a = \frac{x^a}{r}\left(\frac{dk}{dr} + \frac{d\ell}{dr} + \frac{2}{r}\ell\right) = \frac{x^a}{r} s(r) \neq 0 . \tag{6b}$$

Consider the gauge transformed field $\vec{A}' = \vec{A} + \vec{D}\Lambda$, such that $\vec{\partial} \cdot \vec{A}'^a = \frac{x^a}{r}(s(r) + \in(r))$ where $\in(r) \langle\langle s(r)$, i.e., the $\vec{\partial} \cdot \vec{A}'$ surface is infinitesimally close to the $\vec{\partial} \cdot \vec{A} = \frac{x^a}{r} s(r)$ surface. For a non-trivial $\Lambda$ to exit, the proper treatment of the zero mode requires

$$\int_o^R dr\, r^2\, s(r) \in (r) = 0 \tag{7}$$

We regularized the system by considering finite space of large radius R. Since $s(r) = \dfrac{a}{r^2} + br$ and the constants a and b have a specified R behaviour (for $L^2$ behaviour), then $r^2 s(r)$ is a rapidly increasing function. For equation (7) to be satisfied, $\in (r)$ must be fast decreasing and oscillatory. However, from the results of reference (11), it follows that the condition $\in (r) \langle\langle\, s(r)$ cannot be satisfied for all $r$, thus, the only admissible $\in (r)$ is $\in (r) = 0$. This proves that the orbit cannot leave the surface and the horizon.

## IV. THE HAMILTONIAN FORMULATION

The Yang-Mills Hamiltonian, whether in the Coulomb gauge or in the non-linear gauge defined by equation (1;a,b), is given by

$$H = \int d^3x \left\{ \frac{1}{2}\vec{E}^a \cdot \vec{E}^a + \frac{1}{4}F_{ij}^a F_{ij}^a + \frac{1}{2}(\partial_i \alpha^a)^2 \right\}, \tag{8}$$

where $\vec{E}^a = \partial_o \vec{A}^a$. The instantaneous Coulomb term, in both gauges, is given by

$$(\vec{D}\cdot\vec{\partial})^{ab}\alpha^b = \vec{D}^{ab}\cdot\vec{E}^b. \tag{9}$$

The differences between the two gauge-fixing conditions are summarized below:

| Coulomb Gauge | Non-Linear Gauge |
|---|---|
| $\vec{\partial}\cdot\vec{A}^a = 0$ | $(\vec{\partial}\cdot\vec{D})^{ab}(\vec{\partial}\cdot\vec{A}^b) = (\vec{D}\cdot\vec{\partial})^{ab}(\vec{\partial}\cdot\vec{A}^b) = 0$ |
| $\vec{\partial}\cdot\vec{E}^a = 0$ | $(\vec{\partial}\cdot\vec{D})^{ab}(\vec{\partial}\cdot\vec{E}^b) - \epsilon^{abc}\vec{\partial}\cdot[(\vec{\partial}\cdot\vec{A}^b)\vec{E}^c] = 0.$ |

In the Coulomb gauge, for as long as the Fadeev-Popov operator $(\vec{D}\cdot\vec{\partial})$ is non-singular, equation (9) can be solved for the instantaneous Coulomb term $\alpha^a$. The solution can be expanded using the $\dfrac{1}{r}$ potential.

However, if $(\vec{D}\cdot\vec{\partial})$ is singular, then a non-trivial solution for $\alpha^a$ only exists if the zero mode of $(\vec{D}\cdot\vec{\partial})$ is orthogonal to $\epsilon^{abc}\vec{A}^c\cdot\vec{E}^b$. If this condition is imposed, Peccei (see reference 2) showed that the instantaneous Coulomb term, when computed perturbatively, is effectively unchanged. Thus, confinement remains unproven, which agrees with the computation of Greensite.[12]

If we use instead the non-linear gauge, we find that we have a totally different situation. Acting the operator $(\vec{\partial} \cdot \vec{D})$ on equation (9) gives

$$(\vec{\partial} \cdot \vec{D})^{ab} (\vec{D} \cdot \vec{\partial})^{bc} \alpha^c = (\vec{\partial} \cdot \vec{D})^{ab} (\vec{D}^{bc} \cdot \vec{E}^c) \qquad (10)$$

Since $(\vec{\partial} \cdot \vec{D})^{ab} (\vec{D} \cdot \vec{\partial})^{bc}$ is singular with $\vec{\partial} \cdot \vec{A}^a$ as the zero mode, for equation (10) to have a non-trivial solution, we must have

$$\int d^3 x (\vec{\partial} \cdot \vec{A}^a)(\vec{\partial} \cdot \vec{D})^{ab} (\vec{D}^{bc} \cdot \vec{E}^c) = 0 \qquad (11)$$

and indeed this is true because an integration by parts will yield the gauge-fixing condition. Thus unlike in the Coulomb gauge, the consistency condition is trivially satisfied in the non-linear gauge.

Equation (10) can be solved perturbatively using the Greens function for the operator $(\nabla^2)^2$ which is a linear potential. If we consider static sources given by $j_o^a$, equation (10) will be augmented by the term $(\vec{\partial} \cdot \vec{D})^{ab} j_o^b$ (the modification of equation (11) still holds). Pertubative expansion of the "Coulomb" term will yield the usual Coulomb term $\int d^3 x d^3 y \, j_o^a(\vec{x}) \frac{1}{|\vec{x} - \vec{y}|} j_o^a(\vec{y})$.

However, higher order effects will yield "confining" terms because of the linear potential. The lowest order "confining" potential is given by $\int d^3 x d^3 y \, \epsilon^{abc} j_o^a(\vec{x}) |\vec{x} - \vec{y}| \cdot (\vec{\partial} \cdot \vec{A}^b)_y j_o^c(\vec{y})$. Higher orders terms are even more confining.

Finally, we point out the difference between the effective potentials (static fields) in the Coulomb and non-linear gauge.

$$V_{eff} = \int d^3 x \left\{ \frac{1}{2} (B_i^a)^2 + \frac{1}{2} (\partial_i \alpha_a)^2 \right\} + tr \, \ell n \left\{ \frac{(\vec{\partial} \cdot \vec{D})}{\theta} \right\} \qquad (12)$$

From equation (9), the Coulomb term is given by $(\vec{D} \cdot \vec{\partial})^{ab} \alpha^b = 0$. The first term is the classical contribution while the second term is the quantum effect. The $tr \, \ell n (\vec{\partial} \cdot \vec{D})$ is for the Coulomb gauge while $tr \, \ell n \theta$, where $\theta$ is the "Fadeev-Popov" given by equation (13), is for the non-linear gauge,

$$\theta^{ad} = (\vec{D} \cdot \vec{\partial})^{ab} \, (\vec{\partial} \cdot \vec{D})^{bd} - \epsilon^{abc} \left[ \vec{\partial} (\vec{\partial} \cdot \vec{A}^b) \right] \cdot \vec{D}^{cd} . \tag{13}$$

This operator is non-singular in spite of the fact that $(\vec{\partial} \cdot \vec{D})$ is singular. The proof follows simply from first-order perturbation theory.

In the Coulomb gauge, if $(\vec{D} \cdot \vec{\partial})$ is non-singular, $\alpha^a = 0$ and the classical effective potential has minimum at $\vec{B}^a = 0$. The quantum correction $tr \, \ell n \, (\vec{\partial} \cdot \vec{D})$ leads to a vacuum with non-trivial field strength. However the imaginary part of the quantum effective potential makes the non-trivial vacuum unstable and it eventually decays.[13]

If $(\vec{D} \cdot \vec{\partial})$ is singular, (which is valid for fields on the horizon), the instantaneous Coulomb contribution is non-trivial. However, the $\exp[\, tr \, \ell n \, (\vec{\partial} \cdot \vec{D})] = \det (\vec{\partial} \cdot \vec{D})$ is zero. Thus although the classical effective action may have a non-trivial minimum, the contribution of this field configurations to the path-integral is effectively zero because of the determinant.

On the other hand, for the non-linear gauge, both the "instantaneous Coulomb" and the quantum corrections are non-trivial. Firstly, $\alpha^a = \ell_o (\vec{\partial} \cdot \vec{A}^a)$ where $\ell_o$ is a length scale. Secondly, since $\theta$ is non-singular, $\det \theta$ is non-zero and the contribution of these field configurations to the path-integral does not vanish. Thus, the effective potential in the non-linear gauge will give a non-trivial vacuum.

## V. THE NEW DEGREES OF FREEDOM AND THEIR EFFECTIVE DYNAMICS

In this section, we will shift to 4-D Euclidean space-time and consider the covariant version of equations (1;a,b). Equation (1;a,b) is solved by the following:

$$\ell_o^2 D_\mu^{ab} f^b = (A_\mu^a - \partial_\mu \frac{1}{\Box^2} f^a) - t_\mu^a , \tag{14}$$

where the length scale $\ell_o$ is introduced for dimensional reasons. The new vector field $t_\mu^a$ is transverse and must also satisfy another condition (to have the same number of degrees of freedom).

Solving equation (14) for $A_\mu^a$ we find

$$A_\mu^a = \frac{1}{(1+g\ell_o^2 \vec{f} \cdot \vec{f})} \left\{ \delta^{ab} + g\ell_o^2 \, \epsilon^{abc} \, f^c + g^2 \ell_o^4 \, f^a f^b \right\}$$

$$\times (\ell_o^2 \, \partial_\mu f^b + \partial_\mu \frac{1}{\Box^2} f^b + t_\mu^b). \qquad (15)$$

Rescaling the fields, $g\ell_o^2 f^a \rightarrow f^a$, so that the new field $f^a$ is dimensionless, we have

$$A_\mu^a = \frac{1}{(1+\vec{f} \cdot \vec{f})} \left\{ \delta^{ab} + \epsilon^{abc} \, f^c + f^a f^b \right\} \left( \frac{1}{g} \partial_\mu f^b + \frac{1}{g\ell_o^2} \partial_\mu \frac{1}{\Box^2} f^b + t_\mu^b \right)$$

$$(15')$$

To guarantee $\partial_\mu A_\mu^a = f^a$, we must impose

$$\rho^a = \left[ \epsilon^{abc} + \epsilon^{abd} \, f^c f^d + \epsilon^{adc} \, f^d f^b + f^a \, \epsilon^{bcd} \, f^d \right] (\partial_\mu f^b)$$

$$\times \left( \frac{1}{g\ell_o^2} \partial_\mu \frac{1}{\Box^2} f^c + t_\mu^c \right) = 0. \qquad (16)$$

What we have done is to change variables from $A^a$ which satisfies the non-linear gauge condition, to $f^a$ and $t_\mu^a$. The $t_\mu^a$ is not only transverse but also satisfies equation (16). Thus, the number of degrees of freedom remain the same.

In terms of the new degrees of freedom, the path-integral is:

$$PI = \int (df^a)(dt_\mu^a)\,\delta(\partial \cdot t^a)\,\delta(\rho^a) \det \theta \det(1+\vec{f} \cdot \vec{f})\, e^{-S_{YM}}, \qquad (17)$$

where equation (15') is substituted in $\theta$ and $S_{YM}$. Note that the field strength can be expanded as

$$F_{\mu\nu}^a = \frac{1}{g} Z_{\mu\nu}^a(f) + L_{\mu\nu}^a(f;t) + g\, Q_{\mu\nu}^a(f;t) \qquad (18)$$

where $Z_{\mu\nu}^a$ is zeroth order in $t$, $L_{\mu\nu}^a$ is linear in $t$ and $Q_{\mu\nu}^a$ is quadratic in $t$. The action becomes

$$S_{YM} \sim \int d^4x \left\{ \frac{1}{g^2} Z^a_{\mu\nu} Z^a_{\mu\nu}(f) + \frac{2}{g} Z^a_{\mu\nu}(f) L^a_{\mu\nu}(f;t) \right.$$
$$+ \left[ 2Z^a_{\mu\nu}(f) Q^a_{\mu\nu}(f;t) + L^a_{\mu\nu}(f;t) L^a_{\mu\nu}(f;t) \right]$$
$$\left. + 2g\, L^a_{\mu\nu}(f;t) Q^a_{\mu\nu}(f;t) + g^2 Q^a_{\mu\nu}(f;t) Q^a_{\mu\nu}(f;t) \right\}. \qquad (19)$$

The effective dynamics of the new degrees of freedom are read from equation (19). We notice right away the following:

- The pure $f$ term has a typical non-perturbative behaviour.
- The $f$ dynamics is extremely non-linear because of the $(1 + \vec{f} \cdot \vec{f})^{-1}$ term. Furthermore, the $f$ term has four-derivatives and also non-local.
- The vector field $t^a_\mu$ is not only transverse but is also constrained to satisfy equation (16). Also, the vector field is not gauge-invariant.
- Fermions interact in a very non-linear way with the $f$ field as given by $g\bar{\psi}\gamma_\mu T^a A^a_\mu \psi$ with $A^a_\mu$ given by equation (15').

Because of these observations, we conjecture that confinement of quarks and the transverse gluons $(t^a_\mu)$ is due to the $f$ dynamics.

## VI. GENERALIZATION TO HIGHER DERIVATIVES AND NON-LINEAR GAUGES

In section IV, we found that the expansion Greens function used in solving the "instantaneous Coulomb term" is the linear potential. This came about because the gauge-fixing yielded a "Fadeev-Popov" operator that is of order $(\nabla^2)^2$. Thus, the "hint" of confinement seems to be an artifact of gauge-fixing. Consequently, if we include more derivatives in the gauge-fixing, then we should get a more confining potential.

The above arguments are specious for two reasons. First, the arguments in references (10), (11) and Sections II and III of this paper clearly showed the Coulomb gauge is insufficient because the horizons on each of the arbitrary surfaces $\vec{\partial} \cdot \vec{A}^a = f^a(\vec{x})$ are completely ignored. Thus, physical affects associated with this ignored field configurations will be missed completely. And as we have shown in this paper, the physical effects of the missed configurations are primarily non-perturbative.

Secondly, we will now show that the non-linear gauge-fixing condition cannot be generalized to include higher derivatives and higher order non-linear terms. Note that the generalization we will be seeking must contain the Coulomb gauge. Let us consider the following gauge-fixing (note: $\vec{D} \cdot \vec{A} = \vec{\partial} \cdot \vec{A}$)

$$(\vec{\partial} \cdot \vec{D})^2 \, (\vec{\partial} \cdot \vec{A}) = 0 \tag{20}$$

This gauge-fixing has three regimes:

(a) $\vec{\partial} \cdot \vec{A} = 0$;    the Coulomb surface

(b) $\vec{\partial} \cdot \vec{A} = f(\vec{x})$;  $(\vec{\partial} \cdot \vec{D})f = 0$;  the non-linear gauge considered in this paper.

(c) $\vec{\partial} \cdot \vec{A} = f(\vec{x})$;  $(\vec{\partial} \cdot \vec{D})f = k(\vec{x}) \neq 0$;  $(\vec{\partial} \cdot \vec{D})k(\vec{x}) = 0$;  the new regime.

We will now show that the regime defined by (c) is inconsistent. To express $\vec{A}$ in terms of $k(\vec{x})$ and $f(\vec{x})$ (to derive the analogue of equation (15')), we first have to solve the equation $(\vec{\partial} \cdot \vec{D})f = k(\vec{x})$. But since $k(\vec{x})$ is the zero mode of $(\vec{\partial} \cdot \vec{D})$, we cannot find an $(L^2)$ $f$ because the zero mode is the source. Thus, the gauge-fixing defined by equation (20) is not consistent. And in a similar way, we can argue the inconsistency of all gauge conditions of the form $(\vec{\partial} \cdot \vec{D})^n \, (\vec{\partial} \cdot \vec{A}) = 0$, for $n > 3$.

The other types of gauge-fixing conditions that we can consider are

$$(\nabla^2)^n \, (\vec{\partial} \cdot \vec{A}) = 0 \,, \tag{21a}$$

$$(\vec{D}^+ \cdot \vec{D})^n \, (\vec{\partial} \cdot \vec{A}) = 0 \,, \tag{21b}$$

and other combinations of $\nabla^2$, $(\vec{D}^+ \cdot \vec{D})$, and $(\vec{\partial} \cdot \vec{D})$ acting on $\vec{\partial} \cdot \vec{A}$. Equation (21a) is simply equivalent to the Coulomb gauge because the $\nabla^2$ is a positive-definite operator. The same thing is true with $\vec{D}^+ \cdot \vec{D}$ because it is equal to $U^+ \nabla^2 U$ where $U = P \exp\left\{i \int_{-\infty}^{\vec{x}} \vec{A} \cdot d\vec{z}\right\}$. Thus, generalizations which combine $\nabla^2$, $(\vec{D}^+ \cdot \vec{D})$ and $\vec{\partial} \cdot \vec{D}$ can either be reduced to the Coulomb gauge, the non-linear gauge or are inconsistent.

## CONCLUSION

This paper had argued the importance of generalizing the Coulomb gauge to the non-linear gauge defined by equation (1;a,b). We have also shown that the gauge-fixing can no longer be generalized any further. And best of all, we had shown that non-perturbative physics is found in the non-linear regime of the non-linear gauge.

To conclude, we will speculate on an elegant proof of confinement arising from the non-linear regime.  Consider the pure $f$ term of equations (18) and (19).  The relevant path-integral is

$$\int (df^a)\det(1 + \vec{f} \cdot \vec{f})\det \tilde{\theta} \exp\left\{-\frac{1}{g^2}\int d^4x\, Z^a_{\mu\nu}Z^a_{\mu\nu}\right\}, \qquad (22a)$$

where

$$Z^a_{\mu\nu} = F^a_{\mu\nu}(A^a_\mu = \text{given by (15') with } t^a_\mu = 0), \qquad (22b)$$

$$\tilde{\theta} = \theta\,(A^a_\mu = \text{given by (15') with } t^a_\mu = 0). \qquad (22c)$$

Suppose we can transform from $f^a \rightarrow \phi^a = \phi^a(f)$, which may be a non-local transformation, such that

$$\frac{1}{4}(Z^a_{\mu\nu})^2 = (\Box^2\phi^a + \frac{\delta V(\phi)}{\delta\phi^a})^2, \qquad (23a)$$

and

$$\det\left[(\Box^2\delta^{ab} + \frac{\delta^2 V}{\delta\phi^a\delta\phi^b})\frac{\delta\phi^b}{\delta f^c}\right] = \det(1 + \vec{f} \cdot \vec{f})\det\tilde{\theta}. \qquad (23b)$$

Then the complicated dynamics of the $f^a$'s is equivalent to a scalar field $\phi^a$ (with action $\int d^4x\left[\frac{1}{2}(\partial_\mu\phi^a)^2 + V(\phi^a)\right]$) in a random magnetic field.  If this is true, then the Parisi-Sourlas mechanism[14] is valid.  Confinement is proven via dimensional reduction because of the existence of a supersymmetry.  The non-perturbative aspect of QCD$_4$ is equivalent to a two-dimensional scalar theory.

Note that equation (23a) is plausible because $(Z^a_{\mu\nu})^2 \approx (\partial f)^4$.  Furthermore, the potential $V(\phi)$ and $\phi^a = \phi^a(f)$ are to be determined from equations, (23;a,b).  From the general forms of these equations, we cannot rule out the existence of the transformation.  Unfortunately, the author was not able to show, up to the present, the Parisi-Sourlas mechanism.

ACKNOWLEDGEMENT

This research was supported in part by the Natural Sciences Research Institute.

REFERENCES

1. V.N. Gribov, Nucl. Phys. B139 (1978), 1.
2. R. Peccei, Phys. Rev. D17 (1978), 1097.
3. K. Fujikawa, Prog. of Theor. Physics 61 (1979) 627.
4. P. Hirchfeld, Nucl. Phys. B152 (1978) 37.
5. C. Bender, T. Eguchi and H. Pagels, Phys. Rev. D17 (1978) 1086.
6. R. Cutkosky, Phys. Rev. Lett. 51 (1983) 538.
7. T. Maskawa and H. Nakajima, Prog. of Theor. Phys. 60 (1978) 1526.
8. G. Dell'Antonio and D. Zwanziger, Comm. in Math. Physics 138 (1991) 291 and other related works cited in the references.
9. M.A. Semionov-Tian-Shansky and V.A. Franke, Report of the Steklov Institute (Leningrad) 1982.
10. J.A. Magpantay, Prog. of Theor. Physics 91 (1994) 573.
11. J.A. Magpantay and E. Cuansing, Jr., Modern Physics Letters A11 (1996) 87.
12. J. Greensite, Physical Review D18 (1978) 3842.
13. G.K. Savvidy, Phys. Letters 71B (1977) 133.
14. G. Parisi and N. Sourlas, Phys. Rev. Lett. 43 (1979) 744.

# Time-Ordered Products and Exponentials

C.S. Lam[*]

Department of Physics, McGill University
3600 University St., Montreal, QC, Canada H3A 2T8

**Abstract**

I discuss a formula decomposing the integral of time-ordered products of operators into sums of products of integrals of time-ordered commutators. The resulting factorization enables summation of an infinite series to be carried out to yield an explicit formula for the time-ordered exponentials. The Campbell-Baker-Hausdorff formula and the nonabelian eikonal formula obtained previously are both special cases of this result.

## 1   Introduction

It is a great pleasure to dedicate this article to Hiroshi Ezawa on the occasion of his 65th birthday. I am priviledged to have known him for 35 years, and am proud to see that he has such an illustrious career. I wish him and his wife Yoshiko the very best, healthy and a long life that will last for at least another 35 years.

True to the theme of this Workshop I will talk about a mathematical method that has applications in quantum mechanics. This is a method that could simplify computations of (integrals of) time-ordered products and time-ordered exponentials $U$. These quantities are ubiquitous in quantum mechanics. In one form or another they describe the time-evolution opreator of quantum systems, the non-integrable phase factor (Wilson line) of Yang-Mills theories, and the scattering amplitudes of perturbation diagrams. The method relies on a *decomposition formula*, which expresses $U$ in terms of more primitive quantities $C$, the time-ordered commutators.

I shall concentrate in what follows to the description of this and other related formulas. There is certainly no time for the proof and very little to illustrate the applications. For those I refer the readers to the literature [1, 2, 3, 4, 5].

## 2   Preliminaries

Let $H_i(t)$ be operator-valued functions of time. No attempts will be made to discuss domains and convergence problems of these operators. Let $[s] = [s_1 s_2 \cdots s_n]$

be a permutation of the $n$ numbers $[12 \cdots n]$, and $S_n$ the corresponding permutation group. We define the *time-ordered product* $U[s]$ to be the integral

$$U[s] = U[s_1 s_2 \cdots s_n] = \int_{R[s]} dt_1 dt_2 \cdots dt_n H_{s_1}(t_{s_1}) H_{s_2}(t_{s_2}) \cdots H_{s_n}(t_{s_n}) \qquad (1)$$

taken over the hyper-triangular region $R[s] = \{T \geq t_{s_1} \geq t_{s_2} \geq \cdots \geq t_{s_n} \geq T'\}$, with operator $H_{s_i}(t_{s_i})$ standing to the left of $H_{s_j}(t_{s_j})$ if $t_{s_i} > t_{s_j}$. The average of $U[s]$ over all permutations $s \in S_n$ will be denoted by $U_n$:

$$U_n = \frac{1}{n!} \sum_{s \in S_n} U[s]. \qquad (2)$$

The *decomposition formula* expresses $U_n$ in terms of the *time-ordered commutators* $C[s] = C[s_1 s_2 \cdots s_n]$. These are defined analogous to $U[s_1 s_2 \cdots s_n]$, but with the products of $H_i$'s replaced by their nested multiple commutators:

$$C[s] = \int_{R[s]} dt_1 dt_2 \cdots dt_n$$
$$[H_{s_1}(t_{s_1}), [H_{s_2}(t_{s_2}), [\cdots, [H_{s_{n-1}}(t_{s_{n-1}}), H_{s_n}(t_{s_n})] \cdots ]]]. \qquad (3)$$

For $n = 1$, we define $C[s_i] = U[s_i]$. Similarly, the operator $C_n$ is defined to be the average of $C[s]$ over all permutations $s \in S_n$:

$$C_n = \frac{1}{n!} \sum_{s \in S_n} C[s]. \qquad (4)$$

It is convenient to use a 'cut' (a vertical bar) to denote products of $C[\cdots]$'s. For example, $C[31|2] \equiv C[31]C[2]$, and $C[71|564|2|3] \equiv C[71]C[564]C[2]C[3]$. Given a sequence $[s]$ of numbers with $s \in S_n$, its *cut sequence* $[s]_c$ is obtained from $[s]$ by inserting cuts at the appropriate places. A cut is inserted after $s_i$ iff $s_i < s_j$ for all $i < j$. In other words, we should proceed from left to right and put a cut after the smallest number encountered. The first cut is therefore put after the number '1'; the second after the smallest number to the right of '1', etc. For example, $[5413267]_c = [5431|2|6|7]$, and $[1267453]_c = [1|2|67453]$.

# 3 General Decomposition Formula

The main formula [5] states that

$$n!U_n = \sum_{s \in S_n} U[s] = \sum_{s \in S_n} C[s]_c. \qquad (5)$$

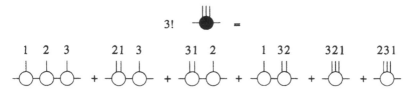

Figure 1: The decomposition of $3!U_3$ in terms of $C[s]$'s.

For illustrative purposes here are explicit formulas for $n = 1, 2, 3$, and 4:

$$1!U_1 = C[1]$$
$$2!U_2 = C[1|2] + C[21]$$
$$3!U_3 = C[1|2|3] + C[21|3] + C[31|2] + C[1|32] + C[321] + C[231]$$
$$
\begin{aligned}
4!U_4 = \ & C[1|2|3|4] + C[321|4] + C[231|4] + C[421|3] + C[241|3] + C[431|2] \\
+ \ & C[341|2] + C[1|432] + C[1|342] + C[21|43] + C[31|42] + C[41|32] \\
+ \ & C[21|3|4] + C[31|2|4] + C[41|2|3] + C[1|32|4] + C[1|42|3] + C[1|2|43] \\
+ \ & C[4321] + C[3421] + C[4231] + C[3241] + C[2341] + C[2431] \qquad (6)
\end{aligned}
$$

Using a filled circle with $n$ lines on top to indicate $n!U_n$, and an open circle for $C[s]$, these formulas can be expressed graphically as shown in Fig. 1.

## 4   Special Decomposition Formula

Great simplification occurs when all $H_i(t) = H(t)$ are identical, for then $U[s]$ and $C[s]$ depend only on $n$ but not on the particular $s \in S_n$. In that case all $U[s] = U_n$ and all $C[s] = C_n$, and the decomposition theorem becomes [5]

$$U_n = \sum \xi(m_1 m_2 \cdots m_k) C_{m_1} C_{m_2} \cdots C_{m_k},$$

$$\xi(m_1 m_2 \cdots m_k) = \prod_{i=1}^{k} \left[ \sum_{j=i}^{k} m_j \right]^{-1} \qquad (7)$$

The sum in the first equation is taken over all $k$, and all $m_i > 0$ such that $\sum_{i=1}^{k} m_i = n$. The quantity $\xi(m_1 \cdots m_k)^{-1}$ is just the product of the number of numbers to the right of every cut (times $n$). Note that it is *not* symmetric under the interchange of the $m_i$'s. It is this asymmetry that makes the formulas for $K_n$ in (12) rather complicated.

We list below this special decompositions up to $n = 5$:

$$
\begin{aligned}
1!U_1 = \ & C_1 \\
2!U_2 = \ & C_1^2 + C_2 \\
3!U_3 = \ & C_1^3 + 2C_2C_1 + C_1C_2 + 2C_3 \\
4!U_4 = \ & C_1^4 + 6C_3C_1 + 2C_1C_3 + 3C_2^2 + 3C_2C_1^2 + 2C_1C_2C_1 + C_1^2C_2 + 6C_4 \\
5!U_5 = \ & C_1^5 + 24C_4C_1 + 6C_1C_4 + 12C_3C_2 + 8C_2C_3 + 12C_3C_1^2 + 6C_1C_3C_1 \\
& +2C_1^2C_3 + 8C_2^2C_1 + 4C_2C_1C_2 + 3C_1C_2^2 + 4C_2C_1^3 + 3C_1C_2C_1^2 \\
& +2C_1^2C_2C_1 + C_1^3C_2. \qquad (8)
\end{aligned}
$$

Figure 2: The decomposition of $U_3$ in terms of $C_m$'s.

The graphical expression for $U_3$ is given in Fig. 2.

# 5 Exponential Formula for Time-Ordered Exponentials

The time-ordered exponential

$$
\mathcal{U} \;=\; T\left(\exp\left(\int_{T'}^{T} H(t)dt\right)\right) = \sum_{n=0}^{\infty} U_n \tag{9}
$$

can be computed from the time-ordered products $U_n$. The factorization character in (7) and (8) suggests that it may be possible to sum up the power series $U_n$ to yield an explicit exponential function of the $C_n$'s. This is indeed the case.

## 5.1 Commutative $C_i$'s

First assume all the $C_{m_i}$ in (7) commute with one another. Then it is possible to show that [5]

$$
\begin{aligned}
\mathcal{U} \;&=\; \sum_{n=0}^{\infty} U_n \\[2mm]
&=\; \prod_{j=1}^{\infty}\sum_{m=0}^{\infty} \frac{1}{j^m m!} C_j^m \\[2mm]
&=\; \exp\left[\sum_{j=1}^{\infty} \frac{C_j}{j}\right]. \tag{10}
\end{aligned}
$$

## 5.2 General Exponential Formula

In general the $C_j$'s do not commute with one another so the exponent in (10) must be corrected by terms involving commutators of the $C_j$'s. The exponent $K$ can be computed by taking the logarithm of $U$ [5]:

$$
\begin{aligned}
U \;&=\; 1+\sum_{n=1}^{\infty} U_n = \exp[K] \equiv \exp\left[\sum_{i=1}^{\infty} K_i\right], \\[2mm]
K \;&=\; \ln\left[1+\sum_{n=1}^{\infty} U_n\right] = \sum_{\ell=1}^{\infty} \frac{(-)^{\ell-1}}{\ell}\left[\sum_{n=1}^{\infty} U_n\right]^{\ell}. \tag{11}
\end{aligned}
$$

The resulting expression must be expressible as (multiple-)commutators of the $C$'s. In other words, only commutators of $H(t)$, in the form of $C_m$ and their commutators, may enter into $K$. This is so because in the special case when $H(t)$ is a member of a Lie algebra, $\mathcal{U}$ is a member of the corresponding Lie group and so $K$ must also be a member of a Lie algebra.

By definition, $K_i$ contains $i$ factors of $H(t)$. Calculation for the first five gives [5]

$$
\begin{aligned}
K_1 &= C_1 \\
K_2 &= \frac{1}{2}C_2 \\
K_3 &= \frac{1}{3}C_3 + \frac{1}{12}[C_2, C_1] \\
K_4 &= \frac{1}{4}C_4 + \frac{1}{12}[C_3, C_1] \\
K_5 &= \frac{1}{5}C_5 + \frac{3}{40}[C_4, C_1] + \frac{1}{60}[C_3, C_2] + \frac{1}{360}[C_1, [C_1, C_3]] + \\
&\quad + \frac{1}{240}[C_2, [C_2, C_1]] + \frac{1}{720}[C_1, [C_1, [C_1, C_2]]]
\end{aligned}
\tag{12}
$$

$K_n$ consists of $C_n/n$, plus the compensating terms in the form of commutators of the $C$'s. By counting powers of $H(t)$ it is clear that the subscripts of these $C$'s must add up to $n$, but beyond that all independent commutators and multiple commutators may appear. For that reason it is rather difficult to obtain an explicit fomula valid for all $K_n$, if for no other reason than the fact that new commutator structures appear at every new $n$. It is however very easy to compute $K_n$ using (11) with the help of a computer. This is actually how $K_5$ was obtained.

Moreover, when we stick to commutators of a definite structure, their coefficients in $K$ can be computed. For example , the coefficient of the commutator term $[C_m, C_n]$ in any $K_{m+n}$ can be shown to be

$$
\eta_2 = \frac{n - m}{2mn(m + n)}.
\tag{13}
$$

See Ref. [5] for similar formulas for multiple commutators.

# 6    Applications

These formulas can be applied to mathematics and physics in various ways, depending on our choice of $H_i(t)$ and the integration interval $[T', T]$. If we choose the interval to be $[T', T] = [0, 2]$, and the operator $H(t)$ to be $P$ for $t \in [1, 2]$ and $Q$ for $t \in [0, 1]$, then $\mathcal{U} = \exp(P)\exp(Q)$, $C_1 = P + Q$, $C_{m+1} = (ad\ P)^m \cdot Q/m!$, and eqs. (11) and (12) lead to [5] the *Baker-Campbell-Hausdorff* formula

$$
\begin{aligned}
\exp(P)\cdot\exp(Q) &= \exp[K_1 + K_2 + K_3 + K_4 + K_5 + \cdots] \\
K_1 &= P + Q \\
K_2 &= \frac{1}{2}[P,Q] \\
K_3 &= \frac{1}{12}[P,[P,Q]] + \frac{1}{12}[Q,[Q,P]] \\
K_4 &= -\frac{1}{24}[P,[Q,[P,Q]]] \\
K_5 &= -\frac{1}{720}[P,[P,[P,[P,Q]]]] - \frac{1}{720}[Q,[Q,[Q,[Q,P]]]] \\
&\quad + \frac{1}{360}[P,[Q,[Q,[Q,P]]]] + \frac{1}{360}[Q,[P,[P,[P,Q]]]] \\
&\quad + \frac{1}{120}[P,[Q,[P,[Q,P]]]] + \frac{1}{120}[Q,[P,[Q,[P,Q]]]], \quad (14)
\end{aligned}
$$

The case when $[P,Q]$ commutes with $P$ and $Q$ is well known. In that case all $K_n$ for $n \geq 3$ are zero. Otherwise, up to and including $K_4$ this formula can be found in eq. (15), §6.4, Chapter II of Ref. [6].

By choosing the interval to be $[T',T] = [-\infty,\infty]$, and the operators to be $H(t) = \exp(ip \cdot k_i t)V_i$, we obtain a *nonabelian eikonal formula* [1, 3, 5] useful in physics. In that case $U_n$ is the $n$th order tree amplitude for an energetic particle with four-momentum $p^\mu$ to emit $n$ bosons with momenta $k_i^\mu \ll p^0$ via vertex factors $V_i$. The decomposition formula (5) can then be interpreted as a repackaging of the tree amplitude into terms in which interference patterns in the spacetime and the internal quantum number variables are explicitly displayed. Such (destructive) interferences lead to cancellations, and the formula can be conveniently used to demonstrate the cancellations necessary for the self consistency of baryonic amplitudes in large-$N_c$ QCD [1, 2]. It can also be used to obtain a simple understanding as to why gluons reggeize in QCD but photons do not reggeize in QED [3, 4].

# References

[*] Email: Lam@physics.mcgill.ca

[1] C.S. Lam and K.F. Liu, *Nucl. Phys.* **B483** (1997) 514.

[2] C.S. Lam and K.F. Liu, *Phys. Rev. Lett.* **79** (1997) 597.

[3] Y.J. Feng, O. Hamidi-Ravari, and C.S. Lam, *Phys. Rev. D* **54** (1996) 3114.

[4] Y.J. Feng and C.S. Lam, *Phys. Rev. D* **55** (1997) 4016.

[5] C.S. Lam, to be published.

[6] N. Bourbaki, *Lie Groups and Lie Algebras*, (Hermann, 1975).

# GAUGE SYMMETRY BREAKING THROUGH SOFT

# MASSES IN SUPERSYMMETRIC GAUGE THEORIES

NORISUKE SAKAI*

*Department of Physics, Tokyo Institute of Technology*
*Oh-okayama, Meguro, Tokyo 152-8551, Japan*

**Abstract**     After reviewing the nonperturbative effects in supersymmetric theories, we have explained our recent study for the effects of soft breaking terms in $N = 1$ supersymmetric gauge theories. For $N_f < N_c$, we include the dynamics of the non-perturbative superpotential and use the original (s)quark and gauge fields. For $N_f > N_c + 1$, we formulate the dynamics in terms of dual (s)quarks and a dual gauge group $SU(N_f - N_c)$. The mass squared of squarks can be negative triggering the spontaneous breakdown of flavor and color symmetry. The general condition for the stability of the vacuum is derived. We determine this breaking pattern, derive the spectrum, and argue that the masses vary smoothly as one crosses from the Higgs phase into the confining phase exhibiting the complementarity. This talk is based on a work in collaboration with Eric D'Hoker, Department of Physics and Astronomy, University of California at Los Angeles, and Yukihiro Mimura, Institute for Cosmic Ray Research, University of Tokyo.

## 1. Introduction

All the available experimental data at low energies ($E < 100\text{GeV}$) can be adequately described by the standard model with $SU(3) \times SU(2) \times U(1)$ gauge group. The three different gauge coupling constants originate from the three different interactions, namely, strong, weak and electormagnetic interactions. The standard model has many parameters which have to be measured by experiments. There are also other conceptually unsatisfactory points as well. For instance, the electric charge is found

*e-mail: nsakai@th.phys.titech.ac.jp

to be quantized in nature, but this phenomena is just an accident in the standard model.

The three interactions described by the three different gauge groups can be truly unified into a single gauge group if we choose a simple gauge group to describe all three interactions. This is realized by the grand unified theories proposed by Georgi and Glashow [1]. The unification energy $M_G$ is very large compared to the electroweak mass scale $M_W$ [2]

$$\frac{M_W^2}{M_G^2} \approx \left(\frac{10^2}{10^{15}}\right)^2 \approx 10^{-26} \tag{1.1}$$

Even if one do not accept the grand unified theories, one is sure to accept the existence of gravitational interactions. The mass scale of the gravitational interactions is given by the Planck mass $M_{Pl}$

$$\frac{M_W^2}{M_{Pl}^2} \approx \left(\frac{10^2}{10^{19}}\right)^2 \approx 10^{-34} \tag{1.2}$$

Now we have a problem of how to explain these extremely small ratios between the mass squared $M_W^2$ to the fundamental mass squared $M_G^2$ or $M_{Pl}^2$ in eq.(1.1) or eq.(1.2). This problem is called the **gauge hierarchy problem**.

When we say **explain**, we mean that it should be given a symmetry reason. This principle is called the naturalness hypothesis [3], [4]. More precisely, the system should acquire higher symmetry as we let the small parameter going to zero. The examples of the enhanced symmetry corresponding to the small mass parameter are

$$
\begin{aligned}
m_{J=1/2} \to 0 &\quad \Leftrightarrow \quad \text{Chiral symmetry} \\
m_{J=1} \to 0 &\quad \Leftrightarrow \quad \text{Local gauge symmetry}
\end{aligned}
\tag{1.3}
$$

The electroweak mass scale $M_W$ originates from the vacuum expectation value $v$ of the Higgs field. The scale of $v$ in turn comes from the quadratic term of the higgs potential, namely the (negative) mass squared of the Higgs scalar $\varphi$. Therefore we need to give symmetry reasons for the vanishing Higgs scalar mass in order to explain the gauge hierarcy problem.

Classically the vanishing mass for scalar filed does give rise to an enhanced symmetry called scale invariance. However, it is well-known that the scale invariance cannot be maintained quantum mechanically. Therefore we have only two options to explain the gauge hierarchy problem.

1. Technicolor model [5]

   We can postulate that there is no elementary Higgs scalar at all. The Higgs scalar in the standard model has to be provided as a composite field at low energies. This option requires nonperturbative physics already at energies of the order of TeV. It has been rather difficult to construct realistic models which pass all the test at low energies especially the absence of flavor-changing neutral currrent. Models with composite Higgs scalar are called Technicolor models.

2. Supersymmetry [6] [7]

   Another option is to postulate a symmetry between Higgs scalar and a spinor field. Then we can postulate chiral symmetry for the spinor field to make it massless. The Higgs scalar also becomes massless because of the symmetry beween the scalar and the spinor. This symmetry between scalar and spinor is called supersymmetry [8]. Contrary to the Technicolor models, we can construct supersymmetric models which can be treated perturbatively up to extremely high energies along the spirit of the grand unified theories.

Since we have proposed the supersymmetric grand unified theories in 1981, ten years of vigorous experimental progress has finally gave us enough experimental data to discriminate between models with and without supersymmetry. If we extrapolate from the data compiled from the LEP experiments at CERN, the renormalization group running with the particle content of the minimal supersymmetric model provides an excellent agreement of all three coupling constants at a point [9]. On the other hand, the three coupling constants do   not converge at a point if we use the particle content of nonsupersymmetric model. This fact can be regarded as an indirect evidence for the supersymmetric unified models.

Supersymmetric theories have now become a standard theory to solve the gauge hierarchy problem and to unify all the forces in nature. Recent advances to understand the dynamics of supersymmetric gauge theories haveprovided a clear picture of nonperturbative effects [10] – [14].

It has been anticipated that these features are not restricted to $N = 1$ supersymmetric gauge theories, but should survive – at least in a qualitative way – to nonsupersymmetric gauge theories. The models in which this proposal is perhaps most easily verified are those where supersymmetry is spontaneously broken (through the introduction of an additional sector of fields [15] – [16]) or those where soft, explicit supersymmetry breaking terms are added as a perturbation on the gauge dynamics [17]. The latter scheme of supersymmetry breaking was used in the original proposal of supersymmetric grand unified theories [6], [18] and provides a general framework for the usual formulation of the Minimal Supersymmetric Standard Model [19].

In a series of papers, it was argued that the addition of perturbative, soft supersymmetry breaking mass terms, with $m^2 \geq 0$, essentially preserves the qualitative picture of the dynamics derived for $N = 1$ supersymmetric QCD (SQCD) [20]. An effective low energy theory is used in terms of color singlet meson and (for $N_f \geq N_c$) baryon fields, appropriate for the confining phase, and the effects of the non-perturbative superpotential of Affleck, Dine and Seiberg [11] are included.

More recently we have investigated $N = 1$ supersymmetric QCD, again with soft supersymmetry breaking mass terms added, but this time with $m^2 < 0$ for at least some of the squark fields [21]. In the present paper, we will briefly report our findings. For general values of the soft supersymmetry breaking mass terms with $m^2 < 0$, the Hamiltonian will become unbounded from below, thus destabilizing the entire theory. For certain simple ranges of the masses, however, we show that stable vacua exist through a balance between the soft supersymmetry breaking mass terms, the quartic $D^2$ term for the squark fields and ( for $N_f < N_c$ ) the non-perturbative effective potential. We argue that in these solutions, flavor as well as color symmetry are spontaneously broken through the vacuum expectation value of the squarks. We consider the most general soft supersymmetry breaking terms respecting $R$-symmetry, for simplicity.

For the analysis of spontaneous flavor and color symmetry breakdown when $N_f < N_c$, we formulate the dynamics in terms of the fundamental quark superfields instead of in terms of meson superfields, in contrast with the analysis in ref.[20]. This choice appears more natural when dealing with the theory in the Higgs phase, rather than in the confining phase. In fact, the original calculation of the nonperturbative superpotential was justified precisely by considering gauge symmetry breakdown due to the vacuum expectation values of squark fields [11].

For small values of the soft supersymmetry breaking mass terms, we expect the confining and Higgs phases to be smoothly connected to one another, with matching low energy spectra. We find that this principle of complementarity between the confining and Higgs phases [22] can indeed be satisfied in these theories. To do so however, it is necessary to consider the nonminimal kinetic terms for color singlet fields in the confining phase. This is in contrast to the minimal kinetic term advocated in ref.[20].

For large values of the soft supersymmetry breaking mass terms, we expect the semi-classical spectrum for the Higgs phase, derived in terms of the fundamental quark superfields, to remain reliable. Thus, we shall calculate the semi-classical spectrum for this model for all ranges of soft breaking mass terms. For $N_f > N_c + 1$, we will use the dual variables of ref.[12], which are more appropriate to describe the spontaneous symmetry breakdown of flavor and (dual) color symmetry.

Finally, we note that there is a simple extension of standard SQCD, obtained by gauging also an additional anomaly-free $U(1)_X$ symmetry, and adding a Fayet-Illiopoulos $D$-term [15] for the corresponding gauge multiplet. (The simplest case would be where this $U(1)_X$ is just baryon number symmetry, but any anomaly-free $U(1)_X$ would do.) No explicit supersymmetry breaking terms are added; instead, a supersymmetric mass term is included in the superpotential, which stabilizes the vacuum. In this model, supersymmetry is broken spontaneously, and soft mass terms with $m^2 < 0$ automatically arise from the Fayet-Illiopoulos $D$-term. This model provides an economical realization of some of the effects of soft supersymmetry breaking mass terms generated directly by spontaneous breakdown of supersymmetry. It will be discussed in a companion paper. A different model has been analyzed which obtains the soft breaking terms from the spontaneous breakdown of supersymmetry [23].

## 2. Dynamics for $N_f < N_c$

In this Section, we shall consider supersymmetric Yang-Mills theory with gauge group $SU(N_c)$ and $N_f$ flavors of squarks and quarks (with $N_f < N_c$), transforming under the representation $N_c \oplus \bar{N}_c$ of $SU(N_c)$. This theory is the natural supersymmetric extension of QCD, and will be referred to as SQCD. The corresponding chiral superfields

$$\hat{Q}_a{}^i \qquad \hat{\bar{Q}}_i{}^a \qquad a = 1, \cdots, N_c; \qquad i = 1, \cdots, N_f \, , \qquad (2.1)$$

contain the squark fields $Q$ and $\bar{Q}$ and the left-handed quark fields $\psi_Q$ and $\psi_{\bar{Q}}$ respectively. There is a natural color singlet meson chiral superfield $\hat{T}$, defined by

$$\hat{T}_i{}^j = \hat{\bar{Q}}_i{}^a \hat{Q}_a{}^j \qquad (2.2)$$

with scalar components $T_i{}^j$. Superfields are denoted by a cap on the scalar components.

As a starting point we consider classical massless SQCD whose Lagrangian $\mathcal{L}_0$ is determined by $SU(N_c)$ gauge invariance, by requiring that the superpotential for the quark superfields vanish identically :

$$\mathcal{L}_0 = \int d^4\theta \, \text{tr}\{\hat{Q}^\dagger e^{2g\hat{V}}\hat{Q} + \hat{\bar{Q}}e^{-2g\hat{V}}\hat{\bar{Q}}^\dagger\} + \frac{1}{2}\int d^2\theta \, \text{tr}WW + \frac{1}{2}\int d^2\bar{\theta} \, \text{tr}\bar{W}\bar{W} \quad (2.3)$$

This theory has a global symmetry, $G_f = SU(N_f)_Q \times SU(N_f)_{\bar{Q}} \times U(1)_B \times U(1)_R$, with $R$-charges ( baryon number) are given by $1 - N_c/N_f$ (1) for $Q$, and $1 - N_c/N_f$ (−1) for $\bar{Q}$.

Exact nonperturbative results in supersymmetric gauge theories can be given for the $F$-type term which is a chiral superspace integral of a superpotential $W_{NP}$ [11] of the quark superfields $\hat{Q}$ and $\hat{\bar{Q}}$ given as follows

$$\int d^2\theta W_{NP}(\hat{Q},\hat{\bar{Q}}) = (N_c - N_f)\Lambda^{3+2N_f/(N_c-N_f)} \int d^2\theta \, (\det\hat{\bar{Q}}\hat{Q})^{-1/(N_c-N_f)} \quad (2.4)$$

We choose to break supersymmetry explicitly, by adding to the Lagrangian $\mathcal{L}_0$ soft supersymmetry breaking terms [17] for the quark supermultiplet.

For the sake of simplicity, we shall add to the Lagrangian only soft supersymmetry breaking squark mass terms and neglect effects due to gaugino masses and supersymmetric flavor masses. Generic mass squared for squark and antisquarks are given by matrices $M_Q^2$ and $M_{\bar{Q}}^2$

$$\mathcal{L}_{sb} = -\{\text{tr}QM_Q^2Q^\dagger + \text{tr}\bar{Q}^\dagger M_{\bar{Q}}^2\bar{Q}\} \quad (2.5)$$

As we remarked above, when $M_Q^2$ and $M_{\bar{Q}}^2$ are proportional to the identity matrix, the global flavor symmetry is unchanged : $SU(N_f)_Q \times SU(N_f)_{\bar{Q}} \times U(1)_B \times U(1)_R$.

When either $M_Q^2$ or $M_{\bar{Q}}^2$ is not positive definite, we expect the pattern of symmetry breaking to be substantially different. Global flavor symmetry should be spontaneously broken, and $Q$ and/or $\bar{Q}$ should acquire non-vanishing vacuum expectation values. These non-zero vacuum expectation values, in turn, are expected to break color $SU(N_c)$ and give mass to some of the gauge particles through the Higgs mechanism. This is the so-called Higgs phase.

According to standard lore, (originally derived from lattice gauge theory) the confining and Higgs phases are smoothly connected to one another in at least some region of parameter space [22]. There should be a one to one correspondence between the observables in both phases, suggesting that – in principle – color singlet meson fields could still be used to describe the dynamics of the Higgs phase. In practice, however, a formulation in terms of colored fields appears more suitable instead. Indeed, physical free quarks and certain massive gauge bosons are expected to appear in the low energy spectrum, and it is unclear how to represent these degrees of freedom in terms of meson variables. Thus, we shall use the original squark $Q$, $\bar{Q}$, quark $\psi_Q$, $\psi_{\bar{Q}}$, and gauge boson and fermion fields as physical variables at low energy. Without SUSY breaking soft masses, the vacuum is known to runaway. For generic matrices $M_Q^2$ and $M_{\bar{Q}}^2$, we find the stability condition for the vacuum as

$$m_{Q_i}^2 + m_{\bar{Q}_j}^2 \geq 0 \qquad (2.6)$$

for any pair of $i, j = 1, \cdots N_f$.

For simplicity, we explicitly analyze only the case where the mass squared for all $Q$'s and $\bar{Q}$'s are equal to $-m_Q^2$ and $m_{\bar{Q}}^2$ respectively, thus preserving the entire global symmetry $SU(N_f)_Q \times SU(N_f)_{\bar{Q}} \times U(1)_B \times U(1)_R$. After a somewhat lengthy analysis, we find that there is a unique minimum of the potential at which the first $N_f \times N_f$ block of $Q$ and $\bar{Q}$ are nonvanishing and are proportional to the identity

$$\langle 0|Q|0\rangle = \begin{pmatrix} Q_0 I_{N_f} \\ 0 \end{pmatrix} \qquad\qquad \langle 0|\bar{Q}|0\rangle = \begin{pmatrix} \bar{Q}_0 I_{N_f} & 0 \end{pmatrix} \qquad (2.7)$$

The minimum conditions are

$$
\begin{aligned}
0 &= (\gamma - 1)Q_0^{-2\gamma}\bar{Q}_0^{-2\gamma} + \gamma Q_0^{-2-2\gamma}\bar{Q}_0^{2-2\gamma} - \frac{g^2}{2\gamma}(Q_0^2 - \bar{Q}_0^2) + m_Q^2 \\
0 &= (\gamma - 1)Q_0^{-2\gamma}\bar{Q}_0^{-2\gamma} + \gamma Q_0^{2-2\gamma}\bar{Q}_0^{-2-2\gamma} + \frac{g^2}{2\gamma}(Q_0^2 - \bar{Q}_0^2) - m_{\bar{Q}}^2
\end{aligned} \qquad (2.8)
$$

We have obtained the spectra in this vacuum. As anticipated, the scalar particles contain massless Nambu-Goldstone boson corresponding to the spontaneous breakdown of $U(1)_R$ and $SU(N_f)$ global symmetry.

## 3. Complementarity between confining and Higgs phase

According to the complemetarity argument, one should be able to relate our mass spectra calculated in terms of colored elementary fields with the result in ref. [20]

where the meson fields $T$ are used as fundamental variables in the low energy effective theory. They have assumed a standard minimal kinetic term for the meson fields, because the Kähler potential can not be constrained by holomorphy. Consequently their results on mass spectra differed from ours, although overall qualitative picture was the same.

On the other hand, if we take large VEV for squark fields, we should be in a nearly perturbative region. Therefore there should be the correct Kähler potential for the color singlet meson which correspond to our choice of perturbative kinetic term for squarks. We indeed find that a similar analysis as in ref. [20] with the Kähler potential for meson

$$K[T] = 2\,\mathrm{tr}\left(T^\dagger T\right)^{\frac{1}{2}} \tag{3.1}$$

reproduces our mass spectra correctly.

This shows that the complementarity is also valid in supersymmetric situation, and the higher order terms in Kähler potential are important when some of the fields acquire VEV.

## 4. Dynamics for $N_f > N_c + 1$

We have also analyzed the case of $N_f > N_c + 1$ using the dual description. Qualitaive picure is similar to the previous case of $N_f < N_c + 1$.

The dual description for the gauge group $SU(N_c)$ and $N_f$ flavors of quarks and antiquarks (with $N_f > N_c + 1$) has a gauge group $SU(\tilde{N}_c)$ with $N_f$ flavors, where $\tilde{N}_c = N_f - N_c$. The elementary chiral superfields in the dual theory are dual quark $\hat{q}$ and meson $\hat{T}$ superfields, $\hat{q}^a{}_i$ $\hat{\tilde{q}}^i{}_a$ $\hat{T}^i{}_j$
$a = 1, \cdots, \tilde{N}_c;$ $i, j = 1, \cdots N_f$ .

Since $\hat{q}$, $\hat{\tilde{q}}$ and $\hat{T}$ are effective fields, their kinetic terms need not have canonical normalizations; in particular, they can receive nonperturbative quantum corrections. Thus, we introduce into the (gauged) Kähler potential for $\hat{q}$, $\hat{\tilde{q}}$ and $\hat{T}$ normalization parameters $k_q$ and $k_T$ as follows

$$K[\hat{q}, \hat{\tilde{q}}, \hat{T}, \hat{v}] = k_q \mathrm{tr}(\hat{q}^\dagger e^{2\tilde{g}\hat{v}}\hat{q} + \hat{\tilde{q}}e^{-2\tilde{g}\hat{v}}\hat{\tilde{q}}^\dagger) + k_T \mathrm{tr}\hat{T}^\dagger\hat{T}. \tag{4.1}$$

Here, we denote by $\hat{v}$ the $SU(\tilde{N}_c)$ color gauge superfield, and by $\tilde{g}$ the associated coupling constant. (Pure gauge terms will not be exhibited explicitly.) In principle, these normalization parameters are determined by the dynamics of the underlying microscopic theory. Furthermore, it has been pointed out in [12] that a superpotential coupling $q$, $\bar{q}$ and $T$ should be added as follows

$$W = \hat{\bar{q}}^a{}_i \hat{T}^i{}_j \hat{\bar{q}}^j{}_a. \tag{4.2}$$

We add soft supersymmetry breaking terms to the Lagrangian for the dual quark and meson supermultiplets. For simplicity we shall assume that $R$-symmetry is maintained so that neither $A$-terms nor gaugino masses are present in the Lagrangian.

When the eigenvalues of $M_q^2$, $M_{\bar{q}}^2$ and $M_T^2$ can take generic positive or negative values, the scalar potential may be unbounded from below. A necessary condition for which the potential is bounded from below is that $M_T^2$ be a positive definite matrix. This is because there is no quartic term of $T$.

The D terms vanish when the vacuum expectation values are given by:

$$\langle 0|q|0\rangle = \begin{pmatrix} q_1 & & & \\ & \ddots & & 0 \\ & & q_{\tilde{N}_c} & \end{pmatrix}, \qquad \langle 0|\bar{q}|0\rangle = \begin{pmatrix} \bar{q}_1 & & & \\ & \ddots & & \\ & & \bar{q}_{\tilde{N}_c} & \\ & & & 0 \end{pmatrix}, \tag{4.3}$$

with the combinations $|q_i|^2 - |\bar{q}_i|^2$ independent of $i$.

If we set the squark masses to be zero, the space where $|q_i|^2$ is independent of $i$ and $\bar{q} = 0$ is a subspace of the moduli space of vacua. If we insist on flavor symmetric mass squared matrix and on having a negative eigenvalue, we are forced to have a potential unbounded from below. In fact, in the next subsection, we shall establish more generally that to have a potential bounded from below, we must have

$$m_1^2 + \cdots + m_{\tilde{N}_c}^2 \geq 0, \tag{4.4}$$

where $m_i^2$ are eigenvalues of the matrix $M_q^2$ or $M_{\bar{q}}^2$, and they are set to be $m_1^2 \leq m_2^2 \leq \cdots \leq m_{N_f}^2$.

Therefore we consider the simplest stable situation, where the $n$ eigenvalues of $M_q^2$ is negative and same, while all the others are positive or zero. The $n$ should be smaller than $\tilde{N}_c$. For simplicity we shall also assume that the soft supersymmetry breaking positive mass squared terms for squarks have a flavor symmetry $SU(N_f -$

$n)_Q \times SU(N_f)_{\bar{Q}}$. As a result, the $N_f - n$ positive eigenvalue of $M_q^2$ are all the same, while the $N_f$ eigenvalues of $M_{\bar{q}}^2$ are the same : $M_{\bar{q}}^2 = m_{\bar{q}}^2 I_{N_f}$. We also assume $M_T^{2i}{}_j{}^k{}_l = m_T^2 \delta_j^i \delta_l^k$.

After all, we find that there are only two solutions for possible minimum, described as follows.

1. Only $q_1, \cdots, q_n \neq 0$, while $q_i = 0$, $i > n$ and $\bar{q}_i = 0$ for all $i$. The values of $q_1, \cdots, q_n$ are the same. We call the common value as $q_0$. The value of $q_0$ and the potential in this configuration are given by

$$q_0^2 = \frac{2}{\tilde{g}^2 \tilde{\gamma}} m_{q_1}^2, \qquad\qquad V = -\frac{n}{\tilde{g}^2 \tilde{\gamma}} m_{q_1}^4. \qquad (4.5)$$

where

$$\tilde{\gamma} = \frac{\tilde{N}_c - n}{\tilde{N}_c}. \qquad (4.6)$$

2. Only $q_1, \cdots, q_n \neq 0$ and $\bar{q}_1, \cdots \bar{q}_n \neq 0$, while $q_i = \bar{q}_i = 0$, $i > n$. The values of $q_0 \equiv q_1 = \cdots = q_n$ and $\bar{q}_0 \equiv \bar{q}_1 = \cdots \bar{q}_n$ are then given by

$$\begin{pmatrix} q_0^2 \\ \bar{q}_0^2 \end{pmatrix} = \frac{1}{\tilde{\gamma}\tilde{g}^2 - \frac{1}{k_T}} \begin{pmatrix} \frac{1}{2}\tilde{\gamma}\tilde{g}^2 k_T(m_{q_1}^2 - m_{\bar{q}}^2) + m_{\bar{q}}^2 \\ \frac{1}{2}\tilde{\gamma}\tilde{g}^2 k_T(m_{q_1}^2 - m_{\bar{q}}^2) - m_{q_1}^2 \end{pmatrix} \qquad (4.7)$$

Given the fact that $q_1^2$ and $\bar{q}_1^2$ must be positive, this expression yields a solution only when the following condition is satisfied

$$\frac{1}{2}\tilde{\gamma}\tilde{g}^2 k_T(m_{q_1}^2 - m_{\bar{q}}^2) \geq m_{q_1}^2. \qquad (4.8)$$

The value of the potential at the stationary point is given by

$$V = -\frac{n}{4}\frac{\tilde{\gamma}k_T\tilde{g}^2}{\tilde{\gamma}\tilde{g}^2 - \frac{1}{k_T}}\left(m_{\bar{q}}^2 - \frac{\frac{1}{2}\tilde{\gamma}\tilde{g}^2 - \frac{1}{k_T}}{\frac{1}{2}\tilde{\gamma}\tilde{g}^2}m_{q_1}^2\right)^2 - \frac{n}{\tilde{\gamma}\tilde{g}^2}m_{q_1}^4 \qquad (4.9)$$

Therefore, whenever condition (4.8) is satisfied, solution 2 is the absolute minimum of the potential and describes the true ground state. If condition (4.8) is not satisfied, solution 1 is the absolute minimum.

In our model, the Lagrangian has a global $SU(N_f - n)_Q \times SU(n)_Q \times U(1)_Q \times SU(N_f)_{\bar{Q}} \times U(1)_B \times U(1)_R$ symmetry. When the coupling $\tilde{g}$ is too weak, the condition

(4.8) is not satisfied, solution 1 is the absolute minimum, and flavor symmetry is not broken.

On the other hand, when the gauge coupling is strong, the condition (4.8) is satisfied. Therefore solution 2 is the absolute minimum, and flavor symmetry is spontaneously broken to $SU(N_f - n)_Q \times SU(N_f - n)_{\tilde{Q}} \times SU(n)_V \times U(1)_V \times U(1)_{B'} \times U(1)_{R'}$, where $SU(n)_V$ is the diagonal subgroup of $SU(n) \subset SU(\tilde{N}_c)$ and $SU(n)_Q \times SU(n)_{\tilde{Q}}$. The spontaneous breaking of the global symmetry induces spontaneous breaking of color gauge symmetry $SU(\tilde{N}_c) \rightarrow SU(\tilde{N}_c - n)$. One interesting feature of the present case is that the gauge symmetry is broken without chiral symmetry breaking. We find Nambu-Goldstone bosons for spontaneous breakdown of $SU(N_f)/SU(N_f - n)$.

# References

[1] H. Georgi and H. Glashow, *Phys. Rev. Lett.* **32** (1974) 438.

[2] H. Georgi, H. Quinn and S. Weinberg, *Phys. Rev. Lett.* **33** (1974) 451.

[3] M. Veltman, *Acta Phys. Pol.***B12** (1981) 437.

[4] G. 't Hooft, in Recent Developments in Gauge Theories, Cargèse summer school 1979 p.135.

[5] L. Susskind, *Phys. Rev.***D20** (1979) 2619; S. Weinberg, *Phys. Rev.***D19** (1979) 1277; **D13** (1976) 974; S. Dimopoulos, and L. Susskind, *Nucl. Phys.* **B155** (1979) 237; E. Eichten and K. Lane, *Phys. Lett.* **B90** (1980) 125.

[6] N. Sakai, Z. f. Phys. **C11** (1981) 153.

[7] S. Dimopoulos and H. Georgi, Nucl. Phys. **B193** (1981) 150.

[8] J. Wess and J. Bagger, Supersymmetry and Supergravity, Princeton University Press, (1992).

[9] U. Amaldi et. al. *Phys. Lett.***B281** (1992) 374.

[10] G. Veneziano and S. Yankielowicz, *Phys. Lett.* **113** (1982) 231; T. Taylor, G. Veneziano and S. Yankielowicz, *Nucl. Phys.* **B218** (1983) 493.

[11] J. Affleck, M. Dine and N. Seiberg, *Phys. Rev. Lett.* **51** (1983) 1026, ibid. *Nucl. Phys.* **B241** (1984) 493.

[12] N. Seiberg, *Phys. Rev.* **D49** (1994) 6857, ibid. *Nucl. Phys.* **B435** (1995) 129, hep-th.9408013.

[13] K. Intriligator, R. Leigh, and N. Seiberg, *Phys. Rev.* **D50** (1994) 1092; K. Intriligator and N. Seiberg, *Nucl. Phys.* **B431** (1994) 551, *Nucl. Phys.* **B447** (1995) 125, hep-th.9509066; M. Douglas and S. Shenker, *Nucl. Phys.* **B447** (1995) 271; S. Elitzur, A. Forge, A. Giveon and E. Rabinovici, *Phys. Lett.* **B353** (1995) 79.

[14] N. Seiberg and E. Witten, *Nucl. Phys.* **B426** (1994) 19, ibid. *Nucl. Phys.* **B431** (1994) 484; C. Vafa and E. Witten, *Nucl. Phys.* **B431** (1994) 3.

[15] P. Fayet and J. Iliopoulos, *Phys. Lett.* **B51** (1974) 461.

[16] L. O'Raifeartaigh, *Nucl. Phys.* **B96** (1975) 331.

[17] L. Girardello and M. Grisaru, *Nucl. Phys.* **194B** (1982) 65; K. Harada and N. Sakai, *Prog. Theor. Phys.* **67** (1982) 1877.

[18] S. Dimopoulos and H. Georgi, *Nucl. Phys.* **B193** (1981) 150.

[19] For a review see for instance, H.P. Nilles, *Phys. Rep.* **C110** (1984) 1; P. Nath, R. Arnowitt, and A. Chamseddine, Applied $N = 1$ Supergravity, the ICTP Series in Theoretical Physics, Vol. I (World scientific) 1984; H. Haber and G. Kane, *Phys. Rep.* **C117** (1985) 75.

[20] O. Aharony, M. Peskin, J. Sonnenschein, and S. Yankielowicz, *Phys. Rev.* **D52** (1995) 6157, hep-th.9509165.

[21] E. D'Hoker, Y. Mimura, and N. Sakai, *Phys. Rev.* **D54** (1996) 7724, hep-th/9603206.

[22] E. Fradkin and S. Shenker, *Phys. Rev.* **D19** (1979) 3682; T. Banks and E. Rabinovici, *Nucl. Phys.* **B160** (1979) 349; S. Dimopoulos, S. Raby, and L. Susskind, *Nucl. Phys.* **B173** (1980) 208.

[23] N. Evans, S.D.H.Hsu, and M. Schwetz, *Phys. Lett.* **B355** (1995) 475; N. Evans, S.D.H.Hsu, M. Schwetz, and S. B. Selipsky, *Nucl. Phys.* **B456** (1995) 205;

# SPONTANEOUS COLLAPSE OF SUPERSYMMETRY

*Dedicated to Professor Hiroshi Ezawa on his 65th birthday*

IZUMI OJIMA

Research Center for Mathematical Sciences
Kyoto University
Kyoto 606-8502, Japan

## 1  Introduction

Supersymmetry (SUSY) has been highlighted as a symmetry to weaken UV divergences by mixing bosons and fermions. BUT, so far we have had *no* evidence for superpartners characteristic to supersymmetric theories in spite of many experimental attempts. And concerning the *fate of SUSY at $T \neq 0°K$*[1, 2, 3], there has been *no agreeable consensus* after more than a decade of discussions! In this talk based upon a joint paper[4] by D.Buchholz and I.O., it is shown that,

> *if generators of SUSY transformations (supercharges) can be defined* in a spatially homogeneous physical state, then this state describes the *vacuum only*.

This means that SUSY is inevitably *broken in any thermal state* and can never be "restored" in any mixed states so as for supercharges to be defined. In contrast to the familiar spontaneous breakdown of usual bosonic symmetries, this is a *complete collapse* of SUSY in thermal states. Also it is shown that spatially homogeneous superthermal ensembles are *never* supersymmetric (for more details see[4]).

Setting of the present talk is just adapted for treating general situations including thermal states with the *only assumption of spatial homogeneity*. We first list up the important points overlooked in the usual discussions:

1) **Necessity for thermodynamic limit**: This is for attaining the homogeneity of infinite-volume systems to eliminate accidental "boundary effects" in finite systems. However, it invalidates the familiar Gibbs formula valid only in a finite volume, because of which it is impossible to interchange thermodynamic limit and spatial integrations in the expression of symmetry generators in terms of current densities.

2) **Necessity for "renormalizing" symmetry generators**

    a) The formula expressing symmetry generators as volume integrals of current densities is *not* valid as it stands, outside the regime of vacua

b) Existence of non-trivial commutants in mixed states implies the possibility to *rescue* generators by "renormalization" and, at the same time, *ambiguities* in their definitions.

As a consequence, for instance, non-vanishing energy in thermal states does *not necessarily* imply spontaneous breakdown of SUSY.
Cf. Time translation itself is *not* broken in a thermal (KMS) state "in spite of" its non-vanishing energy density.

3) **What is symmetry breakdown?**: To answer this question in a precise way, it is necessary to distinguish between *algebraic* transformations and their *generators*. Then, we can clarify the contrast between **spontaneous breakdown vs. spontaneous collapse** in the light of a general relation between symmetries and thermodynamic *phases*, the latter of which are defined as follows:

A. pure and mixed phases / order parameters / cluster property / central decompostion

    a. pure phase: a state with sharp c-number values of macroscopic order parameters, satisfying cluster propery[5] (factorial state with trivial centre)

    b. mixed phase: a state with statistically fluctuating order parameters violating cluster property (centre non-trivial)

**NB** : Don't confuse pure/mixed *phases* with pure/mixed *states*! Any thermodynamic phase, pure or mixed, is a mixed *state*, and, only for $T = 0°K$, a pure phase/state and a mixed phase/state are, respectively, identical.

**NB** : Arbitrary states can be decomposed *uniquely* into pure phases satisfying cluster property (= central decomposition), whereas decompositions into pure states are not unique in general[5].

B. **Patterns of symmetry breaking:**

    i) pure phases with unbroken symmetry

    ii) pure phases with broken symmetry which can be restored by suitable mixing of phases (spontaneous symmetry breakdown [SSB])

    A well-known example of the case ii) is given by the breakdown of rotation invariance in a ferromagnet and its restoration: In a state $\langle \cdot \rangle_\theta$ with sharp direction $\theta$ of magnetization (=order parameter), spatial rotations cannot be defined on the corresponding Hilbert space $\mathcal{H}_\theta$ (except for rotations around the axis $\theta$). However, the mixed phase corresponding to spherical average of these states, $\langle \cdot \rangle = \int d\theta \langle \cdot \rangle_\theta$, is invariant under any infinitesimal rotations $\delta$: $\langle \delta(\cdot) \rangle = 0$. Thus, there exist generators for rotations on an "enlarged" Hilbert space $\int d\theta^{1/2} \mathcal{H}_\theta$ corresponding to $\langle \cdot \rangle$. This is the usual and typical situation in the spontaneous breakdown of bosonic symmetries.

However, there is the *third possibility not known before*:

    iii) **spontaneous collapse** = symmetry *without restoration* in any mixed
    phases.

In the following, we examine the fate of SUSY under situations *outside of vacua
with* $T = 0°K$, paying enough attentions to the above three points 1), 2), and 3),
by introducing, respectively, algebraic formulation of QFT, algebraic formulation of
(super)symmetry transformations, and general definition of "implementability" in
arbitrary states.

## 2 Status of Supercharges

In the usual treatment of SUSY transformation, it is first formulated in terms of
underlying fields, and the invariance of the action under it implies the existence of
the supercurrents $j_{\mu\alpha}(x), j^{\dagger}_{\nu\dot{\beta}}(x)$ and their conservation,

$$\partial^{\mu} j_{\mu\alpha}(x) = \partial^{\nu} j^{\dagger}_{\nu\dot{\beta}}(x) = 0. \tag{1}$$

On the basis of (1), supercharges are defined as the conserved charges generating
the SUSY transformations:

$$Q_\alpha \stackrel{?}{=} \text{“} \int d^3x \, j_{0\alpha}(x) \text{ ”}, \quad Q^{\dagger}_{\dot{\beta}} \stackrel{?}{=} \text{“} \int d^3x \, j^{\dagger}_{0\dot{\beta}}(x) \text{ ”}.$$

While these supercharges $Q_\alpha$, $Q^{\dagger}_{\dot{\beta}}$ appear explicitly in almost all the discussions of
SUSY in the physical literature, the meaning of RHS, and hence, the existence of
these objects is very subtle, depending upon whether SUSY suffers from breakdown
or not. To avoid such subtlety, we here adopt the

**[Algebraic Formulation of SUSY Transformations]**
While we also start from the SUSY transformation of underlying fields (denoted
generically by $\varphi(x)$), we avoid the direct use of such dubious objects as $Q_\alpha, Q^{\dagger}_{\dot{\beta}}$
systematically by extending the former to field algebra $\mathcal{F}$ as an *anti-derivation* [where
$\mathcal{F}$ is a *-algebra generated by polynomials $F = \sum c \, \varphi(f_1)\varphi(f_2) \cdots \varphi(f_n)$ in field operators
$\varphi(f) = \int d^4x \, \varphi(x)f(x)$ smeared by test functions $f$ with compact supports]. This provides
us with an algebraic formulation of SUSY transformation *independent of particular
choice of representations* of field algebra $\mathcal{F}$. For this purpose, we first introduce the
algebraic distinction between Bosons and Fermions in terms of $2\pi$-rotation $\gamma$ of field
operators:

$$\begin{aligned}
\mathcal{F}_\pm &\equiv \{F_\pm \in \mathcal{F}; \, \gamma(F_\pm) = \pm F_\pm\}, \quad \pm : \text{Boson/Fermion}, \\
\mathcal{F} &= \mathcal{F}_+ + \mathcal{F}_-,
\end{aligned}$$

which brings in a $\mathbf{Z}_2$-grading in $\mathcal{F}$: $\mathcal{F}_\pm\mathcal{F}_\pm \subseteq \mathcal{F}_+, \quad \mathcal{F}_\pm\mathcal{F}_\mp \subseteq \mathcal{F}_-$.

SUSY transformations in their algebraic form are now given as *anti-derivations* characterized by "($\mathbb{Z}_2$-)graded" Leibniz rule:

$$\delta_\alpha(F_\pm F) = \delta_\alpha(F_\pm)F \pm F_\pm \delta_\alpha(F), \tag{2}$$

$$\bar{\delta}_{\dot{\beta}}(F_\pm F) = \bar{\delta}_{\dot{\beta}}(F_\pm)F \pm F_\pm \bar{\delta}_{\dot{\beta}}(F), \tag{3}$$

where $F_\pm \in \mathcal{F}_\pm, F \in \mathcal{F}$. We have also

$$\delta_\alpha(\mathcal{F}_\pm) \subset \mathcal{F}_\mp, \quad \bar{\delta}_{\dot{\beta}}(\mathcal{F}_\pm) \subset \mathcal{F}_\mp, \tag{4}$$

and their hermitian conjugation property:

$$\delta_\alpha(F_\pm)^\dagger = \mp \bar{\delta}_{\dot{\alpha}}(F_\pm^\dagger). \tag{5}$$

We denote spacetime translations of $F = \sum c\,\varphi(f_1)\,\varphi(f_2)\cdots\varphi(f_n) \in \mathcal{F}$ by

$$\alpha_x : F \mapsto \alpha_x(F) = F(x) = \sum c\,\varphi(f_{1,x})\,\varphi(f_{2,x})\cdots\varphi(f_{n,x}), \quad \text{with } f_{i,x}(y) = f_i(y - x).$$

Writing $x = (x_0,\ \boldsymbol{x}) \in \mathbf{R}^4$ in a chosen Lorentz frame, we also denote

   time translation: $\alpha_{x_0} \equiv \alpha_{(x_0,0)}, \quad \alpha_{x_0}(F) \equiv F(x_0)$,
   spatial translation: $\alpha_{\boldsymbol{x}} \equiv \alpha_{(0,x)}, \quad \alpha_{\boldsymbol{x}}(F) \equiv F(\boldsymbol{x})$.

In the above notation, **Fundamental relation of SUSY** is given by

$$\bar{\delta}_{\dot{1}} \circ \delta_1 + \delta_1 \circ \bar{\delta}_{\dot{1}} + \bar{\delta}_{\dot{2}} \circ \delta_2 + \delta_2 \circ \bar{\delta}_{\dot{2}} = 4\delta_0, \tag{6}$$

where $\delta_0 = -id\alpha_{x_0}/dx_0|_{x_0=0}$ is the infinitesimal time translation and $\circ$ denotes composition of maps on $\mathcal{F}$.

**Remark:** This gives a mathematically sound formulation, meaningful independently of specific states, to the familiar relation $\sum_{\alpha=1}^{2}\{Q_\alpha, Q_{\dot{\alpha}}^\dagger\} = 4H$ involving supercharges which may *not* exist.

Now, the statement to be proved is: *If SUSY is unbroken in an arbitrary state $\langle \cdot \rangle$ invariant under spatial translations, then $\langle \cdot \rangle$ is a vacuum.*

**Remarks:**

1. Spatial homogeneity of $\langle \cdot \rangle$ [$\langle \alpha_x(\cdot) \rangle = \langle \cdot \rangle$ for $\forall x \in \mathbf{R}^3$] implies

$$\langle F \rangle = \langle F(\boldsymbol{x}) \rangle = \frac{1}{|V|} \int_V d^3\boldsymbol{y}\, \langle F(\boldsymbol{y}) \rangle \tag{7}$$

   for $\forall$bounded spatial domain $V \subseteq \mathbf{R}^3$ ($|V|$: volume).

2. Omitting for simplicity the symbol $\pi$ for a representation of the field algebra, we express the *GNS representation*[5, 6] of $\langle \cdot \rangle$ as $\langle F \rangle = \langle 1| F |1\rangle$ & $\mathcal{H} \equiv \overline{\mathcal{F}|1\rangle}$.

3. What does "unbroken SUSY" mean? : See (b) below.

4. What does "vacuum" mean? : See (c) & (d) below.

The proof goes in the following steps:

(a) Verification of Bose-Fermi superselection rule,

(b) Definition of "unbroken" SUSY $\Rightarrow$ equivalence of "unbroken" in the weakest sense (= implementability) and the strongest one (=invariance),

(c) Derivation of energy positivity from "unbroken" SUSY ($\rightarrow$ vacuum),

(d) Violation of energy positivity in thermal equilibrium (=KMS state).

(a) **Bose-Fermi superselection rule** is *unbroken in any spatially homogeneous state* $\langle \cdot \rangle$, i.e., $\langle \gamma(\cdot) \rangle = \langle \cdot \rangle$, or equivalently

$$\langle F_- \rangle = 0 \quad \text{for } F_- \in \mathcal{F}_-.$$

Proof) For $\forall$fermionic $F_- \in \mathcal{F}_-$, spatial homogeneity, Eq.(7), of $\langle \cdot \rangle$ implies

$$\langle F_- \rangle = \langle F_-(\boldsymbol{x}) \rangle = \frac{1}{|V|} \int_V d^3\boldsymbol{x} \, \langle 1| F_-(\boldsymbol{x}) |1\rangle. \tag{8}$$

Then, from local anticommutativity of fermionic fields, we obtain

$$\| \frac{1}{|V|} \int_V d^3\boldsymbol{x} F_-(\boldsymbol{x}) |1\rangle \|^2 + \| \frac{1}{|V|} \int_V d^3\boldsymbol{x} F_-(\boldsymbol{x})^\dagger |1\rangle \|^2$$
$$= \frac{1}{|V|} \int_V d^3\boldsymbol{x} \frac{1}{|V|} \int_V d^3\boldsymbol{y} \, \langle \, \{ F_-(\boldsymbol{x})^\dagger, F_-(\boldsymbol{y}) \} \, \rangle$$
$$\leq \frac{1}{|V|} \int d^3\boldsymbol{z} \, |\langle \, \{ F_-(\boldsymbol{z})^\dagger, F_- \} \, \rangle| \xrightarrow[V \nearrow \mathbb{R}^3]{} 0.$$

(Note that the last integral exists because the anti-commutator $= 0$ for large spatial translations $\boldsymbol{z}$.) Thus, we obtain

$$\lim_{V \nearrow \mathbb{R}^3} \| \frac{1}{|V|} \int_V d^3\boldsymbol{x} F_-(\boldsymbol{x}) |1\rangle \| = 0, \tag{9}$$

and hence, $\langle F_- \rangle = 0$ follows from Eqs.(8), (9).

(b) **Definition of implementability** and the invariance following from it:
*Implementability* of SUSY in a state $\langle \cdot \rangle$ is defined to be the existence of such operators $Q_\alpha, Q_{\dot{\beta}}^\dagger$ (=hermitian conjugate of $Q_\beta$) in $\mathcal{H}$ that they contain $\mathcal{F} |1\rangle$ in their domains and satisfy the relations

$$Q_\alpha F_\pm |1\rangle = \delta_\alpha(F_\pm) |1\rangle \pm F_\pm Q_\alpha |1\rangle, \tag{10}$$
$$Q_{\dot{\beta}}^\dagger F_\pm |1\rangle = \bar{\delta}_{\dot{\beta}}(F_\pm) |1\rangle \pm F_\pm Q_{\dot{\beta}}^\dagger |1\rangle. \tag{11}$$

**Remark:** Here the invariance of $|1\rangle$, $Q_\alpha |1\rangle = Q_{\dot\beta}^\dagger |1\rangle = 0$ is *not* assumed. Also because of ambiguities in the definition of generators due to contributions from commutants in mixed states [see 2)b) in Sec.1], neither $Q_\alpha$ nor $Q_{\dot\beta}^\dagger$ is guaranteed to be translation invariant simply by the commutativity $\delta_\alpha \circ \alpha_x = \alpha_x \circ \delta_\alpha$, $\bar\delta_{\dot\beta} \circ \alpha_x = \alpha_x \circ \bar\delta_{\dot\beta}$ of SUSY transformations with spacetime translations in the algebraic sense. We do not need such property either.

*If SUSY is implementable in $\langle \, \cdot \, \rangle$ in the above sense, it is unbroken in the usual sense,* i.e.,

$$\langle \delta_\alpha(\cdot) \rangle = \langle \bar\delta_{\dot\beta}(\cdot) \rangle = 0.$$

Proof) Since $\delta_\alpha(F_+), \bar\delta_{\dot\beta}(F_+) \in \mathcal{F}_-$ [Eq.(4)] for $F_+ \in \mathcal{F}_+$, Bose-Fermi superselection rule $\langle F_- \rangle = 0$ tells us $\langle \delta_\alpha(F_+) \rangle = \langle \bar\delta_{\dot\beta}(F_+) \rangle = 0$. Thus, it suffices to show for $\forall$fermionic $F_- \in \mathcal{F}_-$

$$\langle \delta_\alpha(F_-) \rangle = \langle \bar\delta_{\dot\beta}(F_-) \rangle = 0.$$

Now, Definition Eqs.(10), (11) of implementability and commutativity between anti-derivation $\delta_\alpha$ and spatial translations $\alpha_x$ imply

$$\langle \delta_\alpha(F_-) \rangle = \frac{1}{|V|} \int_V d^3x \, \langle 1| \delta_\alpha(F_-(x)) |1\rangle$$
$$= \frac{1}{|V|} \int_V d^3x \, \langle 1| (Q_\alpha F_-(x) + F_-(x)Q_\alpha) |1\rangle,$$

similarly to the above (a). By the Cauchy-Schwarz inequality, we obtain

$$|\langle \delta_\alpha(F_-) \rangle| \leq \|Q_\alpha^\dagger |1\rangle\| \cdot \|\frac{1}{|V|} \int_V d^3x \, F_-(x) |1\rangle\|$$
$$+ \|\frac{1}{|V|} \int_V d^3x \, F_-(x)^\dagger |1\rangle\| \cdot \|Q_\alpha |1\rangle\|.$$

In the limit of $V \nearrow \mathbf{R}^3$, $\langle \delta_\alpha(F_-) \rangle = 0$ for $F_- \in \mathcal{F}_-$ follows from Eq.(9). By the same token, we obtain $\langle \bar\delta_{\dot\beta}(F_-) \rangle = 0$, $F_- \in \mathcal{F}_-$.

**Remark :** In the above discussion, we need only the existence of supercharges satisfying Eqs.(10), (11), while cluster property being totally irrelevant. Therefore, the above conclusion holds, irrespective of whether $\langle \, \cdot \, \rangle$ is a *pure or mixed phase*.

### (c) A SUSY-invariant state is a vacuum only

If $\langle \, \cdot \, \rangle$ is supersymmetric, $\langle \delta_\alpha(\cdot) \rangle = \langle \bar\delta_{\dot\beta}(\cdot) \rangle = 0$, and if Bose-Fermi superselection rule holds, $\langle F_- \rangle = 0$ , then it is a vacuum, invariant under spacetime translations and satisfying the relativistic spectral condition (i.e., positivity of energy in all Lorentz frames).

Proof) Fundamental relation of SUSY and $\langle \delta_\alpha(\cdot) \rangle = \langle \bar\delta_{\dot\beta}(\cdot) \rangle = 0$ imply the invariance of $\langle \, \cdot \, \rangle$ under (space-)time translations: $\langle \delta_0(\cdot) \rangle = 0$. Since $\delta_0$ is a derivation satisfying Leibniz rule and $\delta_0(F)^\dagger = -\delta_0(F^\dagger)$ $(\forall F \in \mathcal{F})$, the relation, $P_0 F |1\rangle = \delta_0(F) |1\rangle$ $(\forall F \in \mathcal{F})$, defines a hermitian* operator $P_0$ satisfying $P_0 |1\rangle = \delta_0(1)|1\rangle = 0$ and $[P_0, F] = \delta_0(F) = -idF(x_0)/dx_0|_{x_0=0}$. This is because $\delta_0(F)F_1|1\rangle = \Big(\delta_0(FF_1) - F\delta_0(F_1)\Big)|1\rangle = P_0 F F_1|1\rangle - F P_0 F_1|1\rangle = [P_0, F]F_1|1\rangle$ and because $F_1|1\rangle$ constitutes a dense subset in $\mathcal{H}$.

Now, if $\langle F^\dagger \delta_0(F) \rangle \geq 0$ $(\forall F \in \mathcal{F})$ holds, we can conclude, from $\langle 1| F^\dagger P_0 F |1\rangle = \langle F^\dagger \delta_0(F) \rangle \geq 0$, that $P_0$ *is a positive energy operator satisfying* $P_0|1\rangle = 0$ *and* $|1\rangle$ *is a ground state for* $P_0$. So, we show that $\langle F^\dagger \delta_0(F) \rangle \geq 0$ $(\forall F \in \mathcal{F})$ follows from $[\langle \delta_\alpha(\cdot) \rangle = \langle \bar{\delta}_{\dot\beta}(\cdot) \rangle = 0]$ and the fundamental relation. Since $\forall F \in \mathcal{F}$ can be decomposed as $F = F_+ + F_-$ into the sum of Bose/Fermi operators $F_\pm \in \mathcal{F}_\pm$, and since $\langle F_\pm^\dagger \delta_0(F_\mp) \rangle = 0$ owing to Bose-Fermi superselection rule, it suffices to examine $\langle F_\pm^\dagger \delta_0(F_\pm) \rangle$. In the equation due to the fundamental relation,

$$4\langle F_+^\dagger \, \delta_0(F_+) \rangle = \langle F_+^\dagger \, (\bar{\delta}_{\dot 1} \circ \delta_1 + \delta_1 \circ \bar{\delta}_{\dot 1} + \bar{\delta}_{\dot 2} \circ \delta_2 + \delta_2 \circ \bar{\delta}_{\dot 2})(F_+) \rangle,$$

we rewrite the first term on RHS by using the graded Leibniz rule for anti-derivations $\delta_\alpha, \bar{\delta}_{\dot\beta}$ which satisfy the hermitian conjugation property Eq.(5),

$$\begin{aligned}\bar{\delta}_{\dot 1}(F_+^\dagger \, \delta_1(F_+)) &= \bar{\delta}_{\dot 1}(F_+^\dagger)\,\delta_1(F_+) + F_+^\dagger \, \bar{\delta}_{\dot 1}(\delta_1(F_+)) \\ &= -\delta_1(F_+)^\dagger \, \delta_1(F_+) + F_+^\dagger \, \bar{\delta}_{\dot 1} \circ \delta_1(F_+).\end{aligned}$$

From $\langle \bar{\delta}_{\dot 1}(\cdot) \rangle = 0$, we obtain

$$\langle F_+^\dagger \, \bar{\delta}_{\dot 1} \circ \delta_1(F_+) \rangle = \langle \delta_1(F_+)^\dagger \, \delta_1(F_+) \rangle \geq 0.$$

A similar argument applies to remaining terms, and hence, $\langle F_+^\dagger \, \delta_0(F_+) \rangle \geq 0$. The same result holds if $F_+$ is replaced by $F_- \in \mathcal{F}_-$. Thus, we arrive at

$$\langle F^\dagger \, \delta_0(F) \rangle \geq 0 \qquad \text{for } F \in \mathcal{F}.$$

### $\langle \, \cdot \, \rangle$ is a ground state in any Lorentz frame:
(From the spinorial transformation property of supercurrents $j_{\mu\alpha}$) Lorentz transform of $\delta_\alpha, \bar{\delta}_{\dot\beta}$ under $A \in SL(2, \mathbb{C})$ is given by

$$\delta_\alpha' = A_\alpha{}^\beta \delta_\beta, \quad \bar{\delta}_{\dot\alpha}' = \overline{A_\alpha{}^\beta}\,\bar{\delta}_{\dot\beta} = \bar{A}_{\dot\alpha}{}^{\dot\beta}\bar{\delta}_{\dot\beta}.$$

I.e., a change of Lorentz frame induces a linear transformation of these maps. Thus, if $\langle \delta_\alpha(\cdot) \rangle = \langle \bar{\delta}_{\dot\beta}(\cdot) \rangle = 0$ in one Lorentz frame, they hold in any frame.

Combining this observation with the fundamental relation for Lorentz-transformed $\delta_\alpha', \bar{\delta}_{\dot\beta}', \delta_0'$, one obtains the same result for $P_0'$ and $|1\rangle$. So, $\langle \, \cdot \, \rangle$ is a vacuum.

---

*Making use of temperedness of the underlying fields, one can show that $P_0$ is essentially self-adjoint on its domain of definition.

(d) **Energy spectrum cannot be positive definite in a thermal equilibrium state:**

In spite of possible ambiguities in the definition of generators, one can show that our $\langle \cdot \rangle$ does not describe a thermal equilibrium at a temperature $\beta^{-1} > 0$. To see this, we show that the condition $P_0 \geq 0$ is incompatible with KMS condition characterizing a thermal equilibrium state[7, 5].

For $F(g) = \int dx_0\, g(x_0) F(x_0)$, we have $F(g)|1\rangle = (2\pi)^{1/2} \tilde{g}(P_0) F|1\rangle$ ($\tilde{g}$: Fourier transform of $g$). If supp $\tilde{g} \cap \mathbf{R}_+ = \emptyset$, $F(g)|1\rangle = 0$ follows from $P_0 \geq 0$. If $\langle \cdot \rangle$ satisfies KMS condition at $\beta^{-1}$, then the function $x_0 \mapsto \langle\, F_1 (F_2^\dagger F(g))(x_0)\, \rangle$ $(F_1, F_2 \in \mathcal{F})$ has an analytic continuation to a complex domain $\{z \in \mathbf{C} : 0 < \mathrm{Im}\, z < \beta\}$, whose boundary value at $\mathrm{Im}\, z = \beta$ is given by $x_0 \mapsto \langle\, (F_2^\dagger F(g))(x_0) F_1\, \rangle$. This implies $\langle\, F_1 (F_2^\dagger F(g))(x_0)\, \rangle = \langle 1| F_1 F_2^\dagger(x_0) F(g)(x_0)|1\rangle = 0$ for $\forall x_0 \in \mathbf{R}$, and hence, $0 = \langle\, F_2^\dagger F(g) F_1\, \rangle = \langle 1| F_2^\dagger F(g) F_1 |1\rangle = ((\langle 1| F_2^\dagger) F(g) (F_1 |1\rangle))$. Therefore, $F(g) = 0$ follows from the arbitrariness of $F_1, F_2$. By the same token, $F^\dagger(g) = 0$ follows from the same argument for $F^\dagger(g)$, and hence, we obtain $F(\bar{g}) = F^\dagger(g)^\dagger = 0$.

Hence $F(g) = 0$ for any $g$ whose Fourier transform vanishes in some neighbourhood of the origin. This means that the function $x_0 \mapsto F(x_0)$ is a polynomial in $x_0$. By time translation invariance of $\langle \cdot \rangle$, $\langle\, F(x_0)^\dagger F(x_0)\, \rangle = \langle\, F^\dagger F\, \rangle$, it is constant. Namely, for $F \in \mathcal{F}$ and $x_0 \in \mathbf{R}$, $F(x_0) = F$, which leads us to a trivial dynamics Therefore, $\langle \cdot \rangle$ cannot satisfy KMS condition (except for such a trivial case).

In summary, any spatially homogeneous $\langle \cdot \rangle$ satisfies

1. Bose-Fermi superselection rule: unbroken

2. If generators (= supercharges) of SUSY transformations exist in $\mathcal{H}$, then this state is SUSY invariant : $\langle \delta_\alpha(\cdot) \rangle = \langle \bar{\delta}_{\dot\beta}(\cdot) \rangle = 0$.

3. A state with properties 1.& 2. is only a vacuum at $T = 0^\circ K$.

4. SUSY is always broken in any thermal equilibrium states irrespective of whether they describe pure or mixed phases, and supercharges do not exist there. Thermal effects induce an inevitable spontaneous collapse of SUSY.

## 3  Role of Supertrace

In the usual discussion, the supertrace and a superthermal ensemble are defined, respectively, by $\mathrm{STr} = \mathrm{Tr}(e^{i\pi Q_F}\cdot)$ $s(\,\cdot\,) = \mathrm{STr}(e^{-\beta H}\cdot)$, which may be endangered in infinite systems as noted in the Introduction. Following van Hove[8], we interpret here a superthermal ensemble as a linear functional $s(\,\cdot\,)$ on $\mathcal{F}$ given by a weighted difference $s(\,\cdot\,) = p_b \langle\, \cdot\, \rangle_b - p_f \langle\, \cdot\, \rangle_f$ of states $\langle\, \cdot\, \rangle_b$, $\langle\, \cdot\, \rangle_f$ corresponding to bosonic and fermionic subensembles, and examine its thermodynamic limit. ($p_b$, $p_f$: non-negative weights normalized by $p_b + p_f = 1$)

Someone argues[3] that the behaviour of $s(\,\cdot\,)$ under $\delta_\alpha$, $\bar{\delta}_{\dot\beta}$ provides the appropriate test for spontaneous breakdown of SUSY: If $s(\delta_\alpha(\cdot)) = s(\bar{\delta}_{\dot\beta}(\cdot)) = 0$: SUSY is said to be unbroken, otherwise it is regarded as spontaneously broken. (e.g., van Hove, Fuchs, etc.) Is this correct? Our answer to it is "NO" as is seen in the following statement.

Proposition on supertrace states: *There are only following two alternatives in spatially homogeneous situations*:

i) $s(\,\cdot\,) = 0$ (i.e., trivial case),

*or* ii) $s(\delta_\alpha(\cdot)) \neq 0$ and $s(\bar{\delta}_{\dot\beta}(\cdot)) \neq 0$.

Namely, only supersymmtric superthermal ensemble is 0!!

**Remark** Physical meaning of the above statement: If $s(\,\cdot\,) = 0$, $p_b = p_f$ and $\langle\,\cdot\,\rangle_b = \langle\,\cdot\,\rangle_f$, then "bosonic phase" and "fermionic phase" cannot be distinguished. [Since $\langle\,\cdot\,\rangle_b = \langle\,\cdot\,\rangle_f$ can be a mixture of multiple phases, one cannot conclude the uniqueness of phase from i).]

If $s(\,\cdot\,) \neq 0$, there are two possibilities : Either (i) there are at least two different phases [$\langle\,\cdot\,\rangle_b \neq \langle\,\cdot\,\rangle_f$], or (ii) $\langle\,\cdot\,\rangle_b = \langle\,\cdot\,\rangle_f$ and $p_b \neq p_f$.

It has been argued (e.g. van Hove[8]) that the latter case does not occur in situations of physical interest. However, since this argument is based on the existence of supercharges it is not applicable in the thermodynamic limit. To establish the existence of different phases, one has to carry out further tests on $s(\,\cdot\,)$.

Thus, supertrace may be used to obtain information about the phase structure of supersymmetric theories. Apart from the trivial case $s(\,\cdot\,) = 0$, however, there is no restoration of SUSY at finite temperature even in the sense of supertrace.

Proof) We prove the above statement on $s(\,\cdot\,)$ in two steps. First, we assume that both $\langle\,\cdot\,\rangle_b$ and $\langle\,\cdot\,\rangle_f$ are pure phases. Then, as was explained in Sec.1, they have cluster property, which we use here in a somewhat weaker form:

$$\frac{1}{|V|} \int_V d^3x \; \langle\, F_1(x) F_2 \,\rangle_{b,f} - \langle\, F_1 \,\rangle_{b,f} \langle\, F_2 \,\rangle_{b,f} \xrightarrow[V \nearrow \mathbb{R}^3]{} 0. \tag{12}$$

Assumption of the SUSY invariance $s(\delta_\alpha(\cdot)) = 0$ of $s(\,\cdot\,)$ combined with the graded Leibniz rule Eq.(2) implies for $\forall F_\pm \in \mathcal{F}_\pm$

$$0 = s(\delta_\alpha(F_-(x)\, F_+)) = s(\delta_\alpha(F_-(x))\, F_+ - F_-(x)\, \delta_\alpha(F_+)).$$

Substituting $s(\,\cdot\,) = p_b \,\langle\,\cdot\,\rangle_b - p_f \,\langle\,\cdot\,\rangle_f$, one obtains

$$p_b \,\langle \delta_\alpha(F_-(x))_b\, F_+ - F_-(x)\, \delta_\alpha(F_+)\rangle = p_f \,\langle \delta_\alpha(F_-(x))_f\, F_+ - F_-(x)\, \delta_\alpha(F_+)\rangle. \tag{13}$$

From the commutativity of $\delta_\alpha$ with spatial translations, the spatial mean of (13) and the cluster property (12), we obtain

$$p_b \,(\langle \delta_\alpha(F_-) \rangle_b \langle\, F_+ \,\rangle_b - \langle\, F_- \,\rangle_b \langle \delta_\alpha(F_+) \rangle_b ) =$$
$$= p_f \,(\langle \delta_\alpha(F_-) \rangle_f \langle\, F_+ \,\rangle_f - \langle\, F_- \,\rangle_f \langle \delta_\alpha(F_+) \rangle_f ).$$

From Bose-Fermi superselection rule applied to $\langle \cdot \rangle_{b,f}$ we get $\langle F_- \rangle_{b,f} = 0$, and hence,

$$p_b \langle \delta_\alpha(F_-) \rangle_b \langle F_+ \rangle_b = p_f \langle \delta_\alpha(F_-) \rangle_f \langle F_+ \rangle_f. \tag{14}$$

Now, the normalization condition $\langle 1 \rangle_b = \langle 1 \rangle_f = 1$ implies $p_b \langle \delta_\alpha(F_- \rangle_b) = p_f \langle \delta_\alpha(F_-$
Since $\langle \cdot \rangle_{b,f}$ are thermal states, $\exists F_- \in \mathcal{F}_-$ s.t. $\langle \delta_\alpha(F_-) \rangle_{b,f} \neq 0$. Thus, Eq.(14) implies $\langle \cdot \rangle_b = \langle \cdot \rangle_f$ and $p_b = p_f$, from which $s(\cdot) = 0$.

We consider next the general cases with $\langle \cdot \rangle_{b,f}$ thermodynamic mixed phase. Then, $\langle \cdot \rangle \equiv p_b \langle \cdot \rangle_b + p_f \langle \cdot \rangle_f$ can be decomposed into pure phases $\langle \cdot \rangle_\theta$ by a central decomposition:

$$p_b \langle \cdot \rangle_b = \sum_\theta p_b(\theta) \langle \cdot \rangle_\theta, \tag{15}$$

$$p_f \langle \cdot \rangle_f = \sum_\theta p_f(\theta) \langle \cdot \rangle_\theta, \tag{16}$$

where $p_b(\theta)$, $p_f(\theta)$ are non-negative weights normalized by $\sum_\theta p_b(\theta) = p_b$, $\sum_\theta p_f(\theta) = p_f$.

Similarly to the case of pure phases, we proceed from the assumption of $s(\delta_\alpha(\cdot)) = 0$ to relation (13), where we insert the decompositions (15), (16). Taking a spatial mean of the resulting expression and proceeding to the limit $V \nearrow \mathbf{R}^3$, we obtain, by applying cluster property to each component pure phase $\langle \cdot \rangle_\theta$, the relation

$$\sum_\theta p_b(\theta) \langle \delta_\alpha(F_-) \rangle_\theta \langle F_+ \rangle_\theta = \sum_\theta p_f(\theta) \langle \delta_\alpha(F_-) \rangle_\theta \langle F_+ \rangle_\theta. \tag{17}$$

Replacing in (17) the operator $F_+$ by $(1/|V|) \int_V d^3x \, (\delta_\alpha(F_-)^\dagger)(x) F_+$ and making use of cluster properties of pure phases, we obtain in the limit $V \nearrow \mathbf{R}^3$ the relation

$$\sum_\theta p_b(\theta) |\langle \delta_\alpha(F_-) \rangle_\theta|^2 \langle F_+ \rangle_\theta = \sum_\theta p_f(\theta) |\langle \delta_\alpha(F_-) \rangle_\theta|^2 \langle F_+ \rangle_\theta. \tag{18}$$

Repeating this procedure, one arrives at similar relations involving higher products of expectation values of arbitrary operators $F_+$ in pure phases $\langle \cdot \rangle_\theta$. Then, keeping $F_-$ fixed and varying $F_+$, one sees that the two functionals $\sum_\theta p_{b,f}(\theta) |\langle \delta_\alpha(F_-) \rangle_\theta|^2 \langle \cdot \rangle_\theta$ coincide. Because of the uniqueness of central decomposition this implies for any $\theta$ the equality

$$p_b(\theta) |\langle \delta_\alpha(F_-) \rangle_\theta|^2 = p_f(\theta) |\langle \delta_\alpha(F_-) \rangle_\theta|^2. \tag{19}$$

Since $\langle \cdot \rangle_\theta$ is a thermal state, however, there is some $F_- \in \mathcal{F}_-$ such that $\langle \delta_\alpha(F_-) \rangle_\theta \neq 0$ and hence we obtain $p_b(\theta) = p_f(\theta)$. It then follows from relations (15), (16) that $s(\cdot) = 0$, so only the trivial supertrace is supersymmetric.

Thus we conclude that the supertrace is a device to deduce some (partial) information about phase structure in supersymmetric theories. Yet its behaviour under SUSY transformations does not provide any additional information, in accord with the result of the previous section that SUSY always suffers from a spontaneous collapse in thermal states.

# 4   Conclusions

Our conclusion obtained above is: In every QFT with the fundamental relation of SUSY valid, SUSY suffers inevitably from spontaneous collapse in any spatially homogeneous states other than vacua.

- The above results are verified for spatially homogeneous states in d=4 dimensions. However, extension is easy to arbitrary dimensions and to more complex situations (e.g., asymptotically homogeneous states, spatially periodic states, etc.)

- Moreover, point-like nature and strict local (anti-)commutativity of the underlying fields are not really crucial. Our arguments require only a sufficiently rapid fall-off of the expectation values of (anti-) commutators of underlying field operators for large spatial translations.

- Therefore, an analogous result may be expected to hold in quantum superstring field theory, provided a pertinent formulation of SUSY can be given in that setting.

- Another example of universal breakdown of symmetries due to thermal effects: *Lorentz symmetry*, broken in thermal equilibrium because KMS condition fixes a rest frame[9]. Although there is no Lorentz-invariant *state* related to KMS states [∵ non-amenability of Lorentz group], one can construct a *representation* containing a KMS state, in which a unitary representation of Lorentz group is defined [see[9]]. So, this represents an *intermediate class between* ii)[usual SSB] *and* iii)[collapse] in Sec.1. This situation of spontaneous breakdown of Lorentz symmetry should be clearly distinguished from the spontaneous collapse of SUSY in thermal states, where it is no longer possible to define an action of the symmetry on physical states.

In view of this **vulnerability to thermal effects**, how can SUSY manifest itself in real physical systems ? Predictions on a zero-energy mode in thermal states is of limited value, since this mode need *not* be affiliated with a Goldstino *particle*, but may result from long range correlations between particle-hole pairs. Also, rigorous results on the fate of particle supermultiplets in a thermal environment is still lacking.

For a reliable prediction of the existence of SUSY in physical systems, it seems necessary to show that symmetry properties of the vacuum theory can be recovered from thermal states in the limit of zero temperature.

On the other hand, the possibility that SUSY remains collapsed in this limit, in analogy to some hysteresis effect, may be even more interesting since it could account for the absence of unbroken SUSY in the real world.

Therefore, it would be desirable to clarify which of these two possibilities is at hand in models of physical interest(, as long as one believes in SUSY).

# References

[1] L. Girardello, M. Grisaru and P. Salomonson, *Nucl. Phys.* **B178** (1981) 331; D. Boyanovsky, *Phys. Rev.* **D29** (1984) 743.

[2] J. Fuchs, *Nucl. Phys.* **B246** (1984) 279; Won-Ho Kye, Sin Kyu Kang and Jae Kwan Kan, *Phys. Rev.* **D46** (1992) 1835.

[3] L. van Hove, *Nucl. Phys.* **B207** (1982) 15; T.E. Clark and S.T. Love, *Nucl. Phys.* **B217** (1983) 349.

[4] D. Buchholz and I. Ojima, *Nucl. Phys.* **B498**, Nos.1,2, 228-242 (1997).

[5] O. Bratteli and D.W. Robinson, *Operator Algebras and Quantum Statistical Mechanics*, Vols. 1 and 2 (Springer, Berlin, 1979 and 1981).

[6] R.F. Streater and A.S. Wightman, *PCT, Spin and Statistics and All That* (Benjamin, New York, 1964); R. Haag, *Local Quantum Physics* (Springer, Berlin, 1992[1st ed.], 1996[2nd ed.]).

[7] R. Haag, N.M. Hugenholtz and M. Winnink, *Comm. Math. Phys.* **5** (1967) 215; I. Ojima, *Ann. Phys.* **137** (1981) 1.

[8] L. van Hove, *Fizika* **17** (1985) 267.

[9] I. Ojima, *Lecture Notes in Physics* **176** (K. Kikkawa, et al. eds.) (Springer, Berlin, 1983) pp.161-165; *Lett. Math. Phys.* **11** (1986) 73.

# RELATING RARITA-SCHWINGER AND BARGMANN-WIGNER SPIN-3/2 FIELDS BY PROJECTION TECHNIQUE

## LORENZO C. CHAN

National Institute of Physics, University of the Philippines
Diliman, Quezon City 1101, Philippines

Abstract    Projection operators relating trispinor representations to 5-tensor spinor representations of SO(5) representing spin-3/2 fields are explicitly constructed. Using these projection operators, we are able to analyze the structure of both totally symmetric and mixed symmetric trispinor fields, identifying 4-tensor spinors as the "base" sector of the totally symmetric Bargmann-Wigner field, and the Rarita-Schwinger 4-vector spinors as the "base" sector of mixed symmetric trispinors. The projectors allow us to derived connections relations that augment each base field with auxiliary components to completely specify the 5-tensor spinor representing each type of trispinor. These concepts were applied to obtain field equations and subsidiary equations by working only in the base sectors.

Key Words    Rarita-Schwinger, Bargmann-Wigner, Spin-3/2, Field Equation, Projection, High spin.

## 1. INTRODUCTION

The importance of a systematic theory of high spin fields has been stressed as a result of super-symmetric gauge theories[1,2]. The spin-3/2 field, in particular, plays a

287

key role in supergravity. The simplest traditional descriptions of this field were given by the Bargmann-Wigner (B-W) equations[3],

$$(\gamma \cdot \partial)_{kk'} \psi_{ijk'} = 0 , \qquad (1)$$

based on totally symmetric trispinor fields; and by the Rarita-Schwinger (R-S) equations[4],

$$(\gamma \cdot \partial)_{kk'} \psi^{\alpha}_{k'} = 0 , \qquad (2)$$

based on 4-vector spinor fields satisfying a null matrix constraint,

$$\gamma^{\mu}_{kk'} \psi^{\mu}_{k}(x) = 0 . \qquad (3)$$

The B-W approach has the merit of being systematically generalizable to any spin, working with well-defined irreducible representations of the de Sitter group or SO(5) that clearly identifies with the spin value. However, the degree of freedom of the B-W field is larger than that of the R-S field, and grows very rapidly with increasing spin values. On the other hand, the R-S fields, being reducible vector-spinor representations of SO(4), bear simpler space-time structures. It is therefore wise to consider the possibility of combining the group theoretical simplicity of B-W fields with the structural simplicity of R-S fields. It is of course a well-known fact that the R-S field is contained in the B-W field[5]. However, the difference in the degrees of freedom also suggests that the two fields are not equivalent. There is a need to study the precise relation between these two fields. For this purpose, we shall study both fields in the same footing, namely, considering the 4-vector spinor R-S field as a subfield of a 5-tensor spinor. The B-W symmetric trispinor is also a subfield of this 5-tensor spinor field. We shall show that it is possible to construct a projection operator that projects a 5-tensor spinor field onto the totally-symmetric B-W trispinor field, and another projection operator that projects onto the mixed symmetric trispinor field. The use of these projection operators allows us to discover connection relations between the 4-vector and 4-tensor spinor components of both the symmetric and the mixed symmetric trispinor field, which in turn made it possible to identify the 4-vector spinor as a "base field" for the B-W field, and the 4-tensor spinor as a "base field" for the mixed symmetric trispinor field, thus gaining a detailed understanding of the difference between the R-S and B-W fields. We also illustrate how these connection relations derived from knowledge of the projectors allow us to obtain consistent field equations in the base components of the totally symmetric and mixed symmetric trispinor.

The organization of this paper is as follows: In section 2, we define the trispinor subfields of the 5-tensor spinor field and exhibit corresponding projection operators. In section 3, we derive properties of these projectors. In section 4, we study the tensorial structures of these subfields, discovering "base fields" for each of the trispinor types and connection relations that define "auxiliary" components to these base fields. In section 5, we illustrate the use of these concepts by deriving field equations based on each type of trispinor, by working on their corresponding "base" sectors. Finally, in section 6, we give a discussion of our results.

## 2. SUBFIELDS AND PROJECTION OPERATORS

The Euclidean basis of the Lie algebra $B_2$ of the de Sitter group $SO(5)$ is defined by the commutation rules

$$[\Sigma^{ab}, \Sigma^{cd}] = 2i \, [\delta^{ac}\Sigma^{bd} + \delta^{bd}\Sigma^{ac} - \delta^{ad}\Sigma^{bc} - \delta^{bc}\Sigma^{ad}]. \tag{4}$$

Here, we use early Latin letters a,b,c,..... as 5-dimensional indices. Middle Latin letters shall be reserved for spinor indices while the usual Greek letters for 4-dimensional indices. The fundamental representation of $B_2$ is the Dirac spinor. Bargmann-Wigner fields are totally symmetrized direct products of Dirac spinors that constitute irreducible representations of $B_2$. The B-W bispinor describing spin-1 fields can be exhibited as anti-symmetric rank-2 5-tensors $\psi^{ab}$. For spin-3/2, the B-W trispinor can be exhibited in terms of a rank-2 anti-symmetric 5-tensor spinor $\psi^{ab}{}_k$, *i.e.*, it is a subfield of the direct product of a Dirac spinor with a rank-2 antisymmetric 5-tensor. This rank-2 5-tensor spinor is the tensor spinor of minimal rank containing a B-W trispinor as a subfield. It also contains an $SO(5)$ decomposable mixed-symmetric trispinor subfield. In terms of $SO(4)$ decomposition, the 5-tensor spinor contains a 4-tensor spinor and a 4-vector spinor, the R-S field. It is therefore clear that the relationship between B-W and R-S fields may be studied by working with the rank-2 anti-symmetric tensor spinors of $SO(5)$. The manner in which the B-W field is contained in the tensor spinor field has been well studied[5]. We shall give a brief summary here.

The symmetric trispinor can be written as an expansion over tensor spinors as

$$\psi_{ijk}(x) = 1/2 \, (\Sigma^{ab}C)_{ij}\psi^{ab}{}_k(x) \tag{5}$$

where $\Sigma^{ab}$ is the Dirac spinor representation of $SO(5)$, with

$$\Sigma^{5\mu} = \gamma^\mu \tag{6}$$

and

$$\Sigma^{\mu\nu} = -i\gamma^{\mu}\gamma^{\nu}. \tag{7}$$

The matrix C is the charge conjugation matrix bearing the useful property that $\Sigma^{ab}C$ are symmetric matrices. The inverse relation expressing the components of $\psi^{ab}{}_k$ in terms of trispinor components can be obtained by taking spinor traces:

$$(1/4)\sum_{i,j}[(C^{\dagger}\Sigma^{ab})_{ij}\psi_{ijk}] = (1/4)\,Tr[(1/2)C^{\dagger}\Sigma^{ab}\Sigma^{cd}C]\psi^{cd}{}_k = \psi^{ab}{}_k \tag{8}$$

The B-W subfield as defined in tensor form above satisfies the matrix conditions

$$\gamma^{\mu}\psi^{5\mu}(x) = 0 \quad , \tag{9}$$

$$\Sigma^{\mu\nu}\psi^{\mu\nu}(x) = 0 \tag{10}$$

and

$$\psi^{5\nu} + i\,\gamma^{\mu}\psi^{\mu\nu}(x) = 0 \quad . \tag{11}$$

These conditions guarantee the additional symmetry of the third index and reduce the original 40 degrees of freedom down to the 20 degrees of freedom of the B-W field. The y define uniquely the B-W subfield of the tensor-spinor field.

It is easily verified that the operator

$$P_S = (1/3)I + (1/12)\Sigma^{ab}\Sigma^{pq} \tag{12}$$

acting on the tensor-spinor field,

$$(\psi_S)^{pq}{}_k = (P_S\psi)^{pq}{}_k = [\,(1/3)I + (1/12)\,(\Sigma^{ab}\Sigma^{pq})_{kk'}\,]\,\psi^{ab}{}_{k'}$$

$$= (1/3)\psi^{pq}{}_k + (1/12)\,(\Sigma^{ab}\Sigma^{pq})_{kk'}\,\psi^{ab}{}_{k'} \tag{13}$$

projects it into the B-W subfield S, *i.e.*, equations (9), (10), and (11) are satisfied by $\psi_S$ as defined in equation (13). Unless otherwise stated, all $\Sigma$ matrices shall henceforth be in the spinor representation.

The maximal subfield N of the tensor spinor field satisfying the matrix null conditions, equations (9) and (10), can likewise be defined by a projection operator:

$$
P_N = \begin{cases} I_T - (1/12)\Sigma^{\sigma\tau}\Sigma^{\alpha\beta} \\[2mm] I_V - (1/4)\Sigma^{5\sigma}\Sigma^{5\alpha} \end{cases} \tag{14}
$$

where the upper line acts on the 4-tensor part and the lower line acts on the 4-vector part of the 5-tensor spinor space. Again, direct contractions with $\Sigma^{\sigma\tau}$ and $\Sigma^{5\tau}$ will prove the null conditions. We can easily verify that indeed S is a subfield of N by proving that

$$
P_N P_S = P_S = P_S P_N . \tag{15}
$$

The complement of $P_S$ inside $P_N$, $P_M = P_N \cdot P_S$, therefore projects into the part of the mixed symmetric trispinor that satisfies both matrix conditions, equations (9) and (10). It is easy to verify that this M-subfield is orthogonal to the S subfield, $i.e.$,

$$
P_M P_S = (P_N - P_S) = P_S - P_S = 0 . \tag{16}
$$

Out of the 20 degrees of freedom of the mixed symmetric trispinor, removal of 8 degrees of freedom represented by the two matrix conditions leaves 12 degrees of freedom for this M-subfield. Since the matrix conditions are not SO(5) invariant, the result is that the M subfield is only a representation of SO(4), and not of SO(5). This is not a physical drawback, as we really need representations of the Poincare group.

## 3. PROPERTIES OF PROJECTORS

The projectors defined in the last section are useful in examining the detailed SO(4) structure of each subfield, $i.e.$, we can obtain the linear relationships of their 4-vector spinor $\psi^{5\mu}{}_k$ and the 4-tensor spinor $\psi^{\mu\nu}{}_k$ components. We shall henceforth suppress the spinor index unless necessary. The SO(4) structure can be elucidated by first defining the restrictions of the projectors $P_M$ and $P_S$ to the 4-vector spinor and 4-tensor spinor sectors:

$$
P_N{}^T = P^T P_N P^T = I^T - (1/12)\Sigma^{\sigma\tau}\Sigma^{\alpha\beta} , \tag{17}
$$

$$
P_S{}^T = P^T P_S P^T = (1/3)I^T + (1/12)\Sigma^{\alpha\beta}\Sigma^{\sigma\tau} , \tag{18}
$$

$$P_M{}^T = P^T P_M P^T = (1/3)I^T - (1/12)\Sigma^{\alpha\beta}\Sigma^{\sigma\tau} - (1/12)\Sigma^{\sigma\tau}\Sigma^{\alpha\beta}, \tag{19}$$

$$P_N{}^V = P^V P_N P^V = I^V - (1/4)\Sigma^{5\sigma}\Sigma^{5\alpha}, \tag{20}$$

$$P_S{}^V = P^V P_S P^V = (1/3)I^V + (1/6)\Sigma^{5\alpha}\Sigma^{5\sigma}$$

$$= (2/3)I^V - (1/6)\Sigma^{5\sigma}\Sigma^{5\alpha} = (2/3)\ P_N{}^V, \tag{21}$$

and

$$P_M{}^V = P^V P_M P^V = (2/3)I^V - (1/6)\Sigma^{5\alpha}\Sigma^{5\sigma} - (1/4)\Sigma 5\sigma\Sigma^{5\alpha}$$

$$= (1/3)I^V - (1/12)\Sigma^{5\sigma}\Sigma^{5\alpha} = (1/3)\ P_N{}^V. \tag{22}$$

Here, $P^T$ and $P^V$ are projectors from the 5-tensor field to its 4-tensor and 4 vector sectors, respectively.

We can verify that contractions of $\Sigma^{\alpha\beta}$ with all three tensor sector restrictions vanish and likewise for the contractions of $\Sigma^{5\alpha}$ with the vector sector restrictions. In general, unless projectors are block diagonal in the vector and tensor sectors, their restrictions to these sectors will not be projectors. Thus, only restrictions of $P_N$ are projectors, while those of $P_S$ and $P_M$ are not. However, for the vector sector restrictions, $P_M{}^V$ and $P_S{}^V$ are proportional to the projectors $P_N{}^V$ obeys the properties

$$(P_S{}^V)^2 = (2/3)P_S{}^V \tag{23}$$

and

$$(P_M{}^V)^2 = (1/3)P_M{}^V \quad . \tag{24}$$

Correspondingly, the tensor sector restrictions have

$$(P_S{}^T)^2 = P_S{}^T - (1/3)P_M{}^T \tag{25}$$

and

$$(P_M{}^T)^2 = (2/3)P_M{}^T \quad . \tag{26}$$

showing that only the tensor sector of the M subfield is invariant to projection, while that of the S subfield "leaks out" into the M subfield upon projection. We can in fact show that $P_S{}^T$ and $P_M{}^T$ are not orthogonal,

$$P_M{}^T P_S{}^T = (1/3)P_M{}^T . \qquad (27)$$

These properties indicate that the M subfield is easier to work with.

## 4. TENSORIAL STRUCTURES OF SUBFIELDS

The restrictions of projection operators allow us to unravel structures of the corresponding subfields in the 4-vector and 4-tensor sectors, giving linear relations that must be satisfied by the 4-vector and 4-tensor components of each subfield. We start by exploring a relation in the N subfield induced by the null conditions. Using the identity

$$i\Sigma^{\sigma\mu} = \Sigma^{5\sigma}\Sigma^{5\mu} - \delta^{\sigma}{}_{\mu}I ,$$

we can write

$$\psi_N{}^{p\sigma} = [\Sigma^{5\sigma}\Sigma^{5\mu} - i\Sigma^{\sigma\mu}] \psi_N{}^{p\mu} \qquad (28)$$

The $p = \tau$ tensor component reads

$$\psi_N{}^{\sigma\tau} = i\Sigma^{\sigma\mu}\psi_N{}^{\tau\mu} - \Sigma^{5\sigma}\Sigma^{5\mu}\psi_N{}^{\tau\mu} , \qquad (29)$$

while the p=5 vector component is simplified by the null condition into

$$\psi_N{}^{5\sigma} = -i\Sigma^{\sigma\mu}\psi_N{}^{5\mu} = -i\Sigma^{\sigma d}\psi_N{}^{5d} . \qquad (30)$$

Equations (29) and (30) are general relations in the N subfield that are therefore satisfied by both $\psi_S$ and $\psi^M$.

We are now ready to study properties specific to the S subfield. First, we can use the null condition to rewrite equation (13) as

$$\psi_S{}^{p\sigma} = (1/8) [ \Sigma^{ab}, \Sigma^{p\sigma} ] \psi_S{}^{ab} = (i/2) [\Sigma^{pb}\psi_S{}^{\sigma b} - \Sigma^{\sigma b}\psi_S{}^{pb}] . \qquad (31)$$

For p=5, use of equation (30) simplifies it further to

$$\psi_S^{5\beta} = i\Sigma^{5d}\psi_S^{\beta d} = -i\Sigma^{\beta d}\psi_S^{5d} \quad . \tag{32}$$

Equation (32) establishes a connection relation from the tensor sector of the S subfield to the corresponding vector sector. We can use this relation to simplify the last term in equation (29), getting

$$\psi_S^{\sigma\tau} = i\Sigma^{\sigma\mu}\psi_S^{\tau\mu} + i\Sigma^{5\sigma}\psi_S^{5\tau} = i\Sigma^{\sigma d}\psi_S^{\tau d}.$$

When this result is compared with equation (31), antisymmetry in the tensor indices is confirmed:

$$\psi_S^{\sigma\tau} = i\Sigma^{\sigma d}\psi_S^{\tau d} = -i\Sigma^{\tau d}\psi^{\sigma d} \quad . \tag{33}$$

Equations (32) and (33) are basic linear relations between the 5-tensor components of the S subfield, they can be expressed together in 5-tensor form as

$$\psi_S^{at} = -(i/2)[\Sigma^{ad}\psi_S^{td} - \Sigma^{td}\psi_S^{ad}] = i\Sigma^{ad}\psi_S^{td} = -i\Sigma^{td}\psi_S^{ad} \quad . \tag{34}$$

The significance of equation (32) is that the vector components of the S subfield is fully defined in terms of the tensor components, in other words, the tensor sector can be taken as the "base field" that spans the S subfield or the B-W field. The null condition, equation (10) reduces the degree of freedom of this base sector from 24 down to 20, making it fully equivalent to the B-W field. To work with the B-W field, it is sufficient to work in the tensor sector, using equation (32) as a connection relation to the vector sector to fully describe its 5-tensor content. Working on this base sector, equation (33) loses significance.

The structure of the M subfield is much easier to describe, mainly owing to the simplicity of equation (26). Consider a 5-tensor field $\psi_M = 3P_M\psi_V$, where $\psi_V$ is any pure 4-vector spinor in N, $\psi_V = P_V\psi_N$. Examination of equation (24) reveals that the vector sector of $\psi_M$ is precisely $\psi_V$ itself, $\psi_M^{5\alpha} = \psi_V^{5\alpha}$, the role of the action of $3P_M$ is therefore interpreted as simply to append tensor components

$$\psi_M^{\alpha\beta} = -i[\Sigma^{5\alpha}\psi_M^{5\beta} - \Sigma^{5\beta}\psi_M^{5\alpha}] \tag{35}$$

to any arbitrary 4-vector tensor to complete its definition as a an M subfield of the 5-tensor spinor. In this fashion, it is transparent why we can consider the vector sector as the "base field" of M. It is thus simple to work on the vector sector of M, with equation (35) acting as the connection relations to the tensor sector. It is equally clear why the M subfield should be identified with the R-S field. We note further that the connection relations for M may be inverted to give

$$\psi_M{}^{5\tau} = -(i/2)\Sigma^{5\beta}\psi_M{}^{\tau\beta} \quad \tag{36}$$

In summary, we have shown in this section that the B-W field is equivalent to the S subfield of the 5-tensor spinor field, with the 4-tensor sector taken as the base field. The R-S field is equivalent to the M subfield with the 4-vector sector taken as a base field. We have also given connection relations that expand the base sectors into full 5-tensor spinors of each subfield. In view of the existence of base sectors for each subfield, we suggest that it should be possible to treat each subfield by working within their base sectors, possibly with the help of connection relations if necessary.

## 5. APPLICATIONS TO FIELD EQUATIONS

We shall now illustrate that it is possible to derive consistent field equations for spin-3/2 fields in the base sectors for both the S field and the M field. It is standard practice to impose the conditions of covariance and the Klein-Gordon equation on field components. These requirements are satisfied by multi-spinors if all indices obey Dirac type equation,

$$(\gamma\cdot\partial)\psi + \mu\psi = 0 \,, \tag{37}$$

where $\mu = \pm m$. The sign of $\mu$ may be different for each index. For the symmetric B-W field, we expect $\mu = m$ for all three indices. However, for the M field, we expect different signs. We shall work our the base components of these Dirac type equations for each field. As a standard, we shall adopt the convention that the sign of $\mu$ for the odd (unsymmetrized) spinor index is positive:

$$(\gamma\cdot\partial)_{kk'}\psi^{(ij)k} + m\psi^{(ij)k} = 0 \,,$$

which reads in tensor spinor form as

$$(\gamma\cdot\partial)\psi^{p\lambda} + m\psi^{p\lambda} = 0 \quad .$$

For the S field, we take the tensor sector

$$(\gamma \cdot \partial)\psi^{\alpha\beta} + m\psi^{\alpha\beta} = 0 , \tag{38}$$

and for the M field, we take the vector sector,

$$(\gamma \cdot \partial)\psi^{5\lambda} + m\psi^{5\lambda} = 0 , \tag{39}$$

which is the R-S equation.

We now consider equation (37) applies to the symmetrized index:

$$(\gamma \cdot \partial)_{ii'}\psi^{(ij)k} + \mu\psi^{(ij)k} = 0 .$$

Again we can convert this to 5-tensor spinor form by taking traces upon action with $C^{\dagger}\Sigma^{\rho\lambda}$. The 4-tensor sector,

$$i(\partial^{\sigma}\psi^{5\tau} - \partial^{\tau}\psi^{5\sigma}) + \mu\psi^{\sigma\tau} = 0 , \tag{40}$$

should apply to the S field, while the 4-tensor sector,

$$i\partial^{\tau}\psi^{\lambda\tau} + \mu\psi^{5\lambda} = 0 \quad , \tag{41}$$

should describe the M field. The sign of $\mu$ in equation (40) describing the S field is required by symmetry to be positive.

We now show that consistency between equations (39) and (41) for the M field requires the sign of $\mu$ to be negative, $\mu = -m$. For massive S field, equation (39) leads easily to the Lorentz condition, $\partial^{\lambda}\psi^{\lambda}=0$. Applying the Lorentz condition of the connection relation for M fields, equation (35), now leads to equation (41) with $\mu=-m$. It is now interesting to note that while conventional R-S theory does not contain the Lorentz condition in the massless case, the M field described by equations (39) and (41) together gives the Lorentz condition even for the massless case. This conclusion is most transparent if equation (41) is rewritten purely in terms of the base field by the use of M-field connection relations,

$$m\psi_{M}^{5\lambda} = \gamma^{\lambda}(\partial^{\tau}\psi_{M}^{\tau}) - (\gamma \cdot \partial)\psi_{M}^{5\lambda} . \tag{42}$$

We can therefore view the M theory as the traditional R-S theory with an explicit auxiliary condition that guarantees the Lorentz condition.

The S field will be describe by equations (38) and (40). The sign of $\mu$ in equation (40) for the S field must be positive because of total symmetry. Here, equation (40) plays the role of a subsidiary equation   Unlike the M field, this subsidiary equation cannot be derived from the main field equation, it is a true subsidiary condition. We can apply the commutation rule

$$[\Sigma^{\alpha\beta}, \gamma \cdot \partial] = 2i(\partial^\beta \gamma^\alpha - \partial^\alpha \gamma^\beta) \tag{43}$$

to equation (38) to get

$$\Sigma^{\alpha\beta}(\gamma \cdot \partial)\,\psi_s{}^{\alpha\beta} = (\gamma \cdot \partial)\Sigma^{\alpha\beta}\psi_s{}^{\alpha\beta} + 4i(\partial^\beta \gamma^\alpha - \partial^\alpha \gamma^\beta) = (\gamma \cdot \partial)\Sigma^{\alpha\beta}\psi_s{}^{\alpha\beta} + 4\partial^\alpha \psi_s{}^{s\alpha}.$$

The last term results from using S-connection relations. It now follows from null conditions

$$(I - P_N{}^T)\psi_s{}^{\alpha\beta} = 0 = (I - P_N{}^V)\psi_s{}^{s\alpha} \tag{44}$$

that

$$\partial^\beta \gamma^\alpha \psi_s{}^{\alpha\beta} = 0 = \partial^\alpha \psi_s{}^{s\alpha}. \tag{45}$$

These can be interpreted as the Lorentz conditions for the base and auxiliary components of the S-subfield. The S-connection relations can be used to recast equation (40) into a subsidiary condition that involves only the base field:

$$\partial^\sigma \gamma^\alpha \psi_s{}^{\tau\alpha} - \partial^\tau \gamma^\alpha \psi_s{}^{\sigma\alpha} = m\psi_s{}^{\sigma\tau}. \tag{46}$$

We can verify that the auxiliary vector sector of the M-field is consistent with the R-S theory. First, using equation (40) on the connection relation, equation (32), we get the R-S equation:

$$m\psi_s{}^\beta = -i\ m\gamma^\alpha \psi_s{}^{\alpha\beta} = -\gamma^\alpha(\partial^\alpha \psi_s{}^\beta - \partial^\beta \psi_s{}^\alpha) = -(\gamma \cdot \partial)\psi_s{}^\beta.$$

Furthermore, we can also get equation (41):

$$2m\psi_s{}^\beta + 2i\partial^\alpha \psi_s{}^{\beta\alpha} = i(m\gamma^\alpha + 2\partial^\alpha)\,\psi_s{}^{\beta\alpha} + m\psi_s{}^\beta = i[-\gamma^\alpha(\gamma \cdot \partial) + \partial^\alpha]\psi_s{}^{\beta\alpha} + m\psi_s{}^\beta$$

$$= i(\gamma \cdot \partial)\psi_s{}^\beta + m\psi_s{}^\beta = 0.$$

## 5. DISCUSSIONS AND CONCLUSIONS

We were able to find projection operators that relate trispinor representations of SO(5) to the structurally simpler 5-tensor spinor representations describing spin-3/2 fields. In particular, we showed that the B-W symmetric trispinor field is equivalent to the 20 dimensional 4-tensor spinor field rather than the traditional R-S 4-vector spinor field, in the sense that the tensor sector acts as the "base" sector that defines uniquely the "auxiliary" vector spinor components through connection relations derived from projection operators. We found that on the other hand, the vector spinor sector acts as the "base" sector for the mixed symmetric M-subfield, with corresponding connection relations to supply its own "auxiliary" tensor-spinor components.

We illustrated how these concepts can be applied to get field equations in the base sectors of both the M-field and the S-field. Each field is described by a R-S like equation for the base field, plus an auxiliary field equation equivalent to the Lorentz condition. The S field represents a new formulation of the B-W field. It is expected that generalizations of the projective approach to higher spin fields will be possible.

## ACKNOWLEDGEMENTS

This work is partially supported by the National Research Council of the Philippines, and the Natural Science Research Institute , University of the Philippines. The author wishes to express his gratitude towards Dr. Ichiro Ohba and the Physics Department of Waseda University and the Japanese Society for the Promotion of Science for their hospitality and use of their facilities in writing a portion of this paper.

## REFERENCES

1. Freedman, D. Z., van Nieuwenhuizen, P. and Ferrara, S. (1976) Progress Toward a Theory of Supergravity. *Physical Reviews*, **D13**, 3214.
2. Deser, S. and B. Zumino, B. (1976) Consistent Supergravity. Physics *Letters*, **B62**, 335.
3. Bargmann, V. and Wigner, E. P. (1948) Group Theoretical Discussion of Relativistic Wave Equations, In *Proceedings of the National Academy of Sciences, U. S.*, **34**, 211.
4. Rarita, W. and Schwinger, J. (1941) On a Theory of Particles with Half-Integral Spin, *Physical Reviews*, **60**, 61.
5. Lurie, D. (1968) *Particles and Fields*, pp. 44-51.New York: Interscience Publishers.

# THERMO FIELD DYNAMICS

# A DIRECT TRANSFORMATION BETWEEN IMAGINARY-TIME AND REAL-TIME THERMOFIELD THEORIES INDUCED BY SECTION ROTATIONS IN $\eta$-$\xi$ SPACETIME

YUAN-XING GUI

CCAST (World Laboratory), PO BOX 8730, Beijing 100080, P.R.China *and*
Department of Physics, Dalian University of Technology, Dalian 116023,
P.R.China

YA-JUN GAO

Institute of Mathematics, Dalian University of Technology, Dalian 116023,
P.R.China *and*
Department of Physics, Jinzhou Teacher's College, Jinzhou 121003, Liaoning,
P.R.China*

**Abstract**: Some suitable complex-time transformation is combined organically with the theory of $\eta$-$\xi$ spacetime and a so-called $\eta$-$\xi$ complex-time transformation is constructed, which gives some section rotations in $\eta$-$\xi$ spacetime and induces a direct transformation between the imaginary-time and real-time thermo field theories.

## 1. INTRODUCTION

Since the middle of this century, the methods and techniqus of quantum field theory have been generalized and used to the case of finite temperature, a few of thermo field theories (TFTs) have been put forth[1−4]. Thanks to these, great progress has been made in the study of thermal phenomenon. The current TFTs can be divided into two classes, i.e. the imaginary-time formalism and the real-time formalism, and the connection between them attracts much interest[5]. The equivalence of various real-time TFTs was proved by Matsumoto *et al.*[6]. However, there has been no satisfactory explanation about the intrinsic connection between imaginary-time and real-time TFTs. To outward seeming, imaginary-time and real-time TFTs are widely different, e.g., the imaginary-time TFT has very obvious imaginary-time periodicity, while a remarkable characteristic of the real-time TFT is the doubling of the degrees of freedom[4]. In Ref.[7], Dolan and Jackiw attempted to obtain a real-time Green's function from the imaginary-time one through an analytical continuation of the imaginary time, the result, however, has some difficults and drawbacks. In fact, only one component of the real-time thermal Green function was obtained there.

---

* Mailing address.

Recently, a new spacetime, called $\eta$-$\xi$ spacetime, was constructed[8−11]. It provides a unified geometrical background for TFTs, and the usual imaginary-time and real-time TFTs are identified, respectively, with the vacuum field theories on the Euclidean section (ES) and the Lorentzian section (LS) of the $\eta$-$\xi$ spacetime[8−12].

In the present paper, we propose a conformal transformation of the complex time variable, an organic combination of which with the theory of $\eta$-$\xi$ spacetime gives section rotations in $\eta$-$\xi$ spacetime. This kind of rotations induce transformations between the field theories on different sections of $\eta$-$\xi$ spacetime. In particular, we obtain the rotation between the ES and the LS in $\eta$-$\xi$ spacetime, which induces a direct transformation between the imaginary- and real-time TFTs. In this transformation, the imaginary-time periodicity and the doubling of the degrees of freedom of the TFTs are automatically preserved. Thus we see that, based on the $\eta$-$\xi$ spacetime background, the imaginary-time formalism of TFT contains more information than previously expected. In addition, since its structure is simpler, we hope the direct transformation between it and the real-time TFT can be used to simplify the studies and calculations of the real-time TFT.

## 2. COMPLEX-TIME TRANSFORMATION AND SECTION ROTATIONS IN $\eta$-$\xi$ SPACETIME

### (1). $\eta$-$\xi$ spacetime [8−11]

The $\eta$-$\xi$ spacetime is a 4-complex-dimensional manifold with complex metric as

$$ds^2 = \frac{1}{\alpha^2(\xi^2 - \eta^2)}(-d\eta^2 + d\xi^2) + dy^2 + dz^2 \tag{2.1}$$

where $\alpha$ is a real constant, $\eta$, $\xi$, $y$, $z$ are complex coordinates. By restricting $\xi$, $y$, $z$ to be real but $\eta = i\sigma$ to be pure imaginary, we obtain the ES of the $\eta$-$\xi$ spacetime, its real metric is

$$ds^2 = \frac{1}{\alpha^2(\xi^2 + \sigma^2)}(d\sigma^2 + d\xi^2) + dy^2 + dz^2. \tag{2.2}$$

Under the coordinate transformation

$$\sigma = \frac{1}{\alpha}e^{\alpha x}\sin\alpha\tau, \quad \xi = \frac{1}{\alpha}e^{\alpha x}\cos\alpha\tau, \tag{2.3}$$

the metric (2.2) becomes

$$ds^2 = d\tau^2 + dx^2 + dy^2 + dz^2. \tag{2.4}$$

When the $\eta$-$\xi$ spacetime is used to study the TFTs, we should take $\alpha = 2\pi/\beta$, here $\beta$ is the inverse temperature. As a result of (2.3), in (2.4) we shall identify $\tau = 0$ with $\tau = \beta$, i.e. $\tau$, $x$, $y$, $z$ take them values in a hypercylinder $S^1 \times R^3$, while coordinates $\sigma$, $\xi$, $y$, $z$ discribe a space homeomorphic to this hypercylnder [8,9,11].

By restricting $\eta$, $\xi$, $y$, $z$ in (2.1) all to be real, we obtain the LS of the $\eta$-$\xi$ spacetime. The singularity on the hypersurfaces $\eta^2 - \xi^2 = 0$ divides the LS into four disjointed parts I, II, III, IV[8−11], each of which is identified with a Minkowski spacetime. For example, if we introduce a coordinate transformation in regions I($\xi > |\eta|$) and II($-\xi > |\eta|$) as

$$\left. \begin{array}{l} \eta = \pm \frac{1}{\alpha} e^{\alpha x} \sinh \alpha t \\ \xi = \pm \frac{1}{\alpha} e^{\alpha x} \cosh \alpha t \end{array} \right\} \quad \begin{array}{l} \text{the upper sign for region I,} \\ \text{the lower sign for region II,} \end{array} \tag{2.5}$$

then (2.1) becomes a Minkowski metric

$$ds^2 = -dt^2 + dx^2 + dy^2 + dz^2. \tag{2.6}$$

It has been shown that[8−12], the zero-temperature field theories on ES and LS of $\eta$-$\xi$ spacetime give, respectively, the results of imaginary- and real-time TFTs in Minkowski formalism.

(2). **Complex-time transformation and section rotations in $\eta$-$\xi$ spacetime**
As well known, in order to discribe finite temperature theory, the time is extended to a complex variable taking values in some region of complex plan, and each complex path according with certain demands leads to a special formalism of TFT[5]. Therefore, if some suitable complex-time transformation is given, it is possible to obtain a transformation between the different formalisms of TFT. Unfortunately, it is difficult to connect the imaginary-time TFT (with obvious imaginary-time periodicity) directly to the real-time TFT (also with the doubling of the degrees of freedom) by using usual transformations. However, we find that when some suitable complex-time transformation is combined organically with the theory of $\eta$-$\xi$ spacetime[8−12], then a direct transformation between the imaginay- and real-time TFTs can be obtained. To this end, we consider the following complex function transformation

$$F: \quad u \longmapsto F(u) = f_1^{-1} \circ f \circ f_1(u), \tag{2.7}$$

where $f$ and $f_1$ are defined as: for an arbitrary complex variable $u$,

$$f(u) = -i \frac{1-u}{1+u}, \qquad f_1(u) = e^{-i\alpha u}, \tag{2.8}$$

and for transformation $f_1^{-1}$ in (2.7), the branch cut is taken to lie in the upper-half complex plan, i.e., $\ln(-1) = -i\pi$ [13]. Letting $w = -i \frac{1 - e^{-i\alpha u}}{1 + e^{-i\alpha u}}$, we have

$$F(u) = \frac{1}{-i\alpha} \ln w = \frac{i}{\alpha} \ln |w| - \frac{1}{\alpha} \arg[w]. \tag{2.9}$$

This is a conformal transformation, and the "arg" is determined by the above-mentioned choice of the branch cut.

From (2.8), we see that the mapping $f_1$ carries the infinite long strip $\{-iu,$ $\text{Im}(-iu) \in [-\beta, 0]\}$ in the $(-iu)$-plan into a whole complex plan (the $f_1(u)$-plan), and carries any path, in $(-iu)$-plan, running from $t$ to $t - i\beta$ ($t$ is real) with non-increasing imaginary part into a simple closed curve moving around the origin of $f_1(u)$-plan. Furthermore, $f$ carries such a simple closed curve into another. In particular, if take the values of $-iu$ in the Matsubara's path, i.e. $u = \tau$ ($0 \le \tau \le \beta$), and denote

$$\frac{1}{\alpha} \ln \left| \frac{1 - e^{-i\alpha\tau}}{1 + e^{-i\alpha\tau}} \right| = \begin{cases} t, & \text{when } 0 < \tau < \beta/2, \\ t', & \text{when } \beta/2 < \tau < \beta, \end{cases} \tag{2.10}$$

then we have

$$F: \tau \longmapsto F(\tau) = \begin{cases} it, & \text{when } 0 < \tau < \beta/2, \\ it' + \beta/2, & \text{when } \beta/2 < \tau < \beta. \end{cases} \tag{2.11}$$

From (2.10) it is evident that $t|_{\tau \to +0} = -\infty$; $t|_{\tau \to \frac{\beta}{2}-0} = +\infty$; $t'|_{\tau \to \frac{\beta}{2}+0} = +\infty$; $t'|_{\tau \to \beta-0} = -\infty$. And when $\tau \in (0, \beta/2)$, $t$ is a monotonically increasing function of $\tau$ (the same ordering of times); When $\tau \in (\beta/2, \beta)$, $t'$ is a monotonically decreasing function of $\tau$ (the inverse ordering of times). These properties are important in the following.

In order to find the connection between (2.7)–(2.11) and $\eta$-$\xi$ spacetime, we introduce the following complex-coordinate transformation:

$$\sigma = \frac{1}{\alpha} e^{\alpha x} \sin \alpha u, \quad \xi = \frac{1}{\alpha} e^{\alpha x} \cos \alpha u, \tag{2.12}$$

which analytically extends the metric (2.2). In coordinate system $u$, $x$, $y$, $z$, the extended metric becomes

$$ds^2 = du^2 + dx^2 + dy^2 + dz^2. \tag{2.13}$$

When $x$, $y$, $z$ are arbitrary real numbers while $u = \tau(0 \le \tau \le \beta)$ (Matsubara's path), then (2.12) and (2.13) give (2.3) and (2.4). Moreover, performing the transformation (2.7)–(2.11) on $u = \tau$, then (2.2), (2.12) and (2.13) are changed, repectively, into (2.1), (2.5) and (2.6). Thus we see that (2.11) transforms the ES into (the regions I and II of) LS in the $\eta$-$\xi$ spacetime. From these, we come to a conclusion that the transformation (2.7)–(2.11) gives section rotations in $\eta$-$\xi$ spacetime. In the following, we call this transformation an $\eta$-$\xi$ complex-time transformation and denote it by $\mathcal{F}$.

Since[8–12] the vacuum field theories on ES and LS of $\eta$-$\xi$ spacetime give the results, respectively, of imaginary-time TFT and real-time TFT [e.g. thermo-field dynamics (TFD)[2]], we expect that the transformation $\mathcal{F}$ can induce a direct transformation between the imaginary- and real-time TFTs. The discussins in the following sections will show that this is just the case.

In the present paper, we focus our attention on the transformations of the mode solutions of free field equation and the associated Green function. Similar to the case of zero-temperature field theory, these mode solutions and Green function play key roles in TFT. Here we deal only with the spinless bose fields, the case of fermi fields will be discussed in the further work.

## 3. TRANSFORMATION OF MODE SOLUTIONS OF FIELD EQUA-TIONS

In this section, we use the transformation $\mathcal{F}$ to obtain directly the mode solutions on LS of $\eta$-$\xi$ spacetime from the mode solutions on ES. Then the Bogoliubov transformation is given naturally.

We consider the free scalar field $\varphi$ (with mass $m$) on ES of $\eta$-$\xi$ spcetime which satisfying the equation

$$\left[\alpha^2(\sigma^2 + \xi^2)(\frac{\partial^2}{\partial \sigma^2} + \frac{\partial^2}{\partial \xi^2}) + \frac{\partial^2}{\partial y^2} + \frac{\partial^2}{\partial z^2} - m^2\right]\varphi = 0. \tag{3.1}$$

In the coordinate system $\tau$, $x$, $y$, $z$ [cf. (2.2)–(2.4)], Eq.(3.1) becomes

$$\left(\frac{\partial^2}{\partial \tau^2} + \frac{\partial^2}{\partial x^2} + \frac{\partial^2}{\partial y^2} + \frac{\partial^2}{\partial z^2} - m^2\right)\varphi = 0. \tag{3.2}$$

Considering the properties of equal right and infinite extension of the spatial coordinates $x$, $y$, $z$, we have the following "energy-momentum" eigen mode solutions

$$\exp\{-\omega_{\mathbf{k}}\tau + i\mathbf{k}\cdot\mathbf{x}\}, \qquad \exp\{\omega_{\mathbf{k}}\tau + i\mathbf{k}\cdot\mathbf{x}\}, \tag{3.3a,b}$$

where $\omega_{\mathbf{k}} = \sqrt{k^2 + m^2} > 0$.

Performing the $\eta$-$\xi$ complex-time transformation on the mode solution (3.3a), we have

$$\mathcal{F}: e^{-\omega_{\mathbf{k}}\tau + i\mathbf{k}\cdot\mathbf{x}} \longmapsto \phi'_{\mathbf{k}} = \delta_{\mathrm{I}} e^{-i\omega_{\mathbf{k}}t + \mathbf{k}\cdot\mathbf{x}} + \delta_{\mathrm{II}} e^{-\frac{1}{2}\omega_{\mathbf{k}}\beta} e^{-i\omega_{\mathbf{k}}t' + i\mathbf{k}\cdot\mathbf{x}}, \tag{3.4}$$

here $\delta_{\mathrm{I}}$ and $\delta_{\mathrm{II}}$ are defined as: $\delta_{\mathrm{I}} = 1$, $\delta_{\mathrm{II}} = 0$ when spacetime variables take their values in the region I of LS, and $\delta_{\mathrm{I}} = 0$, $\delta_{\mathrm{II}} = 1$ in the region II. From (2.7)–(2.12), when $x$, $y$, $z$ are real, then $\phi'_{\mathbf{k}}$ in (3.4) is analytic in the corresponding region of $\eta$-$\xi$ spacetime as long as the *complex* variable $u$ detours the point $u = \beta/2$. Similar to the argument of W.G.Unruh[14], we see that $\phi'_{\mathbf{k}}$ is a positive frequency mode on LS. Owing to the reverse time ordering of $t'$ (cf. Subsection II.2), without loss of generality, here we may take $t' = -t$ in the Minkowski formalism. Normalizing $\phi'_{\mathbf{k}}$ with respect to the Klein-Gordon inner product and denoting the normalized function by $\phi_{\mathbf{k}}$, we have

$$\phi_{\mathbf{k}} = \cosh\theta_\omega \,^{\mathrm{I}}f_{\mathbf{k}} + \sinh\theta_\omega \,^{\mathrm{II}}f^*_{-\mathbf{k}}, \tag{3.5}$$

where $\mathrm{th}\theta_\omega = \exp(-\frac{1}{2}\omega_{\mathbf{k}}\beta)$, and

$$^{\mathrm{I}}f_{\mathbf{k}} = \frac{\delta_{\mathrm{I}}}{\sqrt{2\omega_{\mathbf{k}}V}}\exp\{-i\omega_{\mathbf{k}}t + i\mathbf{k}\cdot\mathbf{x}\}, \quad ^{\mathrm{II}}f_{\mathbf{k}} = \frac{\delta_{\mathrm{II}}}{\sqrt{2\omega_{\mathbf{k}}V}}\exp\{-i\omega_{\mathbf{k}}t + i\mathbf{k}\cdot\mathbf{x}\}. \quad (3.6)$$

Similarly, performing $\mathcal{F}$ on (3.3b) we obtain another mode solution on LS as

$$\widetilde{\phi}_{\mathbf{k}} = \sinh\theta_\omega \, ^{\mathrm{I}}f^*_{-\mathbf{k}} + \cosh\theta_\omega \, ^{\mathrm{II}}f_{\mathbf{k}}. \quad (3.7)$$

$\phi_{\mathbf{k}}$ and $\widetilde{\phi}_{\mathbf{k}}$ were obtained in Refs.[8–11] by a different method, here we emphasize the direct connection between the imaginary- and real-time mode solutions. The general solution $\Phi(t,\mathbf{x})$ can certainly be expanded in terms of the modes (3.5),(3.7) or (3.6), i.e.

$$\begin{aligned}
\Phi &= \sum_{\mathbf{k}}(b_{\mathbf{k}}\phi_{\mathbf{k}} + b_{\mathbf{k}}^+\phi_{\mathbf{k}}^* + \tilde{b}_{\mathbf{k}}\tilde{\phi}_{\mathbf{k}} + \tilde{b}_{\mathbf{k}}^+\tilde{\phi}_{\mathbf{k}}^*) \\
&= \sum_{\mathbf{k}}(a_{\mathbf{k}}\,^{\mathrm{I}}f_{\mathbf{k}} + a_{\mathbf{k}}^+\,^{\mathrm{I}}f_{\mathbf{k}}^* + \tilde{a}_{\mathbf{k}}\,^{\mathrm{II}}f_{\mathbf{k}} + \tilde{a}_{\mathbf{k}}^+\,^{\mathrm{II}}f_{\mathbf{k}}^*).
\end{aligned} \quad (3.8)$$

By the Klein-Gordon inner product we obtain the Bogoliubov transformation:

$$b_{\mathbf{k}} = a_{\mathbf{k}}\cosh\theta_\omega - \tilde{a}_{-\mathbf{k}}^+\sinh\theta_\omega, \qquad \tilde{b}_{\mathbf{k}} = \tilde{a}_{\mathbf{k}}\cosh\theta_\omega - a_{-\mathbf{k}}^+\sinh\theta_\omega. \quad (3.9)$$

Noticing the mode solutions used in Eq.(3.8), we know that

$$a_{\mathbf{k}}|0\rangle_{\mathrm{MI}} = 0, \qquad \tilde{a}_{\mathbf{k}}|0\rangle_{\mathrm{MII}} = 0 \quad (3.10)$$

give the definitions of the vacuum states $|0\rangle_{\mathrm{MI}}$ and $|0\rangle_{\mathrm{MII}}$ of quantum fields, respectively, on "our" Minkowski spacetime (correponding to the region I) and its "mirror" spacetime (corresponding to the region II)[8, 11], while

$$b_{\mathbf{k}}|0\rangle_{\mathrm{LS}} = \tilde{b}_{\mathbf{k}}|0\rangle_{\mathrm{LS}} = 0 \quad (3.11)$$

defines the vacuum state $|0\rangle_{\mathrm{LS}}$ on LS of $\eta$-$\xi$ spacetime. And from (3.9) we see that $|0\rangle_{\mathrm{LS}}$ is just the thermal state for an inertial observer in Minkowski spacetime[2].

## 4. TRANSFORMATION OF GREEN FUNCTIONS

In the present section, we solve the Green function equation on the ES of $\eta$-$\xi$ spacetime, then perform the transformation $\mathcal{F}$ on the solution (which is identifiable with theimaginary-time thermal Green function[8,9,11]) and obtain directly the Green function on LS, the latter is just the real-time thermal Green function in the Minkowski formalism. Here, the $2 \times 2$ matrix structure of real-time thermal Green function is obtained automatically and the process is very simple.

Using the variables $\tau$, $x$, $y$, $z$, the Green function for free bose field on ES of $\eta$-$\xi$ spacetime satisfies

$$\left(\frac{\partial^2}{\partial\tau^2} + \frac{\partial^2}{\partial x^2} + \frac{\partial^2}{\partial y^2} + \frac{\partial^2}{\partial z^2} - m^2\right)G(\tau,\mathbf{x};\tau',\mathbf{x}') = -\delta(\tau-\tau')\delta(\mathbf{x}-\mathbf{x}'), \quad (4.1)$$

and has the form $G(\tau, \mathbf{x}; \tau', \mathbf{x}') = G(\tau - \tau', \mathbf{x} - \mathbf{x}')$. Letting $s = \tau - \tau'$, we have $-\beta \leq s \leq \beta$. Owing to the geometrical structure of the ES[9], the single valued solution of (4.1) is demanded satisfy the periodic condition

$$G|_{\tau=0} = G|_{\tau=\beta} . \qquad (4.2)$$

When $0 < s < \beta$, in terms of the complete system of eigenfunctions of the operator $\partial^2/\partial\tau^2 + \nabla^2 - m^2$ with the condition (4.2), the Green funtion $G$ can be expanded as

$$G(s, \mathbf{x}, \mathbf{x}') = \beta^{-1} \sum_{\mathbf{k}} \sum_{n=-\infty}^{\infty} f_{\mathbf{k}}(\mathbf{x}) f_{\mathbf{k}}^*(\mathbf{x}') \exp\left\{\frac{-2\pi i n s}{\beta}\right\} \left[\left(\frac{2\pi n}{\beta}\right)^2 + \omega_{\mathbf{k}}^2\right]^{-1}, \qquad (4.3)$$

where $f_{\mathbf{k}}(\mathbf{x}) = \frac{1}{\sqrt{V}} \exp\{i\mathbf{k} \cdot \mathbf{x}\}$. In the above series, the part of the sum with repect to $n$ can be expressed by a function, we have (for $0 < s < \beta$)

$$\beta^{-1} \sum_{n=-\infty}^{\infty} \exp\left\{\frac{-2\pi i n s}{\beta}\right\} \left[\left(\frac{2\pi n}{\beta}\right)^2 + \omega_{\mathbf{k}}^2\right]^{-1} = \frac{(2\omega_{\mathbf{k}}^2)^{-1}}{(1 - e^{-\beta\omega_{\mathbf{k}}})} \left[e^{-\omega_{\mathbf{k}} s} + e^{\omega_{\mathbf{k}}(s-\beta)}\right]. \qquad (4.4)$$

Equation (4.4) can be verified by a direct calculation of the Fourier coefficients of the right hand side. Considering (4.2), when $-\beta < s < 0$, we have

$$G(s, \mathbf{x}, \mathbf{x}') = \sum_{\mathbf{k}} f_{\mathbf{k}}(\mathbf{x}) f_{\mathbf{k}}^*(\mathbf{x}') (2\omega_{\mathbf{k}})^{-1} \left(1 - e^{-\beta\omega_{\mathbf{k}}}\right)^{-1} \left[e^{\omega_{\mathbf{k}} s} + e^{-\omega_{\mathbf{k}}(s+\beta)}\right]. \qquad (4.5)$$

Performing the transformation $\mathcal{F}$ on (4.3)–(4.5) and noting the time ording induced by $\tau$ (cf. Subsection II.2), we can directly obtain the real-time thermal Green function of the usual TFT. The independent variables of the $\mathcal{F}$-transformed Green function will take their values in the regions I and II of LS in $\eta$-$\xi$ spacetime. When the independent variables take their values in the region I (resp. region II), we mark the corresponding Green function by subscript 1 (resp. 2). Thus we naturally obtain the $2 \times 2$ matrix form which reflects the property of doubling of the degrees of freedom. To simplify the notation, we denote the real time difference still by $t$ in the $\mathcal{F}$-transformed Green function $G = \{G_{ab}\}$, $(a, b = 1, 2)$. Explicitly, when $t > 0$, performing $\mathcal{F}$ on (4.3) and (4.4), we obtain

$$G_{11}^+(t, \mathbf{x}, \mathbf{x}') = \sum_{\mathbf{k}} f_{\mathbf{k}}(\mathbf{x}) f_{\mathbf{k}}^*(\mathbf{x}') (2\omega_{\mathbf{k}})^{-1} \left(1 - e^{-\beta\omega_{\mathbf{k}}}\right)^{-1} \left[e^{-i\omega_{\mathbf{k}} t} + e^{i\omega_{\mathbf{k}} t} e^{-\omega_{\mathbf{k}}\beta}\right], \qquad (4.6)$$

when $t < 0$, performing $\mathcal{F}$ on (4.5), we have

$$G_{11}^-(t, \mathbf{x}, \mathbf{x}') = \sum_{\mathbf{k}} f_{\mathbf{k}}(\mathbf{x}) f_{\mathbf{k}}^*(\mathbf{x}') (2\omega_{\mathbf{k}})^{-1} \left(1 - e^{-\beta\omega_{\mathbf{k}}}\right)^{-1} \left[e^{i\omega_{\mathbf{k}} t} + e^{-i\omega_{\mathbf{k}} t} e^{-\omega_{\mathbf{k}}\beta}\right]. \qquad (4.7)$$

Thus $G_{11} = \theta(t)G_{11}^+ + \theta(-t)G_{11}^-$ is just the 1-1 component of the real-time thermal Green function in TFD[4].

In order to calculate $G_{12}(t, \mathbf{x}, \mathbf{x}')$, we note the time ordering induced by $\tau$ under $\mathcal{F}$. In the present case, we have $\tau \in (\beta/2, \beta)$, $\tau' \in (0, \beta/2)$ and $\mathcal{F}: s = \tau - \tau' \longmapsto it + \beta/2$, thus (4.3), (4.4) are transformed, by $\mathcal{F}$, into

$$G_{12}(t, \mathbf{x}, \mathbf{x}') = \sum_{\mathbf{k}} f_{\mathbf{k}}(\mathbf{x}) f_{\mathbf{k}}^*(\mathbf{x}')(2\omega_{\mathbf{k}})^{-1} \left(1 - e^{-\beta\omega_{\mathbf{k}}}\right)^{-1} \left(e^{-i\omega_{\mathbf{k}}t} + e^{i\omega_{\mathbf{k}}t}\right) e^{-\frac{1}{2}\omega_{\mathbf{k}}\beta}.$$

(4.8)

This is the 1–2 component of thermal Green function[4].

Similarly, we can obtain another two components of real-time thermal Green function as:

$$G_{21}(t, \mathbf{x}, \mathbf{x}') = G_{12}(t, \mathbf{x}, \mathbf{x}'),$$

(4.9)

$$G_{22}(t, \mathbf{x}, \mathbf{x}') = \theta(t)G_{22}^+(t, \mathbf{x}, \mathbf{x}') + \theta(-t)G_{22}^-(t, \mathbf{x}, \mathbf{x}'),$$

(4.10)

where

$$G_{22}^{\pm}(t, \mathbf{x}, \mathbf{x}') = \sum_{\mathbf{k}} f_{\mathbf{k}}(\mathbf{x}) f_{\mathbf{k}}^*(\mathbf{x}')(2\omega_{\mathbf{k}})^{-1} \left(1 - e^{-\beta\omega_{\mathbf{k}}}\right)^{-1} \left[e^{\pm i\omega_{\mathbf{k}}t} + e^{\mp i\omega_{\mathbf{k}}t}e^{-\omega_{\mathbf{k}}\beta}\right].$$

Here we have noted the reverse ordering of times on the region II of LS and on Minkowski spacetime.

## 5. SUMMARY AND DISCUSSIONS

We have combined a complex-time transformation organically with the theory of $\eta$-$\xi$ spacetime and constructed the so-called $\eta$-$\xi$ complex-time transformation which gives section rotations in $\eta$-$\xi$ spacetime. Particularly, we obtain the rotation between the ES and LS of $\eta$-$\xi$ spacetime, which induces a direct transformation between the imaginary- and real-time TFTs. The mode solutions of free field equation and the associated Green function are discussed in detail, these are important in TFT. The result of this paper also show that, the $\eta$-$\xi$ spacetime provides not only a unified geometrical background for TFTs, but also a theoretic basis for finding direct connections between different formalisms of TFT.

Finally we point out that the imaginary-time TFT, particularly its perturbative expansion have been studied extensively, and their form is, in general, simpler than that of the real-time TFT. Thus the $\mathcal{F}$ transformation obtained in this paper my give a light for simplifying the studies and calculations of the real-time TFT, this needs some further discussions.

## ACKNOWLEDGEMENT

This work was supported by the National Natural Science Foundation of China.

**REFERENCES**

1. Matsubara, T., Prog. Theor. Phys., **14**(1955), 351.
2. Takahash, Y. and Umezawa, H., Collec. Phenom., **2**(1975), 55.
   Umezawa, H. et al., Thermo Field Dynamics and Condensed States, (North-Holland, Amsterdam 1982).
3. Niemi, A. J. and Semenoff, G. W., Nucl. Phys., **B230**[FS10](1984), 181.
4. Landsman, N. D. and Van Weert, Ch. G., Phys. Rep., **145**(1987), 142.
5. Mills, R., Propagators for Many-Particle Systems (Gordon and Breach Science Publishers, N. Y., 1955).
   Marinaro, M., Phys. Rep. **137**(1986), 81.
6. Matsumoto, H., Nakano, Y., and Umezawa, H., J. Math. Phys. **25**(1984), 3076.
   Matsumoto, H., in: Progress in Quantum Field Theory, eds. Ezawa, H. and Kamefuchi, S. (North-Holland, Amsterdam, 1986).
7. Dolan, L. and Jackiw, R., Phys. Rev. **D9**(1974), 3320.
8. Gui, Y. X., Phys. Rev. **D42**(1990), 1988.
9. Gui, Y. X., Phys. Rev. **D46**(1992), 1869.
10. Gui, Y. X., Sci. Sin. **A31**(1988), 1104.
11. Gui, Y. X., Sci. Sin. **A36**(1993), 561.
12. Gui, Y. X., Phys. Rev. **D45**(1992), 697.
13. Birrell, N. D. and Davies, P. C. W., Quantum Fields in Curved Space (Cambridge, 1982).
14. Unruh, W. G., Phys. Rev. **D14**(1976), 870.

CANONICAL QUANTIZATION FOR EXPANDING
GEOMETRY UNIVERSE

ELEONORA ALFINITO and GIUSEPPE VITIELLO
Dipartimento di Fisica, Università di Salerno, 84100
and INFN Gruppo Collegato di Salerno

**Abstract**     We discuss the canonical quantization formalism for the
expanding geometry Universe by using the technique of the doubling of
the degrees of freedom.Thermal properties of inflating Universe are also
discussed.

## 1. INTRODUCTION

We are very pleased to join the celebration of the 65th birthday of Professor Hiroshi
Ezawa. With his systematic and successful explorations of the wonderful world
of Theoretical Physics he has taught us to appreciate and to enjoy the richness of
Quantum Field Theory. In the present paper we try indeed to exploit such a richness
in the theory to produce a canonical formalism for the expanding geometry Universe.

A phase of primordial inflation seems to be required in order to solve some open
problems in General Relativity such as the flatness and horizon problems or the so-
called problem of quantum fluctuations[1-3]. It is therefore interesting to explore the
possibility of constructing in quantum field theory (QFT) the canonical formalism to
treat the non-unitary time evolution implied by inflationary models and in general
by expanding metrics scenarios.

As we will see the full set of unitarily inequivalent representations of the canonical
commutation relations need to be considered for our task. Moreover, an essential
ingredient in our discussion, which will be limited to gravitational wave modes, will
be the doubling of the system degrees of freedom. As a matter of fact, we recover in
the present context, the connection with Thermo Field Dynamics (TFD)[4-6], which
anyway underlies many works in General Relativity and Cosmology (even if not
referred to) since the paper by Israel[7] on TFD of black holes (see also ref. 8, and
for rigorous results in the algebraic approach see ref. 9).

We will also use some results in the quantization method for non-unitary time
evolution proposed by Feshbach and Tikochinsky[10] (see also ref. 11) which turned
out to be relevant in the study of quantum dissipation in QFT[12].

We clarify the physical meaning of the doubling of the system degrees of freedom. As we will show, it is not simply a mathematical device: from the thermal properties perspective the physical interpretation of the doubled degrees of freedom is the one of the thermal bath degrees of freedom; from the point of view of the vacuum structure the one of *holes* of the relic gravitons; from the hamiltonian formalism point of view the one of the *complement* to the inflating system.

We observe that the doubling of the degrees of freedom is intrinsic to the Bogoliubov transformations, so that one deals with a doubled system anytime one works with such transformations. For this reason all the "mixed modes" formalisms (since Parker's work[13]) necessarily involve the algebraic structure of the doubling of the modes.

Expanding metrics also imply time-dependent frequency for the gravitational wave modes, and this leads us to extend the canonical quantization method for non-unitary time evolution so as to include the quantization formalism for parametric oscillator[14]. As one should expect[15] (see also ref. 16), this naturally brings us to squeezed states. The connection with squeezing is on the other hand also expected since the works by Grishchuk and Sidorov[17-19].

## 2. QUANTIZATION IN EXPANDING GEOMETRY

Using a flat time-dependent metric $g_{\mu\nu}(t) = g^0_{\mu\nu}(t) + h_{\mu\nu}(t)$ and the De Donder gauge condition $\partial_\mu h_{\mu\nu} = 0$ the Einstein equations

$$G_{\mu\nu} = R_{\mu\nu} - \frac{1}{2}g_{\mu\nu}R = 0 \qquad (2.1)$$

give

$$\Box\, h_{\mu\nu} = 0. \qquad (2.2)$$

The field $h_{\mu\nu}$ may be then decomposed in harmonic modes $u_k$ obeying the equation:

$$\ddot{u}_k(t) + H\dot{u}_k(t) + \omega_k^2(t)u_k(t) = 0 \qquad (2.3)$$

with

$$\omega_k{}^2(t) = \frac{k^2c^2}{a^2(t)}, \qquad a(t) = a_0 e^{\frac{1}{3}Ht} \quad . \qquad (2.4)$$

In the Minkowski space-time $\omega_k$ is constant in time, but when the Universe expands, $\omega_k$ is time-dependent $\omega_k = \omega_k(t)$. This can be seen as a generalization of the Doppler effect.

The first order derivative term $H\dot{u}_k$ in eq.(2.3) is generally incorporated into the frequency term by using the conformal time variable $\eta$ ;[17-19] such a computational strategy is very useful in the phenomenological approach, however our purpose in this paper is to illustrate the subtleties of the canonical quantization for non-unitary time evolution and therefore we must explicitly take care of the inflattive term in eq.(2.3). In this way the full structure of the state space will be revealed.

In the following, where not strictly necessary, we will omit the $k$-index, remembering that each equation is written down for fixed $k$. In order to perform the canonical quantization of the oscillator (2.3), according to[12] we consider the double oscillator system

$$\ddot{u} + H\dot{u} + \omega^2(t)u = 0 , \tag{2.5}$$

$$\ddot{v} - H\dot{v} + \omega^2(t)v = 0 , \tag{2.6}$$

and in this sense we may speak of a "double Universe". We observe indeed that in the same way as the $u$ oscillator describes the expanding (inflating) metrics, the oscillator $v$ can be associated to the "contracting" ("deflating") metrics. The physical reason to double the degrees of freedom of the inflating system relies in the fact that one must work with closed systems as required by the canonical quantization formalism.

It is interesting to observe that $v = ue^{Ht}$ is solution of (2.6) and that by setting $u(t) = \frac{1}{\sqrt{2}}r(t)e^{\frac{-Ht}{2}}$ and $v(t) = \frac{1}{\sqrt{2}}r(t)e^{\frac{Ht}{2}}$ the system of equations (2.5) and (2.6) is equivalent to the single parametric oscillator $r(t)$ (see also ref. 20)

$$\ddot{r} + \Omega^2(t)r = 0 . \tag{2.7}$$

This further clarifies the meaning of the doubling of the $u$ oscillator: the $u - v$ system is a non-inflating (and non-deflating) system. This is why it is now possible to set up the canonical quantization scheme. It turns out to be convenient to introduce the *canonical* transformations

$$u(t) \;=\; \frac{U(t) + V(t)}{\sqrt{2}}, \qquad v(t) \;=\; \frac{U(t) - V(t)}{\sqrt{2}}. \tag{2.8}$$

We are thus dealing with the decomposition of the parametric oscillator $r(t)$ on the *hyperbolic* plane (i.e. in the pseudo-Euclidean metrics): $r^2(t) = U^2(t) - V^2(t)$. The Hamiltonian for our coupled oscillator system is:

$$\mathcal{H} = \frac{1}{2}p_U^2 + \frac{1}{2}\Omega^2(t)U^2 - \frac{1}{2}p_V^2 - \frac{1}{2}\Omega^2(t)V^2 - \frac{H}{2}(p_U V + p_V U). \tag{2.9}$$

In eq.(2.9) we put $\Omega_k(t) \equiv \left[\left(\omega_k^2(t) - \frac{H^2}{4}\right)\right]^{\frac{1}{2}} > 0$, for any $t$ and any $k$, in order to avoid an over-damped regime; let us denote by $\omega_0$ the lower bound at $t = 0$: $\Omega_k(0) \geq \omega_0 > 0$. Since $\omega_0$ does not depend on $k$, the lower bound condition reflects on $k$ as (cf. (2.4))

$$k \geq k_0 \equiv \frac{a_0}{c}\sqrt{\omega_0^2 + \left(\frac{H}{2}\right)^2} . \tag{2.10}$$

Now, we introduce the annihilation operators:

$$A = \frac{1}{\sqrt{2}}\left(\frac{p_U}{\sqrt{\hbar\omega_0}} - iU\sqrt{\frac{\omega_0}{\hbar}}\right), \qquad B = \frac{1}{\sqrt{2}}\left(\frac{p_V}{\sqrt{\hbar\omega_0}} - iV\sqrt{\frac{\omega_0}{\hbar}}\right), \tag{2.11}$$

and the corresponding creation operators. Their commutation relations are

$$[A, A^\dagger] = 1 = [B, B^\dagger], \qquad [A, B] = 0 = [A^\dagger, B^\dagger], \tag{2.12}$$

and all other commutators equal to zero. In terms of $A$ and $B$ the Hamiltonian (2.9) is finally written as

$$\mathcal{H} = \mathcal{H}_0 + \mathcal{H}_{I_1} + \mathcal{H}_{I_2} \tag{2.13}$$

$$\mathcal{H}_0 = \frac{1}{2}\hbar\Omega_0(t)(A^\dagger A - B^\dagger B) \equiv \hbar\Omega_0(t)\mathcal{C} , \tag{2.14}$$

$$\mathcal{H}_{I_1} = -\frac{1}{4}\hbar\Omega_1(t)\left[\left(A^2 + A^{\dagger^2}\right) - \left(B^2 + B^{\dagger^2}\right)\right] \equiv -\hbar\Omega_1(t)K_1, \tag{2.15}$$

$$\mathcal{H}_{I_2} = i\Gamma\hbar\left(A^\dagger B^\dagger - AB\right) \equiv i\hbar\Gamma(J_+ - J_-) . \tag{2.16}$$

with $\Gamma \equiv \frac{H}{2}$, $\Omega_{0,1} = \omega_0\left(\frac{\Omega^2(t)}{\omega_0^2} \pm 1\right)$ .

The group structure underlying the Hamiltonian (2.13) is the one of SU(1,1); in fact, the operators $K_0, K_1$ and $K_2 = i\frac{1}{4}\left[\left(A^2 - A^{\dagger^2}\right) + \left(B^2 - B^{\dagger^2}\right)\right]$, close the su(1,1) algebra:

$$[K_1, K_2] = -iK_0 \quad , \quad [K_2, K_0] = iK_1 \quad , [K_0, K_1] = iK_2. \tag{2.17}$$

Similarly, the operators $J_+ = A^\dagger B^\dagger$, $J_- = AB$, $J_0 = \frac{1}{2}(A^\dagger A + B^\dagger B + 1)$ close the su(1,1) algebra

$$[J_+, J_-] = -2J_3 \quad , \quad [J_3, J_\pm] = \pm J_\pm. \tag{2.18}$$

$\mathcal{C} \equiv K_0$ is the Casimir operator for the algebra generated by the $J$'s operators.

We also note that

$$[\mathcal{H}_0, \mathcal{H}_{I_2}] = 0 = [\mathcal{H}_{I_1}, \mathcal{H}_{I_2}] , \tag{2.19}$$

which guarantees that under time evolution the minus sign appearing in $\mathcal{H}_0$ is not harmful (i.e., once one starts with a positive definite Hamiltonian it remains lower bounded).

We study the Hilbert space structure in the following section.

## 3. THE VACUUM STRUCTURE

In order to study the eigenstates of the Hamiltonian (2.13) we consider the set $\{|n_A, n_B>\}$ of simultaneous eigenvectors of $A^\dagger A$ and $B^\dagger B$, with $n_A, n_B$ non-negative integers. These are eigenstates of $\mathcal{H}_0$ with eigenvalues $\frac{1}{2}\hbar\Omega_0(t)(n_A - n_B)$ for any $t$. The eigenstates of $\mathcal{H}_{I_2}$ can be written in the standard basis, in terms of the eigenstates of $\left(J_3 - \frac{1}{2}\right)$ in the representation labeled by the value $j \in Z_{\frac{1}{2}}$ of $\mathcal{C}$, $\{|j, m>; m \geq |j|\}$, and $j, m = \frac{1}{2}(n_A \mp n_B)$:

$$\mathcal{C}|j, m> = j|j, m>, \qquad \left(J_3 - \frac{1}{2}\right)|j, m> = m|j, m> . \tag{3.1}$$

By using the $su(1,1)$ algebra (2.18), one can show that the kets $|\psi_{j,m}> \equiv e^{\left(+\frac{\pi}{2}J_1\right)}|j, m>$ satisfy indeed the equation $J_2|\psi_{j,m}> = \mu|\psi_{j,m}>$ with pure imaginary $\mu \equiv i\left(m + \frac{1}{2}\right)$ .[12]

We are left with the discussion of the eigenstates of $\mathcal{H}_{I_1}$. We can "rotate away" [21] $\mathcal{H}_{I_1}$ by using the transformation $\mathcal{H} \to \mathcal{H}' \equiv e^{i\theta(t)K_2}\mathcal{H}e^{-i\theta(t)K_2}$ with $\tanh\theta(t) = -\frac{\Omega_1(t)}{\Omega_0(t)} \leq 1$ for any $k$ and $t$. We obtain

$$\mathcal{H}' \equiv e^{i\theta(t)K_2}\mathcal{H}e^{-i\theta(t)K_2} = \mathcal{H}'_0 + \mathcal{H}_{I_2} \ , \tag{3.2}$$

with

$$\mathcal{H}'_0 = \hbar\Omega(t)(A^\dagger A - B^\dagger B) \qquad \text{and} \qquad [\mathcal{H}'_0, \mathcal{H}_{I_2}] = 0 \ . \tag{3.3}$$

In conclusion, the eigenstates of the Hamiltonian $\mathcal{H}$ at $t$, eq. (2.13), are states of type $e^{-i\theta(t)K_2}|\psi_{j,m} >$ .

The solution to the Schrödinger equation can be given with reference to the initial time pure state $|j, m_0 >$ (see refs. 10, 12). When in particular, the initial state, say at arbitrary initial time $t_0$, ($t_0 = 0$, $\theta(0) = \theta$ for sake of simplicity) is the *vacuum* for $\mathcal{H}'_0$, i.e. $|n_A = 0, n_B = 0 >\equiv |0 >$, with $A|0 >= 0 = B|0 >$ (i.e. $j = 0, m_0 = 0$ for any $k$), the state

$$|0(\theta) >= e^{-i\theta K_2}|0 > , \tag{3.4}$$

at $t_0$ (and for given $k$), is the zero energy eigenstate (the *vacuum*) of $\mathcal{H}_0 + \mathcal{H}_{I_1}$ at $t_0$:

$$(\mathcal{H}_0 + \mathcal{H}_{I_1})|_{t_0}|0(\theta) >= e^{-i\theta K_2}\mathcal{H}'_0|0 >= 0. \tag{3.5}$$

Notice that $\exp(-i\theta K_2)$ can be factorized as the product of two (commuting) single-mode squeezing generators and for this reason we will refer to the state $|0(\theta) >$ as to the squeezed vacuum (at this level actually it is not, strictly speaking, a *squeezed state* since squeezed states are obtained by applying the squeezing generator to a (Glauber-type) coherent state).

We observe that the operators $A$ and $B$ transform under $\exp(-i\theta K_2)$ (for any given $k$) as

$$A \mapsto A(\theta) = e^{-i\theta K_2}Ae^{i\theta K_2} = A\cosh\left(\frac{1}{2}\theta\right) + A^\dagger\sinh\left(\frac{1}{2}\theta\right) \ , \tag{3.6}$$

$$B \mapsto B(\theta) = e^{-i\theta K_2}Be^{i\theta K_2} = B\cosh\left(\frac{1}{2}\theta\right) + B^\dagger\sinh\left(\frac{1}{2}\theta\right) \ . \tag{3.7}$$

These transformations are nothing else than the *squeezing* transformations and preserve the commutation relations (2.12). One has

$$A(\theta)|0(\theta) >= 0 = B(\theta)|0(\theta) > \ . \tag{3.8}$$

Using the commutativity of $J_2$ with $K_2$, we have that the $t$-evolution of the squeezed vacuum $|0(\theta) >$ is given by

$$|0(\theta, t) >= \exp(-i\theta K_2)\exp\left(\frac{-it\mathcal{H}_{I_2}}{\hbar}\right)|0 >= \exp(-i\theta K_2)|0(t) >, \tag{3.9}$$

where $|0(t)>$ is the vacuum state annihilated by $A(t)$, $B(t)$, i.e.

$$A \mapsto A(t) = e^{-i\frac{t}{\hbar}\mathcal{H}_{I_2}} A e^{i\frac{t}{\hbar}\mathcal{H}_{I_2}} = A \cosh{(\Gamma t)} - B^\dagger \sinh{(\Gamma t)} \quad , \tag{3.10}$$

$$B \mapsto B(t) = e^{-i\frac{t}{\hbar}\mathcal{H}_{I_2}} B e^{i\frac{t}{\hbar}\mathcal{H}_{I_2}} = -A^\dagger \sinh{(\Gamma t)} + B_k \cosh{(\Gamma t)} \quad . \tag{3.11}$$

Finally, we have (at finite volume $V$)

$$|0(\theta, t) >= \frac{1}{\cosh{(\Gamma t)}} \exp{(\tanh{(\Gamma t)} J_+(\theta))}|0(\theta) > \quad , \tag{3.12}$$

with $J_+(\theta_k) \equiv A^\dagger(\theta_k) B^\dagger(\theta_k)$, namely the $su(1,1)$ generalized coherent state (a two mode Glauber-type coherent state) with equal numbers of modes $A(\theta_k)$ and $B(\theta_k)$ condensed in it (for each $k$) at each $t$ .[14]
We now note that the vacuum is not stable. In fact

$$< 0(\theta, t)|0(\theta) > \propto \exp{(-t\Gamma)} \to 0 \quad \textit{for large } t, \tag{3.13}$$

i.e. time evolution brings out of the initial-time Hilbert space for large $t$. This is not acceptable in quantum mechanics since there the Von Neumann theorem states that all the representations of the canonical commutation relations are unitarily equivalent and therefore there is no room in quantum mechanics for non-unitary time evolution as the one in (3.13). On the contrary, in QFT there exist infinitely many unitarily inequivalent representations and this suggests to us to consider our problem in the framework of QFT, which we will do in the next section.

## 4. QUANTUM FIELD THEORY FOR EXPANDING GEOMETRY

To set up the formalism in QFT we have to consider the infinite volume limit; however, as customary, we will work at finite volume and at the end of the computations we take the limit $V \to \infty$. The QFT Hamiltonian is introduced as

$$\mathcal{H} = \mathcal{H}_0 + \mathcal{H}_{I_1} + \mathcal{H}_{I_2} \tag{4.1}$$

$$\mathcal{H}_0 = \sum_k \frac{1}{2}\hbar\Omega_{0,\,k}(t)(A_k^\dagger A_k - B_k^\dagger B_k) = \sum_k \hbar\Omega_{0,\,k}(t)C_k \,, \tag{4.2}$$

$$\mathcal{H}_{I_1} = -\sum_k \frac{1}{4}\hbar\Omega_{1,\,k}(t)\left[\left(A_k^2 + A_k^{\dagger^2}\right) - \left(B_k^2 + B_k^{\dagger^2}\right)\right] = -\sum_k \hbar\Omega_{1,\,k}(t)K_{1,\,k} \,,$$
$$\tag{4.3}$$

$$\mathcal{H}_{I_2} = i\sum_k \Gamma_k\hbar\left(A_k^\dagger B_k^\dagger - A_k B_k\right) = i\sum_k \hbar\Gamma_k(J_{+,\,k} - J_{-,\,k}) \,, \tag{4.4}$$

with

$$[A_k, A^\dagger_{k'}] = \delta_{k,\,k'} = [B_k, B^\dagger_{k'}], \qquad [A_k, B_{k'}] = 0 = [A_k^\dagger, B^\dagger_{k'}]. \tag{4.5}$$

We also have now

$$\mathcal{H}'_0 = \sum_k \hbar\Omega_k(t)(A_k^\dagger A_k - B_k^\dagger B_k) \quad . \tag{4.6}$$

The $\theta$-vacuum, at finite volume $V$ is :

$$|0(\theta)> = \prod_{k} \frac{1}{\cosh(\theta_k)} \exp \ (\tanh(\theta_k) j_{k,+})|0> \ , \qquad (4.7)$$

with $j_{k,+} \equiv a_k^\dagger b_k^\dagger = \frac{1}{2}(A_k^{\dagger 2} + B_k^{\dagger 2})$, and $a_k = \frac{1}{\sqrt{2}}(A_k + iB_k)$, $b_k = \frac{1}{\sqrt{2}}(A_k - iB_k)$. For $t \neq 0$:

$$|0(\theta,t)> = \prod_{k} \frac{1}{\cosh(\Gamma_k t)} \exp \ (\tanh(\Gamma_k t) J_{k,+}(\theta))|0(\theta)> \ , \qquad (4.8)$$

with $J_{k,+}(\theta) = A_k^\dagger(\theta) B_k^\dagger(\theta)$.

The normalized state $|0(\theta,t)>$ is also a $su(1,1)$ generalized coherent state. Eq.(3.13) is now replaced by

$$<0(\theta,t)|0(\theta)> = \exp\left(-\sum_{k} \ln \cosh(\Gamma_k t)\right) \ , \qquad (4.9)$$

$$<0(\theta,t)|0(\theta,t)> = 1 \quad \forall t \ , \qquad (4.10)$$

which again exhibit non-unitary time evolution, provided $\sum_k \Gamma_k > 0$:

$$<0(\theta,t)|0(\theta)> \propto \exp\left(-t\sum_{k}\Gamma_k\right) \to 0 \ \ for \ large \ t \ . \qquad (4.11)$$

Use of the customary continuous limit relation $\sum_k \mapsto \frac{V}{(2\pi)^3}\int d^3 \ k$, for $\int d^3 \ k \ \ln \cosh(\Gamma_k t)$ finite and positive, gives in the infinite volume limit

$$<0(\theta,t)|0(\theta)> \xrightarrow[V\to\infty]{} 0 \quad \forall t \ , \qquad (4.12)$$

$$<0(\theta,t)|0(\theta',t')> \xrightarrow[V\to\infty]{} 0 \quad with \ \ \theta' \equiv \theta(t_0'), \ \forall t, t', t_0' \ , \quad t \neq t' \ . \qquad (4.13)$$

At each time $t$ one has

$$A_k(\theta,t)|0(\theta,t)> = 0 = B_k(\theta,t)|0(\theta,t)> \ , \quad \forall t \ , \qquad (4.14)$$

and the number of modes of type $A_k(\theta)$ in the state $|0(\theta,t)>$ is given, at each instant $t$ by

$$n_{A_k}(t) \equiv <0(\theta,t)|A_k^\dagger(\theta) A_k(\theta)|0(\theta,t)> = <0(t)|A_k^\dagger A_k|0(t)> = \sinh^2(\Gamma_k t) \ ; \ (4.15)$$

and similarly for the modes of type $B_k(\theta)$.

We also observe that the commutativity of $\mathcal{C}$ (i.e. $K_0$) with $\mathcal{H}_{I_2}$ (i.e. $J_2$) ensures that the number $(n_{A_k} - n_{B_k})$ is a constant of motion for any $k$ and any $\theta$. Moreover, one can show [4,12] that the *creation* of a mode $A_k(\theta,t)$ is equivalent to the *destruction* of a mode $B_k(\theta,t)$ and vice-versa. This means that the $B_k(\theta,t)$ modes can be interpreted as the *holes* for the modes $A_k(\theta,t)$: the $B$-system can be considered as the sink where the energy dissipated by the $A$-system flows.

Notice that we have used $k$-dependent $\Gamma$ and the relation between the $\Gamma_k$'s and $\Gamma \equiv \frac{H}{2}$ is established noting that in the continuum limit eqs. (4.5) become

$$[A_k, A^{\dagger}_{k'}] = \delta(k - k') = [B_k, B^{\dagger}_{k'}], \quad [A_k, B_{k'}] = 0, \quad [A^{\dagger}_k, B^{\dagger}_{k'}] = 0, \quad (4.16)$$

and that, as well known, the $A_k$ (and $B_k$) operators are not well defined on vectors in the Fock space; for instance $|A_k >\equiv A^{\dagger}_k|0 >$ is not a normalizable vector since from eqs. (4.16) one obtains $< A_k|A_k >= \delta(0)$ which is infinity. As customary one must then introduce wave-packet (smeared out) operators $A_f$ with spatial distribution described by square-integrable (orthonormal) functions

$$f(x) = \frac{1}{(2\pi)^{3/2}} \int d^3 k f(k) e^{i k x}, \quad A_f = \frac{1}{(2\pi)^{3/2}} \int d^3 k A_k f(k) \quad (4.17)$$

with commutators

$$[A_f, A^{\dagger}_g] = (f,g) = [B_f, B^{\dagger}_g], \quad [A_f, B_g] = 0, \quad [A^{\dagger}_f, B^{\dagger}_g] = 0, \quad (4.18)$$

with $(f,g)$ denoting the scalar product between $f$ and $g$. Now $< A_f|A_f >= 1$ and the $A_f$'s are well defined operators in the Fock space where observables have to be realized. In this connection it is interesting to notice that the reality condition on $\Omega(t)$ naturally introduces the infrared cut-off smearing out the operator fields.

In conclusion, we express the $A_f$ number operator as

$$n_{A_f}(t) \equiv< 0(\theta,t)|A^{\dagger}_f(\theta)A_f(\theta)|0(\theta,t) >= \frac{1}{(2\pi)^3} \int d^3 k \sinh^2(\Gamma_k t)|f(k)|^2 \equiv \sinh^2(\Gamma t),$$
$$(4.19)$$

and similarly for the modes of type $B_f(\theta)$ (cf. with eq. (4.15)). Eq. (4.19) specifies the relation between the $\Gamma_k$'s and $\Gamma$.

The structure of $|0(\theta,t) >$ naturally leads us to recognize its thermal properties. This will done in the following section.

## 5. ENTROPY AND FREE ENERGY IN EXPANDING UNIVERSE

The vacuum state $|0(\theta,t) >$ as given by equation (4.8) can be written as

$$|0(\theta,t) >= \exp\left(-\frac{1}{2}S_{A(\theta)}\right)|\mathcal{I}(\theta) >= \exp\left(-\frac{1}{2}S_{B(\theta)}\right)|\mathcal{I}(\theta) > , \quad (5.1)$$

where

$$|\mathcal{I}(\theta) >\equiv \exp\left(\sum_k A^{\dagger}_k(\theta)B^{\dagger}_k(\theta)\right)|0(\theta) >= \exp(-i\theta K_2)|\mathcal{I} > , \quad (5.2)$$

with $|\mathcal{I} >$ the invariant (not normalizable) vector[4]

$$|\mathcal{I} >\equiv \exp\left(\sum_k A^{\dagger}_k B^{\dagger}_k\right)|0 > , \quad (5.3)$$

and

$$S_{A(\theta)} \equiv -\sum_k \left\{ A_k^\dagger(\theta) A_k(\theta) \ln \sinh^2(\Gamma_k t) - A_k(\theta) A_k^\dagger(\theta) \ln \cosh^2(\Gamma_k t) \right\} =$$

$$= \exp(-i\theta K_2) S_A \exp(i\theta K_2) \ . \tag{5.4}$$

Here $S_A$ is given by

$$S_A \equiv -\sum_k \left\{ A_k^\dagger A_k \ln \sinh^2(\Gamma_k t) - A_k A_k^\dagger \ln \cosh^2(\Gamma_k t) \right\} \ . \tag{5.5}$$

$S_{B(\theta)}$ ($S_B$) has the same expression with $B_k(\theta)$ ($B_k$) and $B_k^\dagger(\theta)$ ($B_k^\dagger$) replacing $A_k(\theta)$ ($A_k$) and $A_k^\dagger(\theta)$ ($A_k^\dagger$), respectively. In the following we shall simply write $S(\theta)$ ($S$) for either $S_{A(\theta)}$ or $S_{B(\theta)}$ ($S_A$ or $S_B$). Observing that $A_k^\dagger A_k$ is the number operator for the $k$-mode and that $\sinh^2(\Gamma_k t)$ is the number of modes $A_k$ in the state $|0(t)\rangle$, $S(\theta)$ (or $S$) is recognized to be the entropy.

Moreover, the difference $S_{A(\theta)} - S_{B(\theta)}$ is constant in time (cfr. (3.2)):

$$[S_{A(\theta)} - S_{B(\theta)}, \mathcal{H}] = 0 \tag{5.6}$$

(and, correspondingly, $[S_A - S_B, \mathcal{H}'] = 0$, cfr. (3.2)) . Since the $B$-particles are the holes for the $A$-particles, $S_{A(\theta)} - S_{B(\theta)}$ is in fact the (conserved) entropy for the closed system.

Eqs. (5.1) and (5.4) show that the operational dependence of $\frac{1}{2} S_{A(\theta)}$ (or respectively, $\frac{1}{2} S_{B(\theta)}$) is uniquely on the $A$ ($B$) variables: thus in eq. (5.1) time evolution is expressed solely in terms of the (sub)system $A$ ($B$) with the elimination of the $B$ ($A$) variables. This reminds us of the procedure by which one obtains the reduced density matrix by integrating out bath variables.

For the time variation of $|0(\theta, t)\rangle$ at finite volume $V$, we obtain

$$\frac{\partial}{\partial t} |0(\theta, t)\rangle = -\frac{1}{2} \left( \frac{\partial S(\theta)}{\partial t} \right) |0(\theta, t)\rangle \ . \tag{5.7}$$

Equation (5.7) shows that $\frac{1}{2} \left( \frac{\partial S(\theta)}{\partial t} \right)$ is the generator of time-translations, namely time evolution is controlled by the entropy variations.

This feature reflects indeed correctly the irreversibility of time evolution characteristic of inflattive motion. Inflation implies in fact the choice of a privileged direction in time evolution (*time arrow*) with a consequent breaking of time-reversal invariance.

We therefore ask ourselves whether such statistical features may actually be related to thermal concepts. We know from ref. 12 (which here we closely follow) that this is indeed the case.

For the sake of definiteness, let us consider the $A$-modes alone and introduce the functional (free energy)

$$F_A \equiv \langle 0(\theta, t)| \left( \mathcal{H}'_{0, A(\theta)} - \frac{1}{\beta} S_{A(\theta)} \right) |0(\theta, t)\rangle = \langle 0(t)| \left( \mathcal{H}'_{0, A} - \frac{1}{\beta} S_A \right) |0(t)\rangle \ . \tag{5.8}$$

Here $\mathcal{H}'_{0,A(\theta)} \equiv \sum_k E_k A_k^\dagger(\theta) A_k(\theta)$ and $\mathcal{H}'_{0,A} \equiv \sum_k E_k A_k^\dagger A_k$; $E_k \equiv \hbar\Omega_k(t_0 = 0) - \mu$, with $\mu$ the chemical potential. The stability condition $\frac{\partial F_{A(\theta)}}{\partial \sigma_k} = 0$, $\sigma_k \equiv \Gamma_k t \ \forall k$, assuming $\beta$ a slowly varying functions of $t$, gives $\beta E_k = -\ln\tanh^2(\sigma_k)$. Then we have

$$n_{A_k}(t) = \sinh^2(\Gamma_k t) = \frac{1}{e^{\beta(t)E_k} - 1} \quad , \tag{5.9}$$

which is the Bose distribution for $A_k$ at time $t$ provided we assume $\beta(t)$ to represent the inverse temperature $\beta(t) = \frac{1}{k_B T(t)}$ at time $t$ ($k_B$ denotes the Boltzmann constant). This allows us to recognize $\{|0(\theta, t) >\}$ as a representation of the canonical commutation relations at finite temperature, equivalent with the Thermo Field Dynamics representation $\{|0(\beta) >\}$ of Takahashi and Umezawa[4].

## 6. CONCLUSIONS

We conclude that the system in its evolution runs over a variety of representations of the canonical commutation relations which are unitarily inequivalent to each other for $t \neq t'$ in the infinite-volume limit: the non-unitary character of time evolution implied by expanding geometry is thus recovered, in a consistent scheme, in the unitary inequivalence among representations at different times in the infinite-volume limit.

We observe that the statistical nature of expanding geometry phenomena naturally emerges from our formalism, even though no statistical concepts were introduced a priori (we have seen that the vacuum structure naturally leads to the *entropy operator* as time evolution generator (see eq.(5.7)).

ACKNOWLEDGMENTS

We are grateful to the Directors of the Jagna International Workshop, Drs. C.C. Bernido, M.V.Carpio-Bernido, K. Nakamura and K. Watanabe for giving to us the opportunity to contribute with this paper to the Festschrift for Professor Ezawa.

This work has been partially supported by INFN, by MURST and by a Network supported by the European Science Foundation.

## REFERENCES

1. R. H. Brandenberger, *Rev. Mod. Phys.* **57** (1985) 1.
2. A. H. Guth, *Phys. Rev.* **D23** (1981) 347.
3. E. Calzetta, B L. Hu, *Phys. Rev.* **D52** (1995) 6770.
4. Y. Takahashi and H. Umezawa, *Collective Phenomena* **2** (1975) 55.
5. H. Umezawa, H. Matsumoto and M. Tachiki *Thermo Field Dynamics and Condensed States* (North-Holland Pub. Co., Amsterdam, 1982).
6. H. Umezawa, *Advanced Field Theory: Micro, Macro, and Thermal Concepts* (American Institute of Physics, N. Y., 1993).
7. W. Israel *Phys. Lett.* **57A** (1976) 107.

8. A. E. I. Johansson, H. Umezawa and Y. Yamanaka, *Class. Quantum Grav.* **7** (1990) 385.

9. G. L. Sewell, *Ann. Phys. (N.Y.)* **141** (1982) 201.

10. H. Feshbach and Y. Tikochinsky, *Transact. N.Y. Acad. Sci.*

11. H. Bateman, *Phys. Rev.* **38** (1931) 815.

12. E. Celeghini, M. Rasetti and G. Vitiello, *Ann.Phys.* **215** (1992) 156.

13. L. Parker, *Phys. Rev.* **D183** (1969) 1057.

14. A. M. Perelomov, *Generalized Coherent States and their Applications* (Springer-Verlag, Berlin, 1986).

15. S. M. Barnett and P. L. Knight, *J. Opt. Soc. Am.* **B2** (1985) 467.

16. Jung Kon Kim and Sang Pyo Kim, *J. Korean Phys. So.* **28** (1995) 7.

17. L. P. Grishchuk, *Zh. Eksp. Teor. Fiz.* **67** (1974) 825 [*Sov. Phys. JEPT* **40** (1975) 409.

18. L. P. Grishchuk and Y. V. Sidorov, *Phys. Rev.* **D42** (1990) 3413 and refs. quoted therein.

19. L. P. Grishchuk, H. A. Haus and K. Bergman, *Phys. Rev.* **D46** (1992) 1440.

20. M. Blasone, E. Graziano, O. K. Pashaev and G. Vitiello, *Ann. Phys. (N.Y.)* **252** (1996) 115.

21. A.I. Solomon, *J. Math. Phys.* **12** (1971) 390.

# List of Participants and Contributors

Jiro Arafune
Institute for Cosmic Ray Research
University of Tokyo
3-2-1 Midori-cho Tanashi-shi
Tokyo 188
Japan


Asao Arai
Department of Mathematics
Hokkaido University
Kita 10-jyo, nishi 8-chome
Kita-ku, Sapporo 060
Japan


Huzihiro Araki
Department of Mathematics
Faculty of Science and
    Technology
Science University of Tokyo
2641 Yamazaki, Noda-shi
Chiba-ken 278
Japan


Flordivino Basco
National Institute of Physics
University of the Philippines
1101 Diliman, Quezon City
Philippines


Christopher C. Bernido
National Institute of Physics
University of the Philippines
1101 Diliman, Quezon City
Philippines
    and
Research Center for Theoretical Physics
Central Visayan Institute
6308 Jagna, Bohol
Philippines


M. Victoria Carpio-Bernido
Research Center for Theoretical Physics
Central Visayan Institute
6308 Jagna, Bohol
Philippines


Jinky Bornales
Mindanao State University- Iligan
    Institute of Technology
9200 Iligan City
Philippines


Lorenzo C. Chan
National Institute of Physics
University of the Philippines
1101 Diliman, Quezon city
Philippines

Timothy J. Dennis
National Institute of Physics
University of the Philippines
1101 Diliman, Quezon City
Philippines

Cecile DeWitt-Morette
Department of Physics
The University of Texas at Austin
Austin, Texas 78712-1081
USA

Hiroshi Ezawa
Department of Physics
Gakushuin University
Mejiro, Toshima-ku
Tokyo 171-8588
Japan

Daniel Garcia
Physics Department
De La Salle University
2401 Taft Avenue
1004 Manila
Philippines

Enrico B. Gravador
Mindanao State University - Iligan
  Institute of Technology
9200 Iligan City
Philippines

Jonathan S. Guevarra
Physics Department
De La Salle University
2401 Taft Avenue
1004 Manila
Philippines

Yuan-Xing Gui
Department of Physics
Dalian University of Technology
Dalian 116023
P. R. China

Takeyuki Hida
Department of Mathematics
School of Science and Technology
Meijo University
Nagoya 468
Japan

Masao Hirokawa
Department of Mathematics
Tokyo Gakugei University
Koganei 184-8501
Japan

Takashi Ichinose
Department of Mathematics
Kanazawa University
Kanazawa 920-1192
Japan

John R. Klauder
Department of Physics and
  Mathematics
University of Florida
Gainesville, Florida 32611
U.S.A.

Harry C. S. Lam
Department of Physics
McGill University
3600 University Street
Montreal QC H3A 2T8
Canada

Jiu Qing Liang
Department of Physics
Shanxi University
Taiyuan, Shanxi 030006
P. R. China


Jose A. Magpantay
National Institute of Physics
University of the Philippines
1101 Diliman, Quezon City
Philippines


Ady Mann
Department of Physics
Technion-Israel Institute of Technology
32000 Haifa
Israel


Koichi Nakamura
Division of Natural Science
Meiji University, Izumi Campus
1-9-1 Eifuku Suginami-ku
Tokyo 168
Japan


Toru Nakamura
Department of Mathematics
Sundai Preparatory School
Kanda-Surugadai, Chiyoda-ku
Tokyo 101-8313
Japan


Izumi Ojima
Research Institute for Mathematical
  Sciences
Kyoto University
Kyoto 606-8502
Japan


Yasunori Okabe
Department of Mathematical
  Engineering & Information Physics
Faculty of Engineering
University of Tokyo
Bunkyo-ku, Tokyo 113
Japan


Ruben Quiroga
Physics Department
De La Salle University
2401 Taft Avenue
1004 Manila
Philippines


Norisuke Sakai
Department of Physics
Tokyo Institute of Technology
Oh-okayama, Meguro
Tokyo 152-8551
Japan


Ludwig Streit
Faculty of Physics
University of Bielefeld
D-4800 Bielefeld
Germany
  and
University of Madeira
Colegio dos Jesuitas - Largo
  do Colegio
9000 Funchal
Portugal


Imelda Timonera
Mindanao State University - Iligan
  Institute of Technology
9200 Iligan City
Philippines

Giuseppe Vitiello
Department of Physics
University of Salerno
84100 Salerno
Italy

Ke Wu
Institute of Theoretical Physics
Academia Sinica
Beijing 100080
P. R. China

Shikun Wang
Institute of Applied Mathematics
Academia Sinica
Beijing 100080
P. R. China

Danilo M. Yanga
National Institute of Physics
University of the Philippines
1101 Diliman, Quezon City
Philippines

Keiji Watanabe
Department of Physics
Meisei University
2-1-1 Hodokubo Hino-shi
Tokyo 191-8506
Japan

Kunio Yasue
Research Institute for Informatics
  and Science
Notre Dame Seishin University
2-16-9 Ifuku-cho
Okayama-shi 700
Japan

Frederik W. Wiegel
Center for Theoretical Physics
University of Twente
7500 A E Enschede
The Netherlands

# SCIENTIFIC WORKS
## OF
# PROFESSOR HIROSHI EZAWA

# SCIENTIFIC WORKS OF
# PROFESSOR HIROSHI EZAWA

Professor Ezawa obtained his Bachelor of Science degree in 1955 and his doctoral degree in theoretical physics in 1960, both from the University of Tokyo. Immediately after his doctorate, he was appointed Assistant Professor in the Department of Physics of the University of Tokyo. Three years later, he visited several universities in the USA and was Research Associate at the Department of Physics and Astronomy of the University of Maryland. In 1966, he was Visiting Scientist at the Institute for Theoretical Physics of the University of Hamburg in Germany. Then, in 1967, he joined the faculty of Gakushuin University in Tokyo as Associate Professor of the Department of Physics and became full professor three years later. From 1972-1974 he was a member of the Technical Staff of Bell Labs in Murray Hill, New Jersey, USA. After this stint Professor Ezawa returned to Gakushuin University later serving as Dean of the Faculty of Science for five years from 1987 until 1992.

Professor Ezawa was President of the Japan Physical Society from September 1995 to August 1996, member of the Council for Science and Technology Department of the National Diet Library, and member of the Science Council of Japan.

Aside from his noted works in the area of probability methods in quantum mechanics and quantum field theory, the various books written by Professor Ezawa have been of invaluable help to countless students and researchers.

Professor Ezawa resides in Tokyo with his wife, Yoshiko, and sons, Hajime and Naofumi.

## List of Publications


### 1. Papers

Quantum Statistics of Fields and Multiple Production of Mesons,
  with Y. Tomozawa and H. Umezawa, Nuov. Cim. **5** (1957), 810-841.

Ionization Loss near the Origin of an Electron Pair of Very High Energy,
  with I. Mito, Prog. Theor. Phys. **18** (1957), 437-447.

To Introduce the Impact Parameter into the Analysis of Multiple-Production of
Mesons, Nuov. Cim **11** (1959), 745-759.

Impact Parameter and Perturbation Treatment of the Distant Collision of Nucleons
with Pion Emission,
  with O. Kamei, K. Mori, H. Shimoida and T. Yoneyama, *Proc. of
  Moscow Conf. on Cosmic Ray Physics, 1959*, English version, 274-284, Russian
  version, 276-283.

Green Functions for Elementary Particles,
  with H. Umezawa, Phys. Rev. **116** (1959), 463-464.

Pion-Nucleon Interaction and the Multiple-Production Experiment,
  with O. Kamei. K. Mori, H. Shimoida and T. Yoneyama, Prog. Theor.
  Phys. **25** (1961), 667-683.

Quantum Statistical Analogue of Ward's Identity,
  with K. Watanabe and O. Kamei, Prog. Theor. Phys. **25** (1961), 735-742.

A Formulation for the Bound State Problem,
  with K. Kikkawa and H. Umezawa, Nuov. Cim. **25** (1962), 1141-1166.

An Approach to the Elementarity of Particles,
  with T. Muta and H. Umezawa, Prog. Theor. Phys. **29** (1963), 877-862.

Some Examples of the Asymptotic Fields - In the Sense of Weak Convergence,

Ann. Phys. **24** (1963), 46-62.

A Note on the Van Hove-Miyatake Catastrophe,
Prog. Theor. Phys. **30**(1963), 545-549.

A Perturbation Theory without the Adiabatic Hypothesis,
*Proc. of Eastern U.S. Theor. Phys. Conf., U. of North Carolina,
Oct. 1963*, II. 94-103.

The Representation of Canonical Variables as the Limit of Infinite Space Volume:
the Case of the BCS Model,
J. Math. Phys. **5** (1964), 1078-1090.

Particle-Mixture Theory and Apparent *CP* Violation in *K*-Meson Decay,
with Y.S. Kim, S. Oneda and J.C. Pati, Phys. Rev. Lett. **14** (1965), 673-676.

A Criterion for Bose-Einstein Condensation and Representation of Canonical Commutation Relations,
with M. Luban, Jour. Math. Phys. **8** (1967), 1285-1311.

Spontaneous Breakdown of Symmetries and Zero-Mass States,
with J.A. Swieca, Commun. Math. Phys. **5** (1967), 330-336.

A Renormalization Scheme for the Strong-Coupling $\lambda\phi^4$ Theory,
with K. Nakamura and Y. Yamamoto, Nuov. Cim. **54 A** (1968), 512-515.

Remarks on the Quantum Field Theory in Lattice Space, I and II.
Commun. Math. Phys. **8** (1968), 261-268, **9** (1968), 38-52.

Regge Trajectories in the Wick-Curkosky Model,
with J. Arafune and K. Nakamura, Lett. Nuov. Cim. **2** (1969), 394-398.

Theory of the Kapitza-Dirac Effect,
with H. Namaizawa, Jour. Phys. Soc. Jpn **26** (1969), 458-468.

Numerical Solution of an Anharmonic Oscillator Eigenvalue Problem by Milne's Method,
with K. Nakamura and Y. Yamamoto, Proc. Japan Acad. **46** (1970), 168-172.

Phonons in a Half Space,
Ann. Phys. **67** (1971), 438-460.

Electrons and 'Surfons' in a Semiconductor Inversion Layer,

with T. Kuroda and K. Nakamura, Surface Sci. **24** (1971), 654-658.

Regge Trajectories in the Wick-Cutkosky Model,
     with N. Murai, K. Nakamura, Prog. Theor. Phys. **46** (1971), 909-937.

On the Divergence of Certain Integrals of the Wiener Process,
     with L.A. Shepp and J.R. Klauder. Ann. Inst. Fourier **24** (1974),
     189-193.

A Path Space Picture for Feynman-Kac Averages,
     with J.R. Klauder and L.A. Shepp, Ann. Phys. **88** (1974), 588-620.

Surfons and the Electron Mobility in Silicon Inversion Layers,
     with S. Kawaji and K. Nakamura, Jpn. J. Appl. Phys. **13** (1974),
     126-155.

Vestigial Effects of Singular Potentials in Diffusion Theory and Quantum Mechanics,
     with J.R. Klauder and L.A. Shepp, Jour. Math. Phys. **16** (1975), 783-799.

The Effect of Large Field on the Lifetime of the Forbidden Lines in Quasars,
     with M. Leventhal, J. Phys. B. Atom. Molec. Phys. **8** (1975),
     1824-1830.

Inversion Layer Mobility with Intersubband Scattering,
     Surface Sci. **58** (1976), 25-32.

Einstein's Contribution to Statistical Mechanics, Classical and Quantum,
     Jap. Studies in the Hist. of Sci. No.18 (1979), 27-72.

Many-Body Effects in the Si Metal-Oxide-Semiconductor Inversion Layer: Subband
Structure,
     with K. Nakamura and K. Watanabe, Phys. Rev. **B 22** (1980), 1892-1904.

Stochastic Calculus and Some Models of Irreversible Process,
     with H. Hasegawa, Suppl. Prog. Theor. Phys. No.69 (1980), 41-54.

Gravitational Radiation of a Particle Falling towards a Black Hole,
     with Y. Tashiro, Prog. Theor. Phys. **66** (1981), 1612-1626.

Remarks on a Stochastic Quantization of Scalar Fields,
     with J.R. Klauder, Prog. Theor. Phys. **69** (1983), 664-673.

Fermions without Fermions — The Nicolai Map Revisited,

with J.R. Klauder, Prog. Theor. Phys. **74** (1985), 904-915.

From Infinite Time to Finite Temperature,
in *Quantum Field Theory — Proc. of the Int'l Symp. in Honor of H. Umezawa*, F. Mancini ed., North Holland (1986), 405-414.

Thermo Field Dynamics of Heat Conduction,
in *Progress in Quantum Field Theory*, H. Ezawa and S. Kamefuchi eds., North Holland (1986), 305-324.

Reconstruction of Lagrangian from a given Hamiltonian,
in *Proc. Int'l Conf. on Physics Education*, Japan Soc. for Physics Education (1986), 297.

Towards a Quantum Theory of Heat Conduction,
in *Wondering in the Fields*, Festschrift for Prof. K. Nishijima on the Occasion of his Sixtieth Birthday, K. Kawarabayashi and A. Ukawa eds., World Scientific (1987), 191-206.

Demonstration of Single-Electron Build Up of an Interference Pattern,
with A. Tonomura, J. Endo, T. Matsuda, T. Kawasaki, Am. J. Phys. **57** (1989), 117-120.

Characterization of Thermal Coherent and Thermal Squeezed States,
with A. Mann, K. Nakamura and M. Revzen, Ann. Phys. **209** (1991), 216-230.

Difference in Elastic Scattering Effects on Resonant Tunneling in One Dimension and Three Dimensions,
with Y. Zohta and K. Nakamura, Solid State Commun. **80** (1991), 885-889.

Quantum Field Theory of Thermal Diffusion,
with K. Watanabe and K. Nakamura, in *Thermal Field Theories*, H. Ezawa, T. Arimitsu and Y. Hashimoto ed., North Holland (1991), 79-94.

Thermal Coherent and Thermal Squeezed States,
with A. Mann, K. Nakamura and M. Revzen, in *Thermal Field Theories*, H. Ezawa, T. Arimitsu and Y. Hashimoto eds., North Holland (1991), 201-205.

Feynman Path Integral Approach to Resonant Tunneling,
with Y. Zohta and K. Nakamura, in *Resonant Tunneling in Semiconductors*, L.L. Chang et al. eds., Plenum (1991), 285-295.

Effect of Inelastic Scattering on Resonant Tunneling studied by the Optical Potential
and Path Integrals,
　　with Y. Zohta, J. Appl. Phys. **72** (1992), 3584-3588.

Tomonagas Studien unter Heisenberg in Leipzig,
　　in *Werner Heisenberg als Physiker und Philosoph in Leipzig*,
　　H. Rechenberg Hrsg., Spektrum Akad. Verlag, Heidelberg (1993), 78-86.

Problems of Quantum Control and Measurement,
　　in *Proc. Int'l Workshop on Quantum Control and Measurement*,
　　H. Ezawa and Y. Murayama eds., Elsevier (1993), 3-8.

Quantum Mechanical Representation of Laser Beam Splitting,
　　in *Proc. of Int'l Symp. on Advanced Topics of Quantum Physics- Shanxi*,
　　J.Q. Liang, M.L. Wang and S.N. Qiao eds., Science Press (1993), 342-353.

Particle- and Wave-Descriptions are not Equivalent,
　　in *On Klauder's Path: A Field Trip - Essays in Honor of John R. Klauder*,
　　G.G. Emch, G.C. Hegelfeld and L. Streit eds., World Scientific
　　(1994), 55-61.

Impurity Scattering in Resonance Tunneling,
　　with K. Nakamura and Y. Zohta, in *Proc. Int'l Conf. on Physics and
　　Technology in 1990's*, C.C. Bernido, M.V. Carpio-Bernido and D.M. Yanga
　　eds., Samhang Pisika ng Pilipinas (1994), 1 -19.

Particle and Wave Pictures: Are they Equivalent?
　　in *Field Theory and Collective Phenomena - In Memory of Professor
　　Hiroomi Umezawa, Perugia, Italy, 28-31 May 1992*, S. De Lillo and P. Sodano,
　　F.C. Khanna, G.W. Semenoff eds., World Scientific (1995), 456-469.

Change of Independent Variable in Variational Problems,
　　Int'l Jour. Mod. Phys. B **10** (1996), 1555 - 1561.

Quantization of Fields in a Fabri-Perot Cavity,
　　Physics Essays **9** (1996), 604 - 608. Prof. H. Umezawa Memorial Volume.

Constructive Proof of the Existence of Bound States in One Dimension,
　　Foundation of Physics **27** (1997), 1103 - 1108. Special Issue dedicated
　　to Professor M. Namiki on the Occasion of his Seventieth Birthday.

Diversity and Unity in Physics Research in the Age of Communication,
　　in *Proc. of the 2nd Int'l Conf. on Research and Communication in Physics*,

Physical Society of Japan (1997), 3 - 15.

The Casimir Force from Lorentz's
  K. Nakamura and K. Watanabe, in press.

Thermo Field Dynamics of Irreversible Processes,
  with K. Watanabe and K. Nakamura, in press.

## 2. Books

All but a few are written in Japanese, exceptions being designated either by [E] for English or [G] for German.

Quantum Theory of Scattering,
  in *Quantum Mechanics, Problems with Solutions*, H. Umezawa and
  M. Kotani ed., Syokabo (1959).

*Topics in the Theory of Elementary Particles*,
  H. Umezawa, K. Kawarabayashi and H. Ezawa, Kyoritsu Publ. (1963).

*Textbook of High School Physics*
  M. Nogami, I. Imai, J. Iwaoka, H. Ezawa, K. Kinoshita, M. Kojima,
  M. Kondo, H. Takami and J. Hayashi, Jikkyo Publ. (1963, 1967, 1969).

*Elementary Particles and Cosmic Rays*,
  O. Kamei and H. Ezawa, Introduction-to-Physics Series, Diamond Pub.
  (1964).

Aspects of Physics Education in the U.S.A.
  in *Nature and Her Laws*, Collected Essays on Pedagogy, vol. 7,
  K. Kakiuchi, R. Hirooka and H. Azuma eds., Syogaku-kan Publ. (1968, 1976).
  pp.252 - 271.

Mathematical Structure of Quantum Mechanics.
  *Quantum Mechanics III*, Iwanami Series on Contemporary Physics, vol. 5,
  H. Yukawa, T. Toyoda and K. Inoue eds., Iwanami Shoten Publ. (1972).
  pp. 3 - 180, 229 - 286.

*Fields and Quanta*, Diamond Publ. (1976).

*Who Has Seen the Atom?*, Science-for-Juniors Series,

Iwanami Shoten Publ. (1976, 1981, 1987).

History of Quantum Mechanics (with T. Tsuneto), Quantum Optics (with M. Takat-suji) and Stability of Matter,
  in *Quantum Mechanics in Perspective — On the Occasion of Its 50th Anniversary*, H. Ezawa and T. Tsuneto eds., Iwanami Shoten Publ. (1977, 1987).

Advanced Quantum Mechanics.
  in *Quantum Mechanics I*, Iwanami Series on Contemporary Physics, vol. 3, H. Yukawa and T. Toyoda eds., Iwanami Shoten Publ. (1978). pp. 193 - 429.

Mathematical Structure of Quantum Mechanics.
  in *Quantum Mechanics II*, Iwanami Series on Contemporary Physics, vol. 4, H. Yukawa and T. Toyoda eds., Iwanami Shoten Publ. (1978). pp.249 - 484.

Einstein's Contribution to Statistical Mechanics [E],
  in *Albert Einstein-His Influence on Physics, Philosophy and Politics* P.C.Eichelburg and R.U.Sexl eds., Friedr. Vieweg & Sohn (1979). pp. 69 - 87

Einsteins Beitrag zur statistischen Mechanik [G],
  in *Albert Einstein — Sein Einfluss auf Physik, Philosophie, und Politik*, P.C.Eichelburg und R.U.Sexl hrsg., Friedr. Vieweg & Sohn (1979). pp. 77 - 90

*Dynamics - Problems with Solutions*,
  H. Ezawa, K. Nakamura and Y. Yamamoto, Tokyo Tosho Publ. (1984).

*Proceedings of the International Symposium on Fpundations of Quantum Mechanics in the Light of New Technology* [E],
  S. Kamefuchi, H. Ezawa, Y. Murayama, M. Namiki, S. Nomura, Y. Ohnuki and T. Yajima eds., Physical Society of Japan (1984).

Encyclopedia of Science, 3rd ed.
  T. Iino, H. Inokuti, S. Ueda, H. Ezawa, H. Kawanabe, Y. Taki, K. Toyoshima and Y. Hishino eds., Iwanami Shoten Publ. (1985).

*Progress in Quantum Field Theory* [E],
  H. Ezawa and S. Kamefuchi eds., North Holland (1986).

*Proposals for the 21st Century: Science Education*,
  S. Saito, M. Oda, Y. Murakami, K. Takeuti and H. Ezawa, Nikkei-Science Publ. (1986).

*Dictionary of Physics and Chemistry*, 4th ed.
  R. Kubo, S. Nagakura, H. Inokuti and H. Ezawa eds., Iwanami Shoten Publ.
  (1987).

*Fourier Analysis*, Mathematical Methods for Scientists and Engineers, vol. 6,
  Kodan-sha Publ. (1987, 1989, 1992, 1994, 1996).

*Quantum Field Theory and Statistical Mechanics*,
  H. Ezawa and A. Arai, Nihon-Hyoron-sha Publ. (1988).

Development of Mathematical Physics,
  H. Ezawa and I. Ojima eds., Tokyo Tosho Publ. (1988).

*Modern Physics*,
  Nihon Broadcasting Corporation Press (1988).

*Proceedings of the 3rd International Symposium on Foundations of Quantum
  Mechanics in the Light of New Technology* [E],
  S. Kobayashi, H. Ezawa, Y. Murayama, and S. Nomura eds., Physical Society
  of Japan (1990).

*Thermal Field Theories* [E],
  H. Ezawa, T. Arimitsu and Y. Hashimoto eds., North Holland (1991).

*Yoshio Nishina in Recollections*,
  H. Tamaki and H. Ezawa eds., Misuzu Shobo Publ. (1991).

*Methods of Renormalization Groups*, New Series of Contemporary Physics, vol. 13,
  H. Ezawa, M. Suzuki, K. Watanabe and H. Tasaki, Iwanami Shoten Publ.
  (1994, 1997)

*Groups and their Representations*, Iwanami Series of Applied Mathematics, vol. 11,
  H. Ezawa and K. Shima, Iwanami Shoten Publ. (1994).

*Asymptotic Analysis*, Iwanami Series of Applied Mathematics, vol. 14,
  Iwanami Shoten Publ. (1995).

*Sin-itiro Tomonaga — Life of a Japanese Physicist*,
  Japanese version (1980), ed. by M. Matsui, Misuzu Shobo Publ.;
  English tr.(1995), ed. by H. Ezawa, MYU Publ. (1995).

*Modern Physics*, Asakura Shoten Publ. (1996).

*Science I*, Collected Essays of Torahiko Terada, vol. 5,

Commentary by H. Ezawa, Iwanami Shoten Publ. (1997). pp. 345 - 375.

S. Tomonaga: *Quantum Mechanics and Me*, Collected Scientific Essays,
    H. Ezawa ed., Iwanami Shoten Publ. (1997). Commentary, pp. 387 - 420.

*Dictionary of Physics and Chemistry*, 5th ed.
    S. Nagakura, H. Inokuti, H. Ezawa, H. Iwamura, H. Sato and R. Kubo eds.,
    Iwanami Shoten Publ. (1998).

# Subject Index